Stellingen behorende bij het proefschrift:

Fatigue Crack Growth in Riveted Joints

by Scott Anthony FAWAZ

Statements going with the above doctor's thesis.

1. De overgang van een hoekscheur naar een door de dikte scheur met een scheefstaand elliptisch scheurfront is moeilijk te vangen in een analytisch model. De overgang is in proeven ook moeilijk waar te nemen.

   The transition of a corner crack to a through crack with an oblique, part elliptical crack front in monolithic metals is difficult to model and observe in experiments.

2. Het formuleren van een voorspellingsmodel voor scheurgroei bij het contactvlak tussen twee platen van een geklonken lapnaad vereist experimentele resultaten om het model te toetsen. Omdat er weinig dergelijke gegevens beschikbaar zijn, is het niet verrassend dat er nog geen goed voorspellingmodel is.

   Prediction modeling of crack growth at the faying surface of a riveted lap joint can only be done if we have experimental results to verify the model. Since there is very little data of this sort, it should not be surprising that we do not yet have a good prediction model.

3. De vorm van vermoeiingsscheuren in een geklonken lapnaad hangt af van de belastinsoverdracht in de verbinding. Die afhankelijkheid kan niet worden beschreven zolang wrijving, fretting en estspanningen rondom de klinknagel niet goed worden begrepen.

   The shape of fatigue cracks in riveted lap joints depends on the load transmission in the joint. We will not understand the dependence until we understand the friction, fretting and residual stresses associated with the rivet.

4. Als lapnaden in een vliegtuigromp langere levensduren moeten hebben, dan moet dat zonder klinknagels of door nagels met meer kracht te klinken.

   If we want fuselage lap joints to have a longer fatigue life, we must either remove the rivets or squeeze them harder.

5. Bij de voorspelling van scheurgroei in geklonken lapnaden is het vervangen van het scheefstaande scheurfront door een rechtstaand scheurfront onnodig.

   In predictions on crack growth in riveted joints, there is no reason to simplify and assume the oblique, part elliptical through crack to be a straight crack.

6. Bij een 3D eindige elementen model van een geklonken lapnaad zijn hogere orde elementen in samenhang met contact-elementen essentieel voor een realistische weergave van de belastingsoverdracht in de verbinding en het complexe niet-lineaire vervormingsgedrag.

   In 3D finite element models of riveted joints, higher order solid elements used in conjunction with contact elements are essential for realistic modeling of the load transmission in the joint and complex nonlinear deformation behavior.

7. Om spanningsintensiteitsfactoren nauwkeurig en efficiënt te bepalen met de eindige elementen methode is het een vereiste om dit volledig automatisch te doen verlopen vanaf "mesh" generatie tot K berekening. Het met de hand bewerken van model-parameters opent de deur voor onvoorziene fouten.

   The key to efficient, accurate stress intensity factor solutions via the finite element method is an automated solution scheme from mesh generation to K calculation. Manual manipulation of any of the model parameters opens the door for inadvertent errors.

8. Onderzoekers willen meestal hun mislukkingen niet vastleggen en publiceren, hoewel hun pogingen toch zeer begrijpelijk zijn. Zodoende zullen de mislukkingen waarschijnlijk worden herhaald.

   Often researchers are reluctant to document and publish their failures even though their attempts have been well conceived. This trend ensures that these failures will probably be repeated.

9. Succes in onderzoek vereist een bevlogen benadering. Zonder dat wordt het water naar de zee dragen.

   Success in research activities can only come if you have a passion for the subject, all else is treading water.

10. Onderzoek is niet onderzoek van slechts één persoon. Zelden wordt het doel alleen bereikt.

    The word "research" does not have any "I's"; rarely do we accomplish important matters alone.

11. Weten wat mogelijk is, is weten wat onmogelijk is. Opgeven? Nooit!

    To know what is possible is to know what is impossible. Never quit!

12. Stellingen zijn er niet om ongenoegen over politieke maatregelen kenbaar te maken.

    The "stellingen" is not the forum to air grievances of institutional policy.

Fatigue Crack Growth in Riveted Joints

Fatigue Crack Growth in Riveted Joints

PROEFSCHRIFT

ter verkrijging van de graad van doctor
aan de Technische Universiteit Delft,
op gezag van de Rector Magnificus
Prof. dr. ir. J. Blaauwendraad,
in het openbaar te verdedigen
ten overstaan van een commissie,
door het College van Dekanen aangewezen,
op dinsdag 2 september 1997 te 13.30 uur

door

Scott Anthony FAWAZ

Master of Science in Aeronautical Engineering,
Air Force Institute of Technology, Dayton, Ohio, Verenigde Staten
geboren te Harbor City, Californië, Verenigde Staten

Dit proefschrift is goedgekeurd door de promotoren:
Prof. ir. L. B. Vogelesang
Prof. dr. ir. J. Schijve

Samenstelling promotiecommissie:
Rector Magnificus, voorzitter
Prof. ir. L. B. Vogelesang *(TU Delft, Faculteit der Luchtvaart- en Ruimtevaarttechniek)*
Prof. dr. ir. J. Schijve *(TU Delft, Faculteit der Luchtvaart- en Ruimtevaarttechniek)*
Prof. dr. ir. Th. de Jong *(TU Delft, Faculteit der Luchtvaart- en Ruimtevaarttechniek)*
Prof. dr. ir. H. Tijdeman *(TU Twente, Faculteit Werktuigbouwkunde)*
Dr. ir. A. Vlot *(TU Delft, Faculteit der Luchtvaart- en Ruimtevaarttechniek)*
Dr. J. C. Newman, Jr. *(NASA Langley Research Center)*
Ir. A. U. de Koning *(Nationaal Lucht- en Ruimtevaartlaboratorium, NLR)*

Published and distributed by:

Delft University Press
Mekelweg 4
2628 CD Delft
The Netherlands
Telephone: +31 15 2783254
Fax: +31 15 2781661
E-mail: DUP@DUP.TUDelft.NL

CIP-DATA KONINKLIJKE BIBLIOTHEEK, DEN HAAG

Fawaz, S. A.

Title: subtitle / S. A. Fawaz. – Delft : Delft University Press. – Illustrations.
Thesis Delft University of Technology. – With ref. – With summary in Dutch.
ISBN 90-4-7-1520-3
NUGI 841
Subject Headings: fatigue, riveted joint, fractography, finite element analysis, crack growth predictions

Abstract

The present investigation covers an experimental part and an analytical part based on the fracture mechanics concept. The aim is to extend empirical information on the growth of small and larger cracks in riveted lap splice joints, similar to joints found in operational transport aircraft. Further, it extends the possibilities for fatigue crack growth predictions for small part through cracks to larger through cracks with oblique crack fronts. A characteristic aspect of fatigue of riveted lap joints is the occurrence of crack growth under a complex stress system, which in its simplest form consists of cyclic tension with superimposed cyclic bending due to the eccentricity in the lap joint. In reality, rivet squeezing leads to hole expansion and built-in residual stresses. Rivet tilting in the rivet hole and the contact problems are other complications.

In the empirical part of the investigation a more simple problem was analyzed first, i.e. fatigue crack growth in a multiple-hole sheet specimen loaded under combined tension and bending stress. It turned out that crack growth development for small part-through cracks could be followed by fractographic observations employing marker load cycles in between constant-amplitude loading. The same marking technique was employed for a simple lap joint having two rivet rows with four rivets in each row. The crack growth history could be reconstructed from a crack length of 75 μm to final fracture at 4.5 mm.

In the analytical part, K-solutions are needed. For part through cracks with a quarter elliptical shape the well known Newman-Raju K-solutions are available, which do apply to our open hole specimens. However, after through cracks are obtained they continue to grow with oblique crack fronts due to the combined tension and bending. Since no K-solutions are available for these cracks, the finite-element method and a three dimensional virtual crack closure technique (3D VCCT) were adopted. The 3D VCCT is shown to be invariant to crack plane mesh orientation, which permits a minimal amount of pre-processing of the mesh in order to generate a new K solution. The technique has been verified by comparing its results to K-values of crack shape geometries for which solutions are available in the literature. K-solutions for the crack shapes obtained in the open hole sheet specimen experiments are then calculated and adopted for the prediction of the growth of these cracks. A satisfactory agreement has been obtained. K-values have been calculated for a range of crack depth to crack length ratios (a/c_l = 0.2, 0.3, 0.4, 0.6, 1.0, 2.0), crack depth to sheet thickness ratios (a/t = 1.05, 1.09, 1.13, 1.17, 1.21, 2.0, 5.0, 10.0) and hole radius to sheet thickness ratios (r/t = 0.5, 1.0, 2.0).

The Newman/Raju K-solutions for the part through portion of the fatigue life and the newly calculated K-solutions for the through crack portion have been incorporated into a crack growth prediction scheme. The prediction algorithm not only predicts the fatigue life within 6% of the actual life, but also accurately predicts the crack growth history until just prior to final fracture.

The manufacturing process used in installing the rivet greatly affects the fatigue life behavior of riveted lap joints as shown by Müller. Specifically, the amount of hole expansion, and thus the amount of residual stress around the hole, are proportional to the squeeze force used to expand the non manufactured head. A 3D finite-element model of a three row riveted lap-splice joint is analyzed not only to investigate the change in the stress system at the hole edge due to remote tension and implied secondary bending, but also to determine the local stress field as a result of rivet tilting. Contact elements are used on the faying surface and at contact points between the rivets and adjoining sheets. Initial analyses model a close clearance rivet installation, which does not generate any residual stresses near the rivet holes due to rivet installation. It simulates a rivet driven with a low squeeze force. As indicated by the contact elements, contact is locally lost between the rivet and the bore of the hole (rivet tilting) as the joint is loaded, resulting in a severe stress concentration at the hole edge. On the other hand, for an interference fit rivet installation which does generate residual stresses models a rivet driven with a high squeeze force resulting in the maximum stress being located above the rivet hole. The location of maximum stress for the rivet with and without residual stresses agrees with observations of fatigue crack nucleation at these sites when a low or a high rivet squeeze force is used. The part through cracks introduced in the model shows that ΔK is three times larger for rivets installed with a low squeeze force as compared to a rivet installed with a high squeeze force.

Acknowledgements

One of the characteristics of the Production and Materials Group, Faculty of Aerospace Engineering at TU Delft is that nearly anything is possible. All ideas are given serious thought and consideration whether innovative or mundane. Truly, the limits of what is conceivable lie only in my mind.

I want to give my deepest thanks to my co-promoters, Prof. dr. ir. J. Schijve and Prof. ir. L. B. Vogelesang, whom made it all possible. Prof. Schijve's interest, experience, and ideas made the project continually move forward. I have never met a professor, doctor, or engineer who can explain the most complex technical issues in a manner that a child could understand. In addition to everything I've learned, I hope this is a lesson I can take with me from Delft. The unwavering support given by Prof. Vogelesang was a constant reminder of the team spirit he embodies and promotes in the lab. His sometimes-unbelievable enthusiasm from start to finish made it a pleasure to come to the lab everyday.

A few colleagues deserve a special thanks. Dr. ir. Ad Vlot has always been there for me from my first day in Holland helping with all the moving-in trials and tribulations to giving keen, pointed criticism on the technical issues. Being the first to read my thesis, he has undoubtedly had one of the more difficult tasks. Dr. ir. Müller has been instrumental. His in depth knowledge of the riveting process allowed me to focus my efforts. Without his work, I would have never been able to design test specimens which allowed me to isolate and study the dominant fatigue mechanisms in lap-splice joints. Ir. Jan Hol from the Computational Mechanics Group, Faculty of Aerospace Engineering was always there to help with the computer problems of the day. His knowledge of computers and programming languages kept the K-calculations going.

The technical and support staff in the lab is unequalled. Of all the people in the lab, I have relied most on Frans Oostrum. Without his enduring support (hundreds of hours), doing the fractography work with the scanning electron microscope, the results from the experimental investigations would have been meager. In addition, Frans was responsible for all the optical photography, which made the crack shape measurements possible. Jan Snijder and Berthil Grashof were always there to help with all the fabrication and testing equipment. Their patience in explaining the operating procedures, sometimes repeatedly until I understood, kept me out of the hospital. Additional thanks to Berthil for having the time to troubleshoot my remote computing problems; he always had an idea how to get the job done better and faster. Rob Leonard and Kees Palvast with their expertise in manufacturing methods allowed me to fabricate and test high quality, reproducible test specimens. Special thanks to Hannie van Deventer for helping my family and me adjust to living in a foreign land.

I am grateful for the continual advice and direction of Dr. J. C. Newman, Jr. of the NASA Langley Research Center, USA. His knowledge of computational methods in fracture mechanics, not to mention computer hardware and software resources, made calculating the new K-solutions possible. Ir. Arij de Koning and Carel Lof of the Nationaal Lucht- en Ruimtevaartlaboratorium, NLR were indispensable in the 3D finite element analysis of the lap joint. They have the same "can-do" attitude so appreciated here in Delft. Although my work has focused on monolithic aluminum, Jan Willem Gunnink of Structural Laminates Company made time to share his experience and offer new ideas.

I am also thankful to Colonel Cary Fisher, head of the Engineering Mechanics Department at the United States Air Force Academy for sponsoring my study. Thanks also go to my program managers Lieutenant Colonel Jim Hogan, Major Fernando Conejo, and Major Ralph Tolle for their flexibility and support.

To my officemates, Tjerk de Vries and Pieter-Willem Provo-Kluit, a heartfelt thanks for their help, ideas, and most importantly friendship which has made my stay in Holland perfect.

At last, to Jennifer, Sarah, and Dylan, who endured and made it worth achieving.

Table of Contents

Appendices

List of Figures

Chapter 1

Chapter 2

Chapter 3

Chapter 4

Chapter 5

Chapter 6

Appendices

List of Tables

Chapter 2

Chapter 3

Chapter 4

Chapter 5

Chapter 6

Appendices

List of Symbols

a	Crack depth
A_1	Constants in stress intensity equation
A_i	Constants used in crack opening function
b	½ Width
B	Biaxiality ratio
c	Crack length
C, n	Paris material constants
c_1	Front (faying) surface crack length
c_2	Back (penetrated) surface crack length
C_i	Nodal force weighting functions
COD	Crack opening displacement
da/dN	Crack growth rate
D_o	Rivet diameter before squeezing
E	Modulus of elasticity
f	Prescribed nodal forces
f	Crack opening function
f_w	Finite width correction factor
F_y	Force in the y-direction
G	Strain energy release rate
G_I	Mode I strain energy release rate
h	Height
Hz	Hertz, units of frequency
in	inches
J	J-integral
K	Stress intensity factor
$K(\phi)$	Stress intensity factor as a function of parametric angle ϕ
K(a), K(c)	Stress intensity factor at crack depth and length, respectively
K_{2P}	Stress intensity factor due to wedge loading
K_c	Critical stress intensity factor
K_I	Mode I stress intensity factor
K_{op}	Crack opening stress intensity factor
K_P	Stress intensity factor due to pin loading
K_w	Stress intensity factor due to remote tension
ΔK	Stress intensity factor range
ΔK_{eff}	Effective stress intensity factor range
ΔK_{th}	Threshold stress intensity factor range
kc	kilocycle
kN	kilonewton, units of force
k	Bending factor
K_t	Stress concentration factor
m	meter
mm	millimeter
MPa	Units of stress (force normalized by area)
N	Number of applied load cycles
P	Axial load
p	Rivet row pitch
p, q	Empirically derived material constants
Q	Flaw shape parameter
r	Hole radius
R	Stress ratio
r_D	Maximum value of r used in Force Method

Symbols (cont.)

s	Rivet pitch
S	Global stiffness matrix
t	Sheet thickness
T	Traction
ΔT	Temperature difference
u	Displacement in the x-direction
v	Displacement in the y direction
w	Displacement in the z-direction
w_i	Finite element length along the crack front
W	Width ($W = 2b$)
z_1, z_2	Lower and upper bound of z-distance used in Force Method

Greek Symbols

α	Planar angle used to calculate skew ratio
β	Geometry factor
Δ	Finite element length on each side and normal to the crack front
γ	Load transfer ratio
ϕ	Parametric angle of an ellipse
Φ	Complete elliptical integral of the second kind
π	3.14159
σ	Normal stress
$\sigma_{\theta\theta}$	Normal stress in the tangential direction
$\Delta\sigma_{eff}$	Effective cyclical stress
σ_{brg}	Normal stress due to bearing
σ_o	Remote normal stress
σ_{op}	Crack opening stress
σ_{RSS}	Normal stress due to residual stress system
σ_y	Normal stress in the y-direction
σ_{ys}	Yield stress
θ	Angular measure around a hole
ν	Poisson's ratio

1.

Introduction

Driven by economic pressures, commercial transport aircraft are remaining in service longer than their original design lifetimes. Ensuring the safety of the flying public is of utmost concern and the responsibility of the aircraft manufacturers, airline operators, and airworthiness authorities. The burden cannot be borne by any single organization, but must be shared.[1,2] This magnitude of this burden was thrust to the forefront of the aviation world on 28 April 1988 with the fatigue failure of the Aloha Airlines, Flight 243 Boeing 737-200.[3] The 4.5 m fatigue crack which resulted in roughly 5.5 m of the fuselage to depart in flight was due in part to poor maintenance and cracking at multiple rivet locations in the joint which connects two fuselage skin sheets, known as the longitudinal lap splice joint. The airworthiness authorities in the USA, Federal Aviation Administration, responded to the Aloha incident by bringing together the Air Transport Association (ATA) and Aerospace Industries Association of America (AIA) to form the Airworthiness Assurance Working Group (AAWG).[1] To avoid confusion in the terminology used to describe the various multiple cracking scenarios, the three AAWG definitions, repeated below, are used throughout.

> Widespread Fatigue Damage (WFD): Characterized by the simultaneous presence of cracks at multiple structural details that are of sufficient size and density whereby the structure will no longer meet its damage tolerance requirements; for example, not maintaining required residual strength after partial structural failure.[1]
>
> Multiple Element Damage (MED): A type of WFD characterized by the simultaneous presence of fatigue cracks in similar adjacent structural elements.[1]
>
> Multiple Site Damage (MSD): Characterized by the simultaneous presence of fatigue cracks in the same structural element; for example, fatigue cracks that may coalesce with or without other damage leading to a loss of required residual strength.[1]

The above list is ordered in decreasing effected area; for example, consider a longitudinal lap-splice joint which joins two skin panels along the entire length of the fuselage. WFD describes a state of generalized cracking in a large portion or maybe the entire joint. MED depicts cracking in two or more adjacent bays, and MSD characterizes cracking in only one fuselage bay. MED and WFD can then be thought of as MSD on larger scale. Since MED and WFD are more extreme cases of MSD, this study is limited to the investigation of MSD in that MSD must first be understood and predictable in order to consider the more complex problems of MED and WFD.

Not all of the airframe structure is susceptible to MSD. For MSD to occur, the structural details must be geometrically similar and experience comparable stress levels to maintain a similar crack driving force at each of the probable crack nucleation sites. If the crack driving force is significantly larger for one location in comparison to another, crack nucleation and growth will occur first where the crack driving force is highest. Another product of the AAWG via the Structural Audit and Evaluation Task Group (SAETG) was the airframe manufacturers committee that has identified fuselage structure potentially susceptible to WFD and listed below.[4]

Longitudinal skin joints, frames, and tear straps
Circumferential joints
Aft pressure dome outer ring and dome web splices
Other pressure bulkhead attachment to skin and web attachment
to stiffener and pressure decks
Stringer to frame attachments
Window and door (cutouts) surrounding structures
Over wing fuselage attachments
Latches and hinges of non-plug doors
Skin at runouts of large doublers

Although MSD has been reported in non-fuselage structure; for example, the Lockheed C-5A Galaxy wing or Boeing 707 empennage (Lusaka accident[5]), the fuselage longitudinal lap-splice joints are the prime focus in this effort. For non-fuselage structure susceptible to MSD, see reference [4].

In view of the definitions above, the Aloha case can be classified as MED since the crack grew through nearly nine fuselage bays where each bay is bordered by two circumferential stiffening elements, frames, of approximately ½ meter in width. The Aloha accident is the only catastrophic failure resulting from cracking at multiple rivet holes in the longitudinal lap splice; however, similar cases have been observed in other aircraft.[2,5-11]

As seen by the rather large list of MSD susceptible fuselage elements listed above, much work still needs to be done. This study will only focus on developing a prediction model for fatigue crack growth in the longitudinal lap-splice joint. Prediction methods play an integral role in evaluating the damage tolerance of the aircraft in that they are used to specify inspection intervals for timely fatigue crack detection and to asses the residual strength reduction in the presence of fatigue cracks. Although important, the residual strength concerns are not within the scope of this research. The value of a crack growth prediction model is assessed in a broad sense on whether it increases the safety and decreases the maintenance costs of both old aircraft in service and new aircraft still on the drawing board. A brief review of how the damage tolerance approach is implemented, as outlined in the Damage Tolerance Requirements, DTR, contained in the Federal Aviation Regulation 25 (FAR/AC 25.571), in the civil aircraft fleets illustrates how the crack growth prediction models are used. Figure 1.1 shows a generic crack growth curve where the crack grows from detectable, a_d, to permissible, a_p, to critical, a_c. The DTR specifies that the residual strength cannot fall below a permissible level

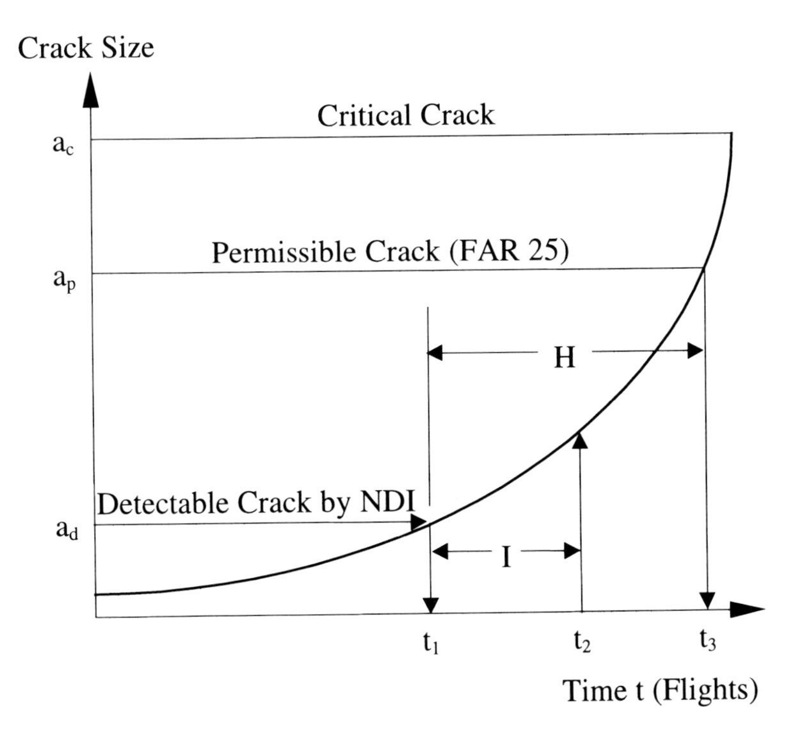

Figure 1.1 Crack Growth Curve[12]

and the crack associated with the permissible residual strength must be detected. The detectable crack size is not constant but depends on the method used for detection; visual, eddy current, ultrasonic, dye penetrant, or X-ray. The permissible crack size is then the largest crack sustainable while maintaining the minimum residual strength as outlined by the DTR. Beyond the permissible crack size, normal operational loading can cause the crack to grow to a critical size resulting in failure of the structural element and possibly catastrophic failure of the aircraft. Assuming the crack is just below the detectable size at t_1 shown in Figure 1.1, and the permissible crack size is reached at t_3 after H flights, inspections must be made at H/2 intervals in order to find and repair the crack. By setting the inspection interval, I, to H/2, the operator has two opportunities to detect and repair the crack before the critical crack size is reached. The designers know the structural details and operational stress levels; therefore, crack growth curves similar to Figure 1.1 must be generated for all fatigue critical locations.

For the operator responsible for an aging fleet, knowing when the MSD cracks may be visible allows more flexibility in scheduling inspections. Also, with the growing trend of flying airplanes beyond their design lifetime, structural life extension programs may be implemented to further extend the aircraft life. In such a scenario, the current "health" of the fuselage can be assessed with a crack growth model that may give cause to modifications or even reskinning the fuselage. For the new aircraft, the crack growth prediction model can be used for damage tolerance assessment and durability analysis of competing designs. For example, assuming two separate joint designs both satisfy the damage tolerance requirements, one joint may be the better design since it is easier to inspect or more damage tolerant.

In general, the crack growth prediction model is just another tool in the designer and maintainer's toolbox to safely and efficiently manufacture and operate the aircraft. Although conservative, the available crack growth prediction models do not accurately represent the MSD cracking scenario for reasons discussed in detail in Chapters 2 and 3. Even though the complex geometry and loading found in a lap-splice joint implies that a MSD crack growth prediction model cannot be developed in the near term, much can be learned about the cracking behavior in lap splice joints. There are several elementary examples:

- Can a MSD cracking scenario be investigated in the laboratory with coupon sized test specimens?
- Why do some cracks nucleate at the top of the rivet hole and others at the hole edge in the net section of the joint?
- What role does rivet tilting play?
- Does the stress state vary between rivets in the same row?
- What are the crack growth rates of MSD cracks?
- Is small crack growth data obtainable?
- Can the crack shape be represented in a known geometric functional form?
- What effect does the local stress state have on the crack shape?
- Is the crack shape changing during its life?
- Does the crack shape affect the fatigue life?
- Are the existing crack growth models sufficient?
- Can the existing crack growth models be improved?

In trying to answer the questions posed above, two main thrusts are pursued. One is the development of stress intensity factor solutions, with the terms K's or SIF's used interchangeably; second, fractographic observations of the fracture surfaces. Before embarking on developing a new prediction algorithm, the existing prediction models and experimental results available in the literature are reviewed in Chapter 2. The experimental program is discussed in Chapter 3 with presentation of the various experimental procedures and results. From the fractography, the crack geometry is discovered for which K solutions are calculated and presented in Chapter 4. The K solutions are then incorporated in a fatigue crack growth, FCG, prediction algorithm described in Chapter 5. No new FCG laws are developed only implementation of the K solutions based on the fractographic findings. In Chapter 6, development and results from a three-dimensional finite element model, FEM, of a lap-splice joint are presented in order to elucidate the more dominant geometry and loading conditions affecting crack growth. Chapter 7 is the final chapter where conclusions of the research and recommendations for continued work are outlined.

[1] Swift, T. WideSpread Fatigue Damage Monitoring-Issues and Concerns, Proc. of 5th International Conference on Aging Aircraft. 16-18 June 1993, Hamburg, Germany.

[2] Goranson, Ulf G. and M. Miller. Aging Jet Transport Structural Evaluation Programs, Proc. of the 15th Symposium of the International Committee on Aeronautical Fatigue, 21-23 June1989, Jerusalem, Isr. West Midlands, UK: EMAS, 1989.

[3] Aircraft Accident Report: Aloha Airlines, Flight 243, Boeing 737-200, N73711, near Maui, Hawaii, April 28, 1988, NTSB/AAR-89/03. Washington DC: U.S. National Transportation Safety Board, 1989.

[4] Goranson, Ulf G. Damage Tolerance Facts and Fiction, 14th Plantema Memorial Lecture. Proc. of the 17th Symposium of the International Committee on Aeronautical Fatigue, 9-11 June1993, Stockholm, Swed. West Midlands, UK: EMAS, 1993.

[5] Wilkenson, G. C. Report on Accident near Lusaka International Airport, Zambia, on 14 May 1977 Boeing 707 321C G-BEBP. Aircraft Accident Report 9/78, Dept. of Trade, Accident Investigation Branch, 1979.

[6] Schijve, J. Comments on the Problem of Multiple-Site-Damage (MSD), B2-91-10. Delft, NL: Faculty of Aerospace Engineering, Delft University of Technology, 1991.

[7] Shiohara, T., M. Nakata, and H. Kumada. Aging Review of the YS-11 Aircraft. Proc. of the 15th Symposium of the International Committee on Aeronautical Fatigue, 21-23 June1989, Jerusalem, Isr. West Midlands, UK: EMAS, 1989.

[8] Nakata, Masahiko, Toshihiko Nishimura, and Kenji Inaba. Damage Toelrance Assessment on the Multi-Site Cracks for the YS-11 Aircraft. Proc. of the 15th Symposium of the International Committee on Aeronautical Fatigue, 21-23 June1989, Jerusalem, Isr. West Midlands, UK: EMAS, 1989.

[9] Shaw, W. J. D. Effects of Prior Fatigue Damage on Crack Propagation Rates in 2024-T351 Aluminum Alloy. Proc. of the 15th Symposium of the International Committee on Aeronautical Fatigue, 21-23 June1989, Jerusalem, Isr. West Midlands, UK: EMAS, 1989.

[10] De la Motte, Eddy B. And Capt. Frank Opalski. MSD: Where We Are and Where We Should Go. Proc. of the 1992 USAF Structural Integrity Program Conference, 30 Nov – 2 Dec 1993, San Antonio, TX, WL-TR-94-4079.

[11] Mayville, R. A. and T. J. Warren. "A Laboratory Study of Fracture in the Presence of Lap Splice Multiple Site Damage," Eds. S. N. Atluri, S. G. Sampath, and P. Tong. Structural Integrity of Aging Airplanes, Springer Series in Computational Mechanics. Berlin: Springer Verlag, 1991.

[12] Broek, D. "The Civil Damage Tolerance Requirements in Theory and Practice." Eds. S. N. Atluri, S. G. Sampath, and P. Tong. Structural Integrity of Aging Airplanes, Springer Series in Computational Mechanics. Berlin: Springer Verlag, 1991.

2.

Background

2.1 Introduction

The fatigue behavior of riveted connections has been investigated over the last half century, although much progress has been made, questions still remain. The distressing Aloha accident has rekindled the interest of fatigue crack growth in riveted joints. The current practice used to ensure flight safety of an aircraft is based on both analytical and experimental methods which are closely linked. In general terms, analysis is used for design and experiments for verifying the design. From a safety point of view, the most important phase of the aircraft development cycle is the full scale static and fatigue tests where the entire, or at least a large portion of the aircraft, is tested to ensure static strength and damage tolerance under the anticipated operational loading. Static or fatigue failures at this stage in development are quite costly to correct; thus, good analytical tools validated by experimental data are a must. The ensuing discussion serves to review and summarize the existing experimental and analytical data related to MSD. The fatigue specimens most widely used can be separated into three groups; open hole, flat unstiffened/stiffened lap-splice joints, and curved stiffened lap-splice joints. Finally, the state of the art analytical tools are surveyed which in turn provides a starting point for the current investigation.

The recent catastrophic fatigue failures in the presence of multiple site damage has spawned a tremendous research effort directed toward an increased understanding of the cause and behavior of multiple site damage. Typical longitudinal lap-splice joints in pressured fuselages, shown in Figure 2.1, are complex in design, manufacture, and fatigue loading conditions. As a result, many researchers have adopted a "bottom-up" research philosophy where the complex structure is decomposed into simpler component parts in an attempt to isolate a particular parameter which may contribute to the MSD situation.

Once the behavior of the simple structure is understood, the structural realism (complexity) can be increased up to full scale testing. A brief description of a generic longitudinal lap-splice joint, shown in Figure 2.2 will aid in the ensuing survey of the existing research results which may be pertinent in developing a fatigue crack growth prediction model.

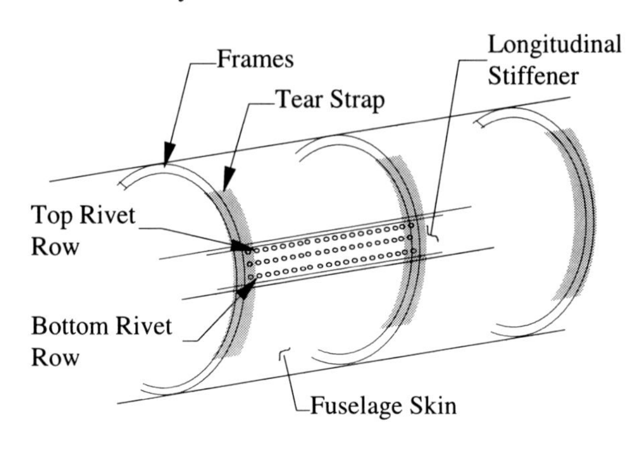

Figure 2.1 Fuselage Longitudinal Lap-Splice Joint

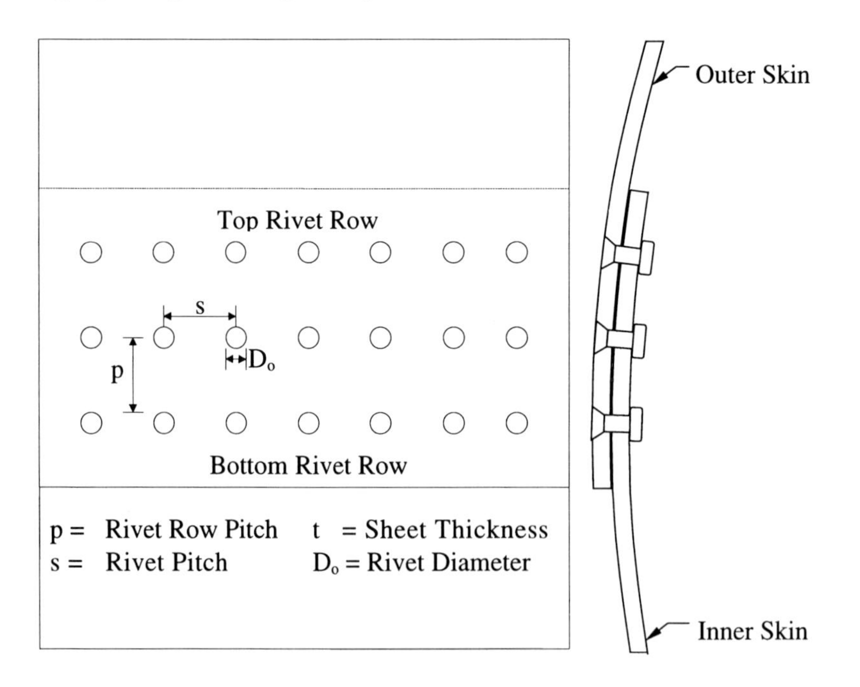

Figure 2.2 Generic Longitudinal Lap Splice Joint

Figure 2.2 could easily suggest that fatigue of a longitudinal lap-splice joint could be simulated in a fatigue test on a simple flat lap joint loaded uniaxially in a testing machine. However, there are some obvious differences:

(i.) In the fuselage, the loading is biaxial.
(ii.) The fuselage skin is curved and loaded by internal pressure.
(iii.) In the fuselage, the hoop stress is not uniformly distributed between the frames. In general, the hoop stress is larger between the frames and approximately constant over a significant number of rivets, see Figure 2.3.

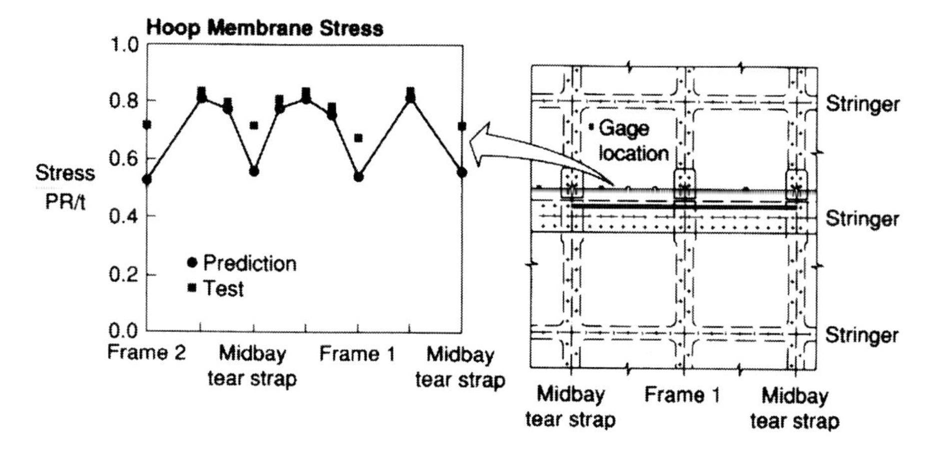

Figure 2.3 Hoop Stress Distribution between Frames[1]

In a multiple row riveted lap joint, the upper row is fatigue critical in the outside skin because of pin loading on the holes and the large bypass load. Moreover, here the fatigue cracks start at countersunk holes. The bottom row is critical in the inner skin, however at the noncountersunk holes. It thus might well be expected that the top row would be more critical than the bottom row. This has been confirmed by some service experience, but in some aircraft the bottom row turned out to be more critical.

Before the MSD phenomenon was recognized, simple lap joint specimens were frequently adopted to obtain S-N curves in order to help the designer assess the allowable stress levels. However, since MSD has become a relevant issue,

Figure 2.4 Multiple Crack Nuclei in the Critical Row of a Lap-Splice Joint

fatigue crack growth in lap joints became of great interest. Also, fractography became highly opportune. Fractographic observations have shown multiple crack nuclei in the critical rivet row after specimen failure as seen in Figure 2.4. In the MSD situation, mutual interactions between cracks and crack coalescence should also be considered. Further, really visible crack growth is observed only during a small fraction of the fatigue life close to final fracture. Thus, there is a case for trying to predict crack growth, especially for small cracks that may still be part through cracks.

The literature then shows that various investigators have tried to investigate and predict multiple crack growth in simple flat specimens, which is discussed in section 2.2.1. At the same time, other investigators have tried to make crack growth observations in riveted lap joints, which is discussed in section 2.2.2. Considerations on different types of specimens, varying from rather simple sheet specimens to more realistic full-scale testing conditions are dealt with in section 2.2.3.

In the early nineties until now, considerable research work on riveted lap joints was carried out in the Structure and Materials Laboratory of the Faculty of Aerospace Engineering of the Delft University of Technology. It turned out that the rivet squeeze force had a large effect on the fatigue behavior of a lap joint. Secondly, it was also found that finite width effects could offer a problem when testing wide panels. These aspects are summarized in section 2.3.

The understanding gained made it clear that an analytical analysis of the fatigue behavior of a riveted lap joint offers considerable problems, amongst them the occurrence of secondary bending due to the eccentricity occurring in a lap joint and the slowly growing small part-through cracks. Analytical aspects and existing prediction models are covered in sections 2.4 and 2.5 in order to define the approach adopted in the present investigation which is discussed in section 2.6.

2.2 Multiple Site Damage Experiments: Observations and Effects

The test results available in the literature, open hole flat coupon, lap-splice joint coupon, flat stiffened panel with lap-splice joint, and curved stiffened panel with lap-splice joint, are discussed in order of increasing structural complexity.

2.2.1 Open Hole MSD Testing

Being the simplest of MSD specimens, many authors have used coupon sized flat aluminum alloy sheet specimens with varying number of holes and cracking scenarios as shown in Table 2.1. The aim of such tests was to primarily determine to what extent a crack could affect the crack growth of adjacent cracks, known as crack interaction. Investigations focusing on determining the residual strength in the presence of MSD have been excluded.

Table 2.1 Open Hole Multiple Site Damage Test Specimen Literature Survey

Researcher	Length (mm)	Width (mm)	Thickness (mm)	Hole Diameter (mm)	Number of Holes	R Ratio	σ_{max} (MPa)
‡ Kobayashi/Shimokawa, 1989[2]	500	150	4.6	6.4	5	?	?
Nishimura, 1990[3]	500	250	4.6	6.4	5	0.1	117.7
‡ Kobayashi/Shimokawa, 1991[4]	750	300	1.0	4.0	3	0.125	96.0
Lehrke/Schöpfel, 1991[5]	-	160	5	?	3	0.1	80
Pártl/Schijve, 1992[6]	1000	400	2.0	4.0	9	0.03	67.0
Dawicke/Newman, 1992[7]	762	304.8	2.3	3.8	10	0.0	71.0
Moukawsher et al., 1992[8-10]	685.8	228.6	2.3	4.0	8 - 11	0.01	40.8 - 82.0
Vermeeren, 1993[11]	300	100	0.8	25.0	1	0.05	120
Brot/Nathan, 1993[12]	300	100	2.5	5.0	6	0.05	96.5
Buhler et al., 1994[13,14]	685.8	228.6	2.3	4.0	8 - 11	0.01	70.0 - 103.4
Rohrbaugh et al., 1994[15]	-	787	2.3	4.8	30	0.1	68.9

? Not Reported

‡ Unpublished, Mitsubishi Heavy Industries, Jap.

All researchers conducted the MSD testing under pure constant amplitude tensile loading with a stress ratio, $R \approx 0$, in view of cabin pressurization being the dominant load condition for a fuselage longitudinal lap-splice joint. The internal cabin pressure is zero when the plane is on the ground and maximum at cruising altitude then returning to zero pressure upon landing. This loading cycle is known as the ground-air-ground, GAG, cycle. The aim of researcher listed in Table 2.1 was to validate crack growth prediction algorithms, except Vermeeren, who fatigued the specimen to obtain naturally occurring cracks before determining the residual strength. All tests showed a definite crack interaction behavior when the cracks are close to one another affecting the crack growth rate and fatigue life. Conversely, when the cracks are not close to one another, the crack growth rate is low resulting in a small effect on the total crack growth life. Pártl and Schijve found the interaction effects are more pronounced when the cracks are closer than ½ the distance between adjacent holes with little effect on the fatigue life. In addition, they discovered the initial locations of the MSD cracks could have an effect on the crack growth later in the crack growth life. A lead crack, either nucleating naturally or simulated, is larger than the MSD cracks. Lehrke and Schöpfel, Dawicke and Newman, Moukawsher et al., Brot and Nathan, and Buhler et al., all noticed crack interaction late in the fatigue life with the size and location of a lead crack dominating the growth of the MSD cracks. The scenario of a lead crack with multiple MSD cracks is one that has been found in service[16] and full scale fatigue test articles;[17] thus receiving ample consideration in the literature.

2.2.2 Flat Stiffened/Unstiffened Lap-Splice Joint MSD Testing

The flat panel, open hole MSD tests reported above, does not consider the bending stress always present in a lap-splice joint. Specifically, as an aircraft is pressurized the in-plane forces resolved from the hoop stress apply an eccentric load in the lap joint. This eccentricity creates a bending moment, usually called secondary bending, and thus creates normal stresses due to (secondary) bending. Depending on the remotely applied stress, the secondary bending is a nonlinear phenomenon because of the out of plane displacements as a result of the eccentric joint geometry. Müller found that the bending stresses can be as large or larger than the normal stress due to the in-plane loading.[18] For brevity, herein the normal stress due to bending and normal stress due to tension are simply referred to as bending and tensile stress, respectively. To investigate the behavior of MSD cracks under combined loading, many researchers have used coupon sized, flat, unstiffened lap-splice joints that are discussed below.

Studying and predicting crack growth in structure subject to combined loading is always an arduous task not only due to the complex test fixtures that are usually required, but also due to the difficulty in quantifying the individual contributions of each load condition; e.g. remote biaxial tension, remote bending, pin loading. In attempt to mitigate the complexity inherent in lap-splice joints, a possible solution is to simplify the joint geometry by using only one rivet. Obviously such a specimen is unrealistic. The load transmission through the rivet is 100% and there is no bypass load. Secondary bending and rivet tilting will be high resulting in a low fatigue resistance. A single rivet lap joint cannot represent an MSD situation; however, this joint geometry has proved useful to study the effects of rivet interference[19] and fretting in a riveted connection.[20,21]

Another attempt at simplifying the lap-splice joint is the 1½ dog-bone specimen initially developed to rate fastener systems during an Advisor Group for Aerospace Research and Development (AGARD) round robin experimental investigation.[22,23] The 1½ dog-bone joint, as shown in Figure 2.5, does lessen the secondary bending as compared to the single rivet lap-splice joint; however, this joint is still inadequate for fatigue crack growth testing according to Palmberg, Segerfröjd, Wang, and Blom who used this joint extensively to investigate crack growth in riveted joints. The problem they could not overcome was crack nucleation and growth from the top of the rivet hole, not at the rivet centerline, i.e. sheet net section, as seen in fuselage lap-splice joints.[24] A detailed discussion of cracking at the top of the rivet hole in 1½ dog bone specimens is presented in section 3.3.3.1.

Efforts in trying to reduce the complexity of the lap-splice joint have had little success. The simplest joint which behaves in a manner similar to fuselage lap-splice joints; i.e., cracking at plate net section and slight rivet tilting, is the asymmetric lap joint shown in Figure 2.6. Leven completed fatigue tests with this specimen geometry and varied the orientation of the rivets, as shown in Figure 2.6 cases A - C. He showed the rivet orientation has a minor effect on both the total stress in the joint and fatigue life for the rivet pitch to rivet diameter (s/D_o) ratios of 3, 4, 6, and 8.[25]

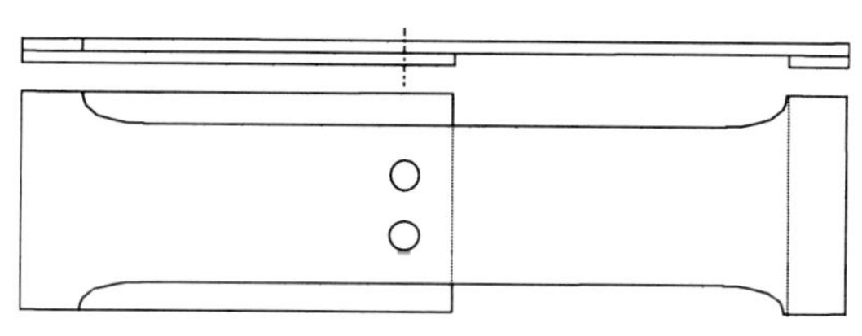

Figure 2.5 1½ Dog Bone Specimen

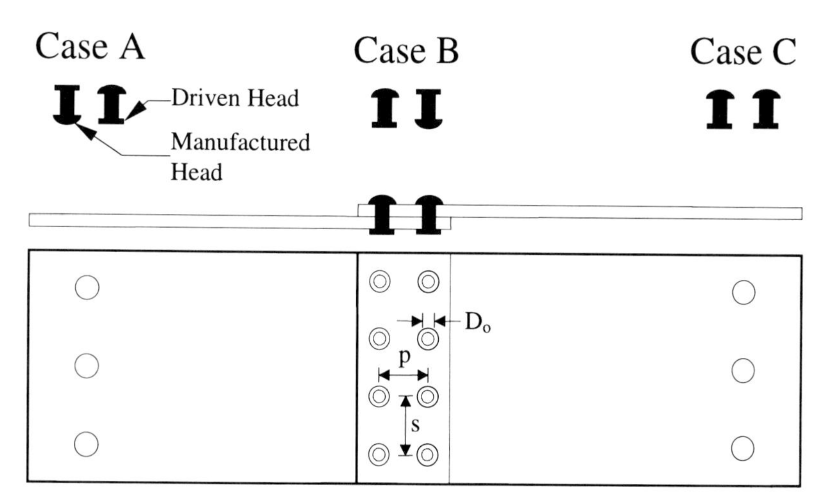

Figure 2.6 Asymmetric Lap-Splice Joint

Investigation of fatigue crack growth in the presence of MSD in flat stiffened or unstiffened lap-splice joints, shown in Table 2.2, has not received as much attention as the open hole MSD research discussed early. This is most likely due to the difficulty in collecting crack growth data. Also, the most immediate concern with regard to fleet safety is to what extent the residual strength of the fuselage has been reduced due to the MSD.

Table 2.2 Flat Stiffened or Unstiffened Lap-Splice Joint Multiple Site Damage Test Specimen Literature Survey

Researcher	Width (mm)	Thickness (mm)	Rivet Diameter (mm)	Number of Rivets/Row (Rivet Rows)	Rivet Pitch (mm)	Rivet Row Pitch (mm)	R Ratio	σ_{max} (MPa)
Hartman et al., 1962[26]	60	0.6	2.4	6 (2)	10.0	7.5	0.02-0.5	118 - 173
Mayville/Warren, 1991[27]	200	1.0	4.0	7 (3)	25.4	25.4	0.1	103
Pelloux et al., 1991[28]	76	1.0	4.0	3(2)	25.4	25.4	0.1	100
Soetikno, 1992[29]	168	1.6	4.8	7 (3)	24.0	24.0	0.1	100,120,150
Wit, 1992[30]	125, 500	1.2	4.0	5 (3), 20 (3)	25.0	25.0	0.05	105
Molent/Jones, 1993[31]	203	1.0	4.0	8 (3)	25.4	25.4	0.05	194
Vlieger, 1994[32]	304.8	1.27	4.0	8 (3)	19, 25.4	25.4	0.1	97, 110
Zhuang et al., 1995[33]	200	1.0	5.0	12 (1)	25.4	‡	0.25	120
Ottens, 1995[34]	20, 40	1.2	3.2	1 (3)	N/A	20.0	0.1	79
Ottens, 1995[34]	140	1.2	3.2	7 (3)	20.0	20.0	0.1	79
Müller[18]	500	1.0	4.0	25 (3)	20.0	20.0	0.05	97

‡ Unique MSD lap joint test fixture.

Hartman, Jacobs, and de Rijk tested two other joint geometries in addition to the one listed in Table 2.2; the dimensions of which were determined by scaling the geometry by 1.66 and 2.66. Although predating the acknowledgment of MSD, the study by Hartman et al. provided verification that the global dimensions of the lap joints with similar geometric ratios does not effect the fatigue life.[26] In other words, they give an initial indication that specimen size effects could be of minor importance. Müller would later show, discussed in section 2.3.2, specimen size could affect fatigue bahavior.

The work of Mayville and Warren concentrated mainly on determining residual strength and developing a link-up criteria; however, they completed one fatigue test where a proof load, 33% larger than the fatigue load, was applied in an attempt to retard or arrest the crack growth.[27] The proof load concept was unsuccessful.

The only researchers to focus on fractographic investigations were Pelloux, Warren, and O'Grady. Their purpose was to determine MSD crack growth rates by fractographic analysis in laboratory specimens and compare them to those seen in-service. Boeing[35] reported rates ranging from 0.13 – 0.5 μm/cycle depending on the airplane and service conditions. Pelloux et al. achieved these rates with their simplified two row, three rivets per row lap-splice joint whose

MSD crack growth rates were 0.2 – 0.5 μm/cycle.[28] They also observed nearly constant crack growth rates through ¼ of the rivet pitch (approximately 6 mm).

Soetikno primarily investigated the fatigue of naturally occurring cracks in lap joints made with fiber metal laminates and used monolithic aluminum, 2024-T3, for comparison.[29] The 2024-T3 lap joint developed cracks in the upper rivet row that began at the faying surface where the tensile stress is maximum and penetrated the free surface to continue growing as a through crack with an oblique front, see Figure 2.7.

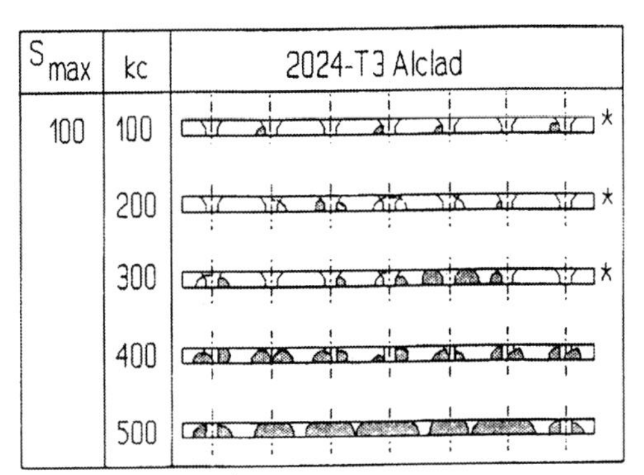

Figure 2.7 MSD in 2024-T3 Alclad 3 Rivet Row Lap Splice Joint

In Figure 2.7, S_{max} is the maximum remote tensile stress and kc is the number of fatigue kilocycles. Note the MSD development commonly seen in lap joints with a uniform stress distribution through the specimen width. Also, the cracks nucleate and grow as part through cracks and maintain a part elliptical crack front throughout their life.

The effect of production processes was reported by Wit who observed a factor of three increase in fatigue life by increasing the expansion of the driven head. The importance of detection of an MSD scenario was also emphasized by Wit since he found that 85% of the fatigue life was spent in crack growth to a visible crack and only 1.3% of the fatigue life remained after the first crack link-up.[30] Upper rivet row cracking was also reported by Molent and Jones. In addition, they found cracks always initiated from the knife-edge condition, if the knife edge condition existed, at the countersunk hole.[31] Although dependent on the joint geometry and stress level, Molent and Jones also noticed the ligament length just prior to link-up was less than 1% of the rivet pitch, 2.0 mm for their joint design.

Comparing uniaxial and biaxial fatigue test results, Vlieger saw lower fatigue lives and crack initiation earlier in the fatigue life for the uniaxial tests. One other difference between the uniaxial and biaxial test results to note is cracking in the uniaxial tests generally was widespread, cracks occurring at many rivets and these cracks did not link-up; whereas, the biaxial tests had cracking from rivets in close proximity to one another which grew simultaneously and linked-up without immediately leading to panel failure.[32] Since fuselage longitudinal lap-splice joints are subject to biaxial loading resulting from cabin pressurization and aerodynamic loading, the task of crack detection might be expedited if the inspector knows to look for groups as opposed to single small cracks.

The local effect of rivet installation was considered by Zhuang, Baird, and Williamson who examined the effect of rivet clamping force on fatigue life.[33] With a specially designed clamping force sensor, they varied the clamping force from 80.0 - 170.0 N and found that as the clamping force increased, so did the fatigue life. Unfortunately, the clamping force could not be correlated to the amount of force used to install a rivet.

A comprehensive experimental effort investigating crack initiation and growth has been completed by Ottens. He attempted to mark the fracture surface by using high stress ratio, R, and overloads, neither of which marked the fracture surface for cracks smaller than 1.0 mm.[34] For the larger cracks, specifically through cracks, the marker bands were visible. Marking the fracture surface is a useful tool in determining crack growth rates for small cracks in addition to reconstructing a crack growth history for cracks that are small and/or difficult to macroscopically measure. A new fracture surface marking technique is introduced in Chapter 3. Ottens also noticed that 30% of the fatigue life is spent nucleating and growing a crack to 0.1 mm, the next 60% of the life is consumed growing the crack through the sheet thickness, and only 10% of the life remains once the cracks are visible.[34] Lastly, he found that the critical row was the bottom row in the formed head side of the sheet. Of the 22 joints Vlieger tested, only one failed as a result of crack growth in the formed head side of the sheet. According to Müller, crack growth in the formed head side of the sheet is related to the force used to install the rivet and will be discussed in more detail later.[18] Supporting similar observations by Wit, Vlieger, and Ottens, Müller found that 85 - 90% of the fatigue life is spent in growing a crack to a visible size, 1 - 2 mm. Concurring with Vlieger's uniaxially loaded narrow panel tests, Müller's wide panel testing also showed cracking at all rivet holes with no grouping of crack locations as Vlieger found in his biaxial panel tests. Müller accomplished the most exhaustive lap joint test matrix in the open literature with the aim of determining which lap joint design and manufacturing parameters govern fatigue performance. Those parameters pertinent to the present investigation are discussed later.

2.2.3 Curved Stiffened Lap-Splice Joint MSD Testing

The most representative test article for MSD short of full scale testing, is a portion of the full-scale fuselage panel. Production and testing costs have limited the number of tests conducted as evident in Table 2.3. Although the list in Table 2.3 is sparse, much of the test data is most likely not reported in the open literature and has only received heightened interest since the Aloha incident in 1988.

Table 2.3 Curved Stiffened Lap-Splice Joint Multiple Site Damage Test Specimen Literature Survey

Researcher	Thickness (mm)	Fuselage Body Type	R Ratio	Maximum Pressure (MPa)
Samavedam et al., 1994[36]	0.9	Narrow	0.105	0.066
Miller et al., 1994[1]	1.0	Narrow	0.0	0.059
Miller et al., 1994[1]	1.6	Wide	0.0	0.059
Gopinath, 1992[37]	1.6	Wide	0.0	0.062

All the test articles listed in Table 2.3 are composed of a fuselage longitudinal lap-splice joint, tear straps, stringers, and frames and are loaded by internal pressure. Samavedam, Thomson, and Jeong analyzed the panel tests performed by Foster-Miller, Inc. (FMI) and found cracking in the upper rivet row at 49% of the fatigue life, much sooner than the flat unstiffened lap joints tests discussed earlier would predict. In 1989 and 1990, the Boeing Commercial Airplane Group built two large pressure test fixtures, one representing a narrow body and the other a wide body.[38] Four narrow and two wide body tests were completed with simulated MSD. Since the cracks originated from saw cuts, the number of cycles until the first detectable crack is unknown. Two of the more notable results from the narrow body tests are that the MSD cracks ahead of a lead crack forces the lead crack to stay in the upper rivet row for a longer period of time as opposed to the crack growing outside of the lap-splice joint in the fuselage skin. Secondly, the panel behavior is independent of the attachment method of the tear straps to the skin.[1] In the wide body tests, the cracks generally grow in the longitudinal direction and didn't curve to continuing growing in the circumferential direction as the narrow body panels did in 2 of the 4 tests.[1] Several other MSD scenarios are presented in reference [1] primarily dealing with residual strength concerns. Full-scale fuselage fatigue test results of the Boeing 747 were reported in reference [37] with non-destructive and destructive tear down inspections conducted by Piascik, Willard and Miller.[39] The first cracks in bay #1 of the four bay test article were observed in the upper rivet row at 38,333 cycles with the first link-up at 58,200 cycles. The crack arrested at the tear straps bordering bay #1 with a final length of 479 mm at the end of the 60,000 cycle test. Interestingly, the first cracks were detected at 64% of the fatigue life, assuming there is little life remaining once the crack has grown to the tear straps, which is later by 15% than the panel test conducted by FMI.

2.3 Multiple Site Damage Fatigue Testing

Described in the preceding sections and illustrated in Figure 2.8 is a common approach in investigating complex structures beginning by testing simple coupon size specimens which are derivatives of the complex structure which can then be modified in each subsequent test series to better represent the complex structure. Not only is the structural configuration simplified in this approach, but also the loading conditions. For the single open hole specimen, the pure tensile loading results in a uniform stress distribution through the sheet thickness and width; however, for all of the joints, secondary bending and load transfer through the rivets must also be considered. Although the bending stress distribution through the thickness of the sheet is assumed to be varying linearly from a maximum at the faying surface to minimum at the free surface; Müller showed the secondary bending is not uniform through the width of the specimen.[18] The asymmetric lap-splice joint with two rivet rows is the simplest joint since the amount of load transfer through each rivet row is known, 50%. For the three rivet row joints, the load transfer through each row is a function of the joint geometry and rivet flexibility (the latter is related to the rivet squeeze force and discussed in the following section). The loading complexity is further increased by the frames and tear straps in a fuselage which locally reduce the hoop stress in the skin as shown in Figure 2.9. In designing test specimens representative of a stiffened structure, care must be taken since the stiffening element design greatly influences the amount of local stress reduction. Furthermore, the manner in which the stiffening elements interact with the sheet material is also important. For example, although the stiffened flat lap-splice joint in Figure 2.8 appears to represent one bay of a fuselage, the simulated frames at the specimen edges locally reduces the stress in the sheet by reducing the secondary bending. However, in a fuselage, the frames decrease the hoop stress by restricting radial expansion of the pressure cabin. From this brief discussion on the complex nature of a pressurized fuselage longitudinal lap-splice joint, the simplification of structural complexity and reduction of loading conditions is paramount to this investigation.

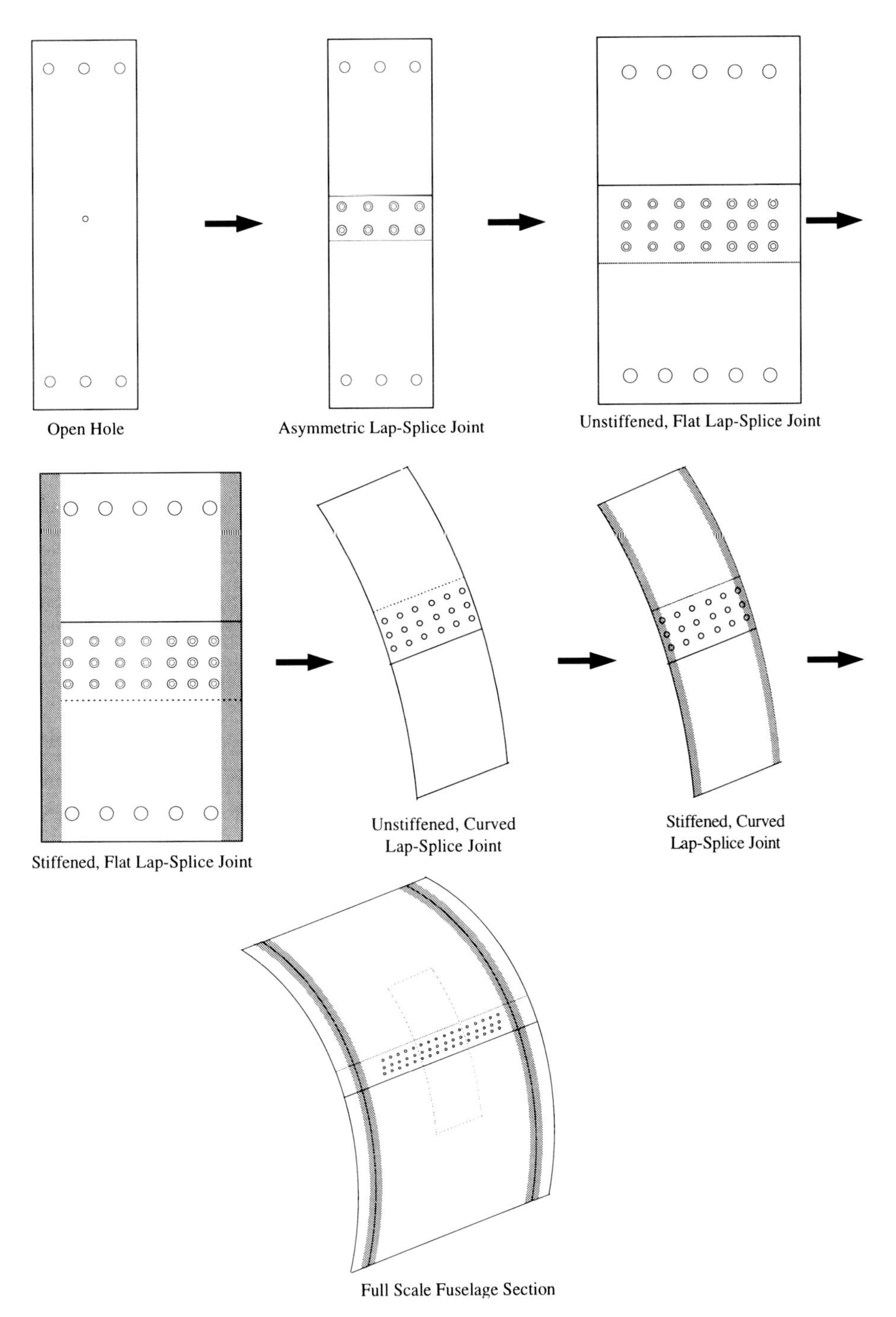

Figure 2.8 Derivatives of Longitudinal Lap-Splice Joint in a Pressurized Fuselage

Except for the single open hole specimen, the riveted joint is the primary focus of all the joints shown in Figure 2.8. Thus, if the riveted joint is viewed as the

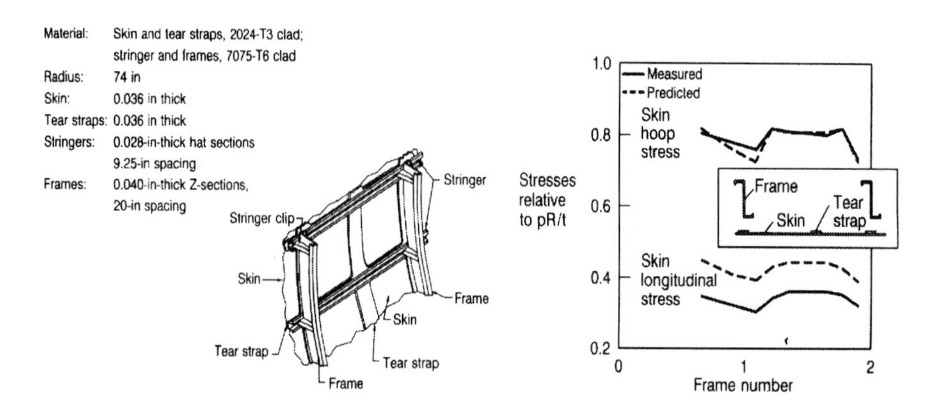

Figure 2.9 Stress Distribution Variation in a Fuselage Bay[42]

complex structure in question, it can also be decomposed into two constituents, the sheet material and rivet. The most prevalent fuselage skin material in use today is aluminum 2024-T3; however, there is a wide range of rivets being used, 2117-T3 (AD), 2017-T31, 2024-T31 (DD), 7050-T73, 7050-T731 (E), and 2219-T81 (KE) according to reference [41]. Aluminum 2024-T3 has been extensively static and fatigue tested and in general has a well characterized and widely documented mechanical and fatigue behavior. The same cannot be said for the aforementioned rivets. The performance of the rivet is directly related to the squeeze force used for it's installation. Müller showed the amount of hole expansion is not only related to the yield stress and work hardening characteristics of the rivet material, but more important a function of the rivet squeeze force.[18] Also affected by the rivet squeeze force is the degree of rivet tilting which has yet to be fully characterized. Since the rivet squeeze force strongly effects the fatigue performance, a low squeeze force should be used to aid in studying the other parameters affecting the fatigue life. Thus, to further reduce the number of variables in the joint design, only one rivet material, 2117-T3, is used with 2024-T3 Alclad sheet. Additional aspects regarding the rivet squeeze force and manufacture of riveted joints are reviewed in the subsequent section.

2.3.1 Effect of Rivet Squeeze Force on Fatigue Crack Growth

Hartman investigated the influence of manufacturing procedures on the fatigue life of 2024-T3 lap-splice joints in 1968. He noticed as the driven head diameter increased so did the fatigue life.[43] Müller extended the work of Hartman by relating the force used to form the driven head, known as the rivet squeeze force, to the driven head diameter. In 2024-T3 lap-splice joint fatigue tests, Müller showed the fatigue life is dependent on the rivet squeeze force realizing a fatigue life improvement of 10 times when the driven head diameter is increased from 1.2 to 1.75 times the rivet shank diameter for most rivet/sheet combinations found in fuselage structure. A high squeeze force has some important consequences. As explained by Slagter in reference [44], the squeezing process of the rivet involves plastic flow of the rivet, causing radial expansion of the shank and yielding in a substantial region of the sheet surrounding the hole. When the squeeze force is released, the rivet and sheet spring back, the radial spring back of the sheet tends to exceed that of the rivet at the hole boundary. The resulting interference fit between the rivet and sheet causes residual stresses in the sheet. Finite element calculations by Slagter confirmed the presence of compressive residual tangential stresses. This pre-stressing of the sheet at the boundary of the hole yields a favorable reduction of the local stress amplitude. In addition, by increasing the squeeze force, the clamping between the two sheets increases. The amount of clamping is considered responsible for the location of crack initiation. A tight clamping can lead to crack initiation outside the hole and to cracking which no longer intersects the hole boundary. Tight clamping makes the hole less critical and fretting fatigue more critical.[45] Another effect of the squeeze force is the shift of the location of the maximum bending stress. Also reported in reference [18], is the variation of the bending stress through the width of a lap joint. Near the rivets the bending stress is higher than between the rivets. With increasing squeeze force, the location of the maximum bending stress near the driven head shifts away from the rivet hole. Moreover, the difference in the maximum bending stress along the width of the lap joint gets larger as the rivet pitch increases. Many other significant contributions to the understanding of lap-splice joints have been made by Müller and can be found in reference [18]. In view of the fatigue life increase possible by varying the rivet squeeze force, great variability in the fatigue life may occur which can be decreased by controlling the rivet squeeze force. With this source of fatigue life variability in check, the remaining causes of variability are sheet material inhomogeneity and rivet hole quality, assuming with today's computer controlled fatigue machines, the testing protocol is undeviating.

2.3.2 Difficulties with Finite Width Effect in Fatigue Testing of Wide Panels

Experience in the aircraft manufacturing industry has shown differences in the fatigue lives of laboratory specimens and full-scale fuselage lap-splice joints.

Commonly, to obtain the same fatigue life on a laboratory specimen as a fuselage, the stress applied to the specimen is on the order of 1.2 times the nominal stress in the fuselage.[18] With the desire to test lap joints which replicate the operational environment of lap joints in fuselage structure, panels on the order of one fuselage bay in width (500 mm), seem to be a logical choice. Moreover, with larger specimens more representative of a fuselage lap-splice joint more rivets are required in the joint resulting in additional crack nucleation locations that can furnish insight into to degree of scatter in crack nucleation period. One unfortunate by-product of wide panel testing is the edge effect, which has been investigated and reported in references [18, 27, 30, 46, 47]. The edge effect is due to a difference in lateral contraction of the two sheets in the overlap region of the joint. As a result, the outer rivets, the rivets at the free edge of the specimen in the width direction, may experience as much as a 10% higher load leading to premature cracking at the edge rivets which in turn prevents development of MSD. For an MSD situation to develop the stress distribution through the width of the lap joint must be nearly homogenous. With MSD being defined earlier as numerous small cracks in a row of rivets in a lap joint, a heterogeneous stress distribution would promote higher crack growth rates in the area of the higher stresses whereby one crack can become dominant.

Several elegant solutions have been used successfully to eliminate the edge effect; however, Müller simply added two zero load transfer rivets at the free edge schematically shown in Figure 2.10. As reported in references [17] and [36] for a lap joint specimen with edge stiffeners and fuselage, respectively, cracks were first detected near the center of the bay where the stress distribution is nearly homogeneous. Away from the center, the tear straps and frames provide additional load paths for reaction of the hoop stress caused by cabin pressurization, thereby locally reducing the skin stresses.

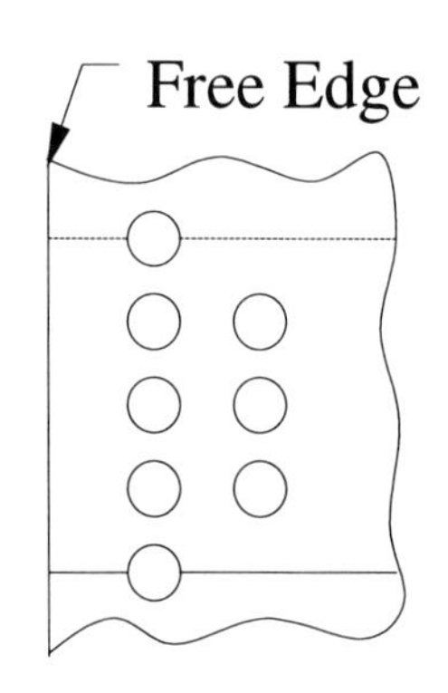

Figure 2.10 Zero Load Transfer Rivets at Free Edge

Another solution to the problems associated with MSD testing was developed by Eastaugh, Simpson, Straznicky, and Wakeman. A uniaxial coupon test specimen for simulating MSD initiation, growth, and link-up was designed in hopes of providing a universal protocol that could be used for proof-of-concept testing.[48] The three rivet row, eight rivets per row lap-splice joint consisted of 2024-T3 alclad sheets with 4 mm 2117-T4 aluminum rivets. Bonding side straps along the joint length eliminated the edge effect. Profile doublers were bonded through the joint width outside the joint overlap region to control the stress distribution through the width of the joint. Both side straps and profile doublers were of the same material as the lap joint sheets. Eastaugh et al. were able to simulate a non aircraft specific MSD scenario; thus, their methodology could be a useful tool in the design process for rating different lap-splice joints.

2.4 Analytical Considerations on the Stress Field

Designing a lap-splice joint that is easy to manufacture and maintain, in addition to satisfying static strength and damage tolerance requirements, is not a simple task. It often has a character of a compromise between the responsible departments, i.e., design, stress, fatigue, and production departments. The engineer performing the damage tolerance analysis must know the geometric details and the local stress field with the former dictating the latter for a given operational load condition. Quantifying the stress field is often complex due to the multi-element interaction between the fuselage skin, frames, tear straps, and longitudinal stiffeners.

The GAG cycle typically results in hoop stresses ranging from 70 to 112 MPa.[49] Cabin pressurization also creates a longitudinal stress that for an unstiffened pressure vessel yields a biaxiality ratio, B, of 0.5. The biaxiality ratio is the ratio between longitudinal and hoop stresses. In a stiffened pressure vessel like a fuselage, the biaxiality ratio is then less than 0.5 due to the longitudinal stiffening elements; i.e., the stringers. However, additional longitudinal stresses are generated as a result of fuselage bending not only due to aircraft weight but also aerodynamic loading, specifically the moment created by the horizontal stabilizer. Depending on the location of the joint with respect to the neutral axis of the fuselage, the loading parallel to the rivet line may be tensile or compressive thereby augmenting or reducing the biaxiality.

The hoop stress is transferred from one skin panel to the next at the longitudinal lap-splice joints through the rivets in the joint. The inherent eccentricity of a single shear joint, which is a common joint for skin splices, creates a local bending moment, secondary bending, in the skin as shown in Figure 2.11. As will be discussed in more detail later, the secondary bending has not been adequately represented in any of the existing lap-splice joint crack growth prediction models even though as already stated, Müller found the secondary bending stresses are largest at the two outer rows in a three rivet row configuration, top row and bottom row, and can be as large or larger than the membrane stresses.[18]

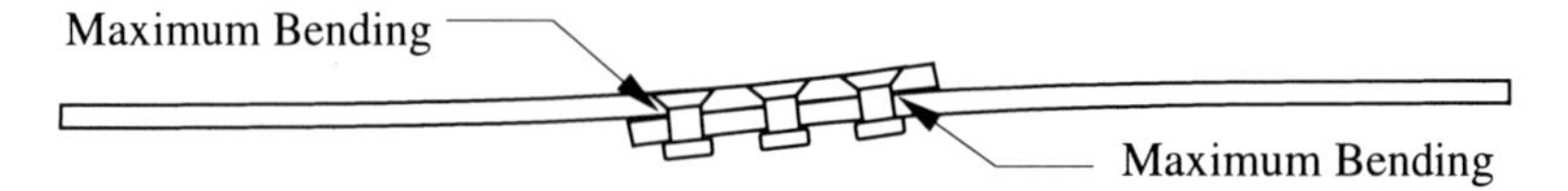

Figure 2.11 Secondary Bending in Lap-Splice Joint

The local stress field is further complicated by the rivets which when installed leave residual stresses around the hole in the plane of the sheet in addition to clamp-up stresses perpendicular to the sheet. The two outer rows in a three rivet row configuration, top row and bottom row, transmit the largest portion of the total load. Beside the load transfer through the rivets, load is also transferred from one sheet to the next by friction. The clamp-up stresses caused by installing the rivet vary with rivet squeeze force; thus more load is transferred by friction for high squeeze force rivet installations. Although having been studied quite extensively, see references [18,20,21,44,50], attempts at quantifying the effects of friction in riveted connections have not been successful. With regard to the hole expansion due to rivet squeezing, the magnitude of the tangential residual stresses and the distance from the hole edge in which they act ultimately affects the cyclical stresses driving crack growth at the hole edge in the net section of the sheet. Again referring to [18], for small rivet squeeze forces the tangential residual stresses at the hole edge are tensile, but become compressive for moderate to large rivet squeeze forces. The rivet flexibility also affects the stress distribution around the hole by governing the contact behavior between the shank of the rivet and bore of the hole. A flexible rivet will lose contact on the side of the hole opposite the applied remote load allowing for the stress concentration at the edge of the hole to develop since the hole is no longer filled, but open to some extent. Although rivet loading as a pin load on the rivet hole has been included in some of the existing prediction models, the effects of residual stresses and rivet flexibility have not. With the multiple loading conditions and complex detailed geometry in longitudinal lap-splice joints, a solution method allowing for superposition may make the task more tractable and is discussed in the following section.

2.5 Existing Lap-Splice Joint Fatigue Crack Growth Prediction Models

In view of complexity of the multiple site damage phenomenon, development of an accurate prediction model seems unlikely. The response of a lap-splice joint to external loading creates difficult to quantify behavior not to mention the statistical nature of the fatigue crack nucleation process. According to Schijve, "Accurate predictions can not be expected."[51] In addition, the recommendation of Swift from the Federal Aviation Administration, USA is

> ". . . research for MSD (should) concentrate on the effects of MSD on lead crack residual strength rather than on crack growth investigations with a view to establishing inspection programs to manage the safety of aircraft in the presence of MSD."[52]

In spite of the poor outlook on prediction routines, research in this area is still progressing. The physics of crack growth in riveted joints is not yet fully understood due to the complexities discussed earlier. Actually, a prediction model should address small crack behavior, crack interaction effects, lead-MSD crack interaction, and cracking in multiple rivet rows. Some researchers, references [53 - 61], have attempted to solve the problem from a global perspective, that is, including the structural elements, fuselage skin, frames, tear straps, and rivets, but excluding the component interaction effects; e.g., rivet flexibility, rivet hole fill, fretting corrosion, friction, rivet clamping, which have yet to be quantified. The application of these various methodologies as a general purpose analysis tool for the prediction of MSD is not yet realized. An alternative philosophy to the global approach is a local approach whereby understanding of each individual phenomenon is obtained and then integrated into a methodology to address the entire problem. The local approach has been adopted by several researchers also, see references [3,6,8,12,13,30,62-67]; however, none of the local approach models have been able to include all of the component interaction effects.

In review of the extensive research conducted, a local approach is adopted here to further increase the understanding of the individual phenomena. Linear Elastic Fracture Mechanics, LEFM, is aptly suited for such a research philosophy. Using the stress intensity factor, K, as a measure of the crack driving force and defined in Eqn. (2.1), with the principles of compounding and superposing of loading conditions and geometry factors, respectively, the effects of the various parameters affecting the MSD situation can be considered in turn.

$$K = \sigma\sqrt{\pi c}\beta \qquad (2.1)$$

where

σ = Applied remote stress

c = Crack length

β = Boundary correction factor

As long as the plastic zone ahead of the crack tip is small, LEFM is applicable. Nilsson and Hutchinson question the suitability of LEFM in view of the large plastic zones that may be present when there are MSD cracks near the lead crack.[68] To further investigate the role of plasticity in cracked stiffened structure, they have developed two elastic-plastic fracture methodologies; one based on LEFM using a damage reduced fracture toughness, and the other, based on a modified Dugdale model. The former model did not produce acceptable results due to the large scale yielding of the panel at the larger lead crack lengths; this method is not discussed further. Although the modified Dugdale model performed well in the stiffened flat panel analyses, further development is necessary before this type of methodology can accommodate the geometric detail and complex stress distribution seen in a lap-splice joint.

2.6 Approach Used in Current Study

As stated previously, the primary goal of this research effort is to develop a prediction algorithm for fatigue crack growth in longitudinal lap-splice joints of a pressurized fuselage. The existing prediction models discussed in the preceding section leave room for additional work in this area. The shortcomings in the existing prediction models relate to the degree of realism in representing the fuselage geometry, rivets, and loading conditions. Since the non-LEFM based models are not as mature as the LEFM models, the former are not discussed further. The most noticeable deficiency in the existing models is lack of verification with lap-splice joint fatigue test crack growth data. Predicting the fatigue life is a necessary but not sufficient requirement in assessing the validity of a prediction model. Furthermore, all of the phenomena related to or affected by the rivet squeeze force have yet to be considered in any prediction model. Another shortcoming is the rather simplistic representation of the secondary bending behavior and its effect on part through and through crack growth. Also, all the stress intensity factors used in the existing models are derived from two-dimensional analyses which omits any consideration of the crack front shape. In view of the bending, quarter elliptical part through cracks and through cracks with oblique crack fronts must be considered.

The focus here is both experimental and analytical with the intent of bridging some of the gaps mentioned above. Experiments are conducted to validate each phase of the prediction algorithm. For example, to verify the boundary correction factors for a finite width plate with a centrally located hole subject to remote tension and bending, a fatigue specimen is designed and tested in order to collect both the crack growth rate data and crack history. With the specimen geometry, loads, and crack history, predictions are made using one of the available crack growth laws, Paris, Forman, or Forman-Newman-de Koning. This procedure is used for each increase in the realism of the joint. Specifically for the riveted joints, a new technique is used which allows for the reconstruction of a large portion of the crack growth curve. Using this technique, the crack growth behavior is investigated as the crack grows from the rivet hole as a part through crack to a through crack ultimately ending in ligament or joint failure. Extensive testing is completed to investigate the secondary bending behavior in riveted connections and its effect on fatigue life and crack shape development. To investigate the three dimensional nature of crack growth in riveted connections, new, three dimensional stress intensity factor solutions are calculated based on the crack shapes and load conditions most commonly seen in lap-splice joints. The influence of the rivet squeeze force on the stress state in the joint and crack growth at the rivet hole edge is studied via a three dimensional finite element analysis which models the complex loading and contact behavior in the joint. The following Chapter reviews the experimental procedures, goals, and results.

2.7 Conclusions

As a result of surveying the literature, the following conclusions are formulated:

- Riveted lap joints are complex joints for geometric reasons with a complicated load transmission that is considerably affected by the riveting technique, especially the rivet squeeze force.
- As a result of the eccentricity in the lap joint, secondary bending cannot be avoided implying any tension stress on the joint will always be accompanied by bending stress (secondary bending). Moreover, as a result of rivet squeezing, a residual stress system is introduced, radial and tangential

stresses around the rivet and clamping stresses between the faying surfaces of the two sheets.

- In general, a riveted lap joint loaded by an approximately homogeneous hoop stress will lead to MSD configurations. For the larger part of the life, the cracks are small and invisible part through cracks with a quarter elliptical crack front. Later when through cracks are present, the crack front will still be curved with an oblique orientation through the sheet thickness.
- Information in the literature on the growth of small cracks and larger cracks in riveted joints is rather limited. However, the information indicates that existing prediction models on fatigue crack growth in these joints are not realistic. There is an obvious need for improvement requiring both analysis to obtain stress intensity factors for complex crack geometries and complex loading conditions. In complementary empirical research, more detailed information on crack shape development under combined tension and bending fatigue loads should be gathered. It is essential to combine these two approaches in view of their mutual significance for progress to be made.

---

[1] Miller, M., M. L. Gruber, K. E. Wilkins, and R. E. Worden. Full Scale testing and Analysis of Fuselage Structure, Proc. of FAA/NASA International Symposium on Advanced Structural Integrity Methods for Airframe Durability and Damage Tolerance, 4-6 May 1994, Hampton, VA, NASA-CP-3274.

[2] Kobayashi, Akira and Toshiyuki Shimokawa. Damage Tolerance Analysis of Multiple Site Cracks, Proc. of the 15th Symposium of the International Committee on Aeronautical Fatigue, 21-23 Jun 1989, Jerusalem, Isr.

[3] Nishimura, Toshihiko, Yoshiharu Hoguchi, Tetsuo Uchimoto, "Damage Tolerance Analysis of Multiple-Site Cracks Emanating from Hole Array," Journal of Testing and Evaluation, 18 (1990): 401-407.

[4] Kobayashi, Akira and Toshiyuki Shimokawa. Fatigue Crack Propagation of 2024-T3 Riveted Plate with Initial Defects Simulating Multi-Site Damages, Proc. of the 16th Symposium of the International Committee on Aeronautical Fatigue, 22-24 May 1991, Tokyo, Jap. Tokyo: Ryoin; West Midlands, UK: EMAS, 1991.

[5] Lehrke, H.-P. and A. Schöpfel. Analysis of Multiple Crack Propagation in Stiffened Sheet, Proc. of the 16th Symposium of the International Committee on Aeronautical Fatigue, 22-24 May 1991, Tokyo, Jap. Tokyo: Ryoin; West Midlands, UK: EMAS, 1991.

[6] Pártl, O. and J. Schijve. Multiple-Site-Damage in 2024-T3 Alloy Sheet, Report LR-660. Delft, NL: Delft University of Technology UP, 1992.

[7] Dawicke, D. S. and J. C. Newman Jr. Analysis and Prediction of Multiple-Site Damage (MSD) Fatigue Crack Growth. NASA-TP-3231, 1992.

---

[8] Moukawsher, E. J., M. A. Neussel, and A. F. Grandt Jr. A Fatigue Analysis of Panels with Multiple Site Damage, Proc. of the 1992 USAF Structural Integrity Program Conference, 1-3 Dec 1992, San Antonio, TX, WL-TR-93-4080.

[9] Moukawsher, E. J., M. A. Neussel, and A. F. Grandt Jr. Analysis of Panels with Multiple Site Damage, Proc. of the AIAA 35th Structures, Dynamics, and Materials Conference, 18-21 Apr 1994, Hilton Head, SC, AIAA Paper No. 94-1459.

[10] Grandt, Jr., A. F. Material Degradation and Fatigue in Aerospace Structures, Proc. of the 2nd Aging Aircraft Conference, 16-20 May 1994, Oklahoma City, OK, AFOSR-TR-94-0756.

[11] Vermeeren, C. A. J. R. Residual Strength Predictions for Fiber Metal Laminates: The R-Curve Approach, Report LR-717. Delft, NL: Delft University of Technology UP. 1993.

[12] Nathan, A. and A. Brot. An Analytical Approach to Multi-Site Damage, Proc. of the 17th Symposium of the International Committee on Aeronautical Fatigue, 9-11 Jun 1993, Stockholm, Swed. West Midlands, UK: EMAS, 1993

[13] Buhler, Kimberley, Allen F. Grandt Jr., and E. J. Moukawsher. Fatigue Analysis of Multiple Site Damage at a Row of Holes in a Wide Panel, Proc. of FAA/NASA International Symposium on Advanced Structural Integrity Methods for Airframe Durability and Damage Tolerance, 4-6 May 1994, Hampton, VA, NASA-CP-3274.

[14] Buhler, Kimberley, Allen F. Grandt Jr., and E. J. Moukawsher. Fatigue Analysis of Multiple Site Damage at a Row of Holes in a Wide Panel, Proc. of the 17th Symposium of the International Committee on Aeronautical Fatigue, 9-11 Jun 1993, Stockholm, Swed. West Midlands, UK: EMAS, 1993.

[15] Rohrbaugh, S. M., D. Ruff, B. M. Hillberry, G. McCabe, and A. F. Grandt Jr. A Probabilistic Fatigue Analysis of Multiple Site Damage, Presented at FAA/NASA International Symposium on Advanced Structural Integrity Methods for Airframe Durability and Damage Tolerance, 4-6 May 1994, Hampton, VA, NASA-CP-3274.

[16] Aircraft Accident Report: Aloha Airlines, Flight 243, Boeing 737-200, N73711, near Maui, Hawaii, April 28, 1988, NTSB/AAR-89/03. Washington DC: U.S. National Transportation Safety Board, 1989.

[17] Piascik, Robert S., Scott A. Willard, and Matthew Miller. The Characterization of WideSpread Fatigue Damage in Fuselage Structure. NASA-TM-109142, 1994.

[18] Müller, Richard Paul Gerhard. An Experimental and Analytical Investigation on the Fatigue Behaviour of Fuselage Riveted Lap Joints, The Significance of the Rivet Squeeze Force, and a Comparison of 2024-T3 and Glare 3. Diss. Delft University of Technology, 1995. Delft:NL, 1995. ISBN 90-9008777-X, NUGI 834.

[19] Swenson, Daniel, Sudhir Gondhalekar, and Dave Dawicke. Analytical Developments in Support of the NASA Aging Aircraft Program with an Application to Crack Growth from Rivets. Proc. 1993 SAE General, Corporate, & Regional Aviation Meeting & Exposition, 18-20 May 1993, Wichita, KS.

[20] Iyer, K., G. T. Hahn, P. C. Bastias, and C. A. Rubin. "Analysis of Fretting Conditions in Pinned Connections." Wear, 181-183 (1995): 524-530.

---

[21] Iyer, K., P. C. Bastias, Ca. A. Rubin, and G. T. Hahn, "Local Stresses and Distortions of a Three-Dimensional, Riveted Lap Joint." Unpublished paper. Department of Mechanical Engineering, Vanderbilt U. 1995.

[22] Van der Linden, H. H., L. Lazzeri, and A. Lanciotti. Fatigue Rated Fastener System in 1½ Dogbone Specimens, NLR-TR-86082. National Aerospace Laboratory of The Netherlands, NL, 1986.

[23] Cook, Robin. Standard Fatigue Test Specimens for Fastener Evaluation, AGARD-AG-304. Neuilly-Sur-Seine, Fr. 1987.

[24] Palmberg, B., G. Segerfrojd, G.-S. Wang, and A. Blom. Fatigue Behaviour of Mechanical Joints: Critical Experiments and Numerical Modeling, Proc. of 18th Symposium of the International Committee on Aeronautical Fatigue, 3-5 May 1995, Melbourne, Austral.

[25] Leven, G. "Fatigue Tests on Riveted Lap Joints." (in German) Zeitschrift für Flugwissenschaften, 10 (1962): 160-167.

[26] Hartman, A., F. A. Jacobs, and P. de Rijk. Tests on the Effect of the Size of the Specimen on the Fatigue Strength of 2024-T4 Alclad Double Row Riveted Single Lap Joints, NLR-M.2104. Amsterdam: National Laboratory of The Netherlands, 1962.

[27] Mayville, R. A. and T. J. Warren. "A Laboratory Study of Fracture in the Presence of Lap Splice Multiple Site Damage," Structural Integrity of Aging Airplanes. Springer-Verlag: Berlin, 1991.

[28] Pelloux, R. A. Warren, and J. O'Grady, "Fractographic Analysis of Initiation and Growth of Fatigue Cracks at Rivet Holes." Eds. S. N. Atluri, S. G. Sampath, and P. Tong. Structural Integrity of Aging Airplanes, Springer Series in Computational Mechanics. Berlin: Springer Verlag, 1991.

[29] Soetikno, T. P. Residual Strength of the Fatigued 3 Rows Riveted GLARE3 Longitudinal Joint. Masters Thesis, Faculty of Aerospace Engineering, Delft University of Technology, Delft, NL, 1992.

[30] Wit, G. P. MSD in Fuselage Lap Joints - Requirements for Inspection Intervals for Typical Fuselage Lap Joint Panels with Multiple Site Damage, LR-697. Delft, NL: Delft University of Technology UP, 1992.

[31] Molent, L. and R. Jones. "Crack Growth and Repair of Multi-Site Damage of Fuselage Lap Joints." Engineering Fracture Mechanics, 44 (1993): 627-637.

[32] Vlieger, H. Results of Uniaxial and Biaxial Tests on Riveted Fuselage Lap Joint Specimens, Proc. of FAA/NASA International Symposium on Advanced Structural Integrity Methods for Airframe Durability and Damage Tolerance, 4-6 May 1994, Hampton, VA, NASA-CP-3274.

[33] Zhuang, W. Z. L., J. P. Baird, and H. M. Williamson. Multi Site Damage and Buckling in Riveted Lap Joints, Proc. of the 18th Symposium of the International Committee on Aeronautical Fatigue, 3- 5 May 1995, Melbourne, Austral. West Midlands, UK: EMAS, 1995.

[34] Ottens, H. H. Multiple Crack Initiation and Crack Growth in Riveted Lap Joint Specimens, Proc. of the 18th Symposium of the International Committee on Aeronautical Fatigue, 3- 5 May 1995, Melbourne, Austral. West Midlands, UK: EMAS, 1995.

[35] Goranson, Ulf G. and M. Miller. Aging Jet Transport Structural Evaluation Programs, Proc. of the 15th Symposium of the International Committee on Aeronautical Fatigue, 21-23 June1989, Jerusalem, Isr. West Midlands, UK: EMAS, 1989.

---

[36] Samavedam, Gopal, Douglas Thomson, and David Y. Yeong. Evaluation of the Fuselage Lap Joint Fatigue and Terminating Action Repair. Proc. of FAA/NASA International Symposium on Advanced Structural Integrity Methods for Airframe Durability and Damage Tolerance, 4-6 May 1994, Hampton, VA, NASA-CP-3274.

[37] Gopinath, K. V. Structural Airworthiness of Aging Boeing Jet Transports - 747 Fuselage Test Program." AIAA 92-1128 (1992):

[38] Maclin, James, R. Performance of Fuselage Pressure Structure, Proc. of the 3rd International Conference on Aging Aircraft and Structural Airworthiness, 19-21 Nov 1991, Washington DC.

[39] Piascik, Robert S., Scott A. Willard, and Matthew Miller. The Characterization of WideSpread Fatigue Damage in Fuselage Structure. NASA-TM-109142, 1994.

[40] United States. Dept. of Defense. Metallic Materials and Elements for Aerospace Vehicle Structures, MIL-HDBK-5G. 1 Nov 1994.

[41] Miller, Matthew, Kevin N. Kaelber, and R. Elaine Worden. Finite Element Analysis of Pressure Vessel Panels. Proc. of International Workshop on Structural Integrity of Aging Airplanes, 31 Mar - 2 Apr 1992, Atlanta, GA, Atlanta: Atlanta Technical Publications.

[42] Hartman, A. The Influence of Manufacturing Procedures on the Fatigue Life of 2024-T3 Alclad Riveted Single Lap Joints, NLR-TR-68072. Amsterdam: National Laboratory of The Netherlands, 1968.

[43] Slagter, Wim Jan. Static Strength of Riveted Joints in Fibre Metal Laminates. Diss. Delft University of Technology, 1994. Delft,NL, 1994. ISBN 90-9007089-3, NUGI 831.

[44] Schijve, J. Multiple-Site Damage Fatigue of Riveted Joints, Report LR-679. Delft, NL: Delft University of Technology P, 1992.

[45] Barneveld, K. W. Biaxial Stresses in Riveted Lap Joints, Part I: Theory. Masters Thesis, Faculty of Aerospace Engineering, Delft University of Technology, Delft, NL, 1994.

[46] van Griensven, B. F. Multiple Site Damage in Flat Uniaxially Loaded and Curved Pressurized Riveted Lap Joints. Masters Thesis, Faculty of Aerospace Engineering, Delft University of Technology, Delft, NL, 1995.

[47] Eastaugh, G. F., D. L. Simpson, P. V. Straznicky, and R. B. Wakeman. A Special Uniaxial Coupon Test Specimen for the Simulation of Multiple Site Fatigue Crack Growth and Link-Up in Fuselage Skin Splices, AGARD-CP-568. Neuilly-Sur-Seine, Fr. 1995.

[48] Niu, Michael, C. Y. Airframe Structural Design, Practical Design Information and Data on Aircraft Structures. Hong Kong: Conmilit Press LTD, 1988.

[49] Fung, C. P. and J. Smart, "An Experimental and Numerical Analysis of Riveted Single Lap Joints," Proceedings of the Institute of Mechanical Engineers, Part G, Journal of Aerospace Engineering, Vol. 208, 1994, pp. 79-90.

[50] Schijve, J. Multiple-Site Damage Fatigue of Riveted Joints, Report LR-679. Delft, NL: Delft University of Technology P, 1992.

[51] Swift, T. WideSpread Fatigue Damage Monitoring-Issues and Concerns, Proc. of 5th International Conference on Aging Aircraft. 16-18 June 1993, Hamburg, Germany.

[52] Atluri, S. N. and Pin Tong. "Computational Schemes for Integrity Analyses of Fuselage Panels in Aging Airplanes." Eds. S. N. Atluri, S. G. Sampath, and P. Tong. Structural Integrity of Aging Airplanes, Springer Series in Computational Mechanics. Berlin: Springer Verlag, 1991.

[53] Park, J. H., T. Ogiso, and S. N. Atluri. "Analysis of Cracks in Aging Aircraft Structures, with and Without Composite-Patch Repairs." Computational Mechanics 10 (1992): 169-201.

[54] Park, J. H. and S. N. Atluri, "Fatigue Growth of Multiple Cracks Near a Row of Fastener Holes in a Fuselage Lap Joint," Computational Mechanics 13 (1993): 189-203.

[55] Singh, Ripudaman, Jai H. Park, and Satya N. Atluri. "Growth of Multiple Cracks and Their Linkup in a Fuselage Lap Joint." AIAA 32 (1994): 2260-2268.

[56] Park, Jai H., Ripudaman Sigh, Chang R. Pyo, and Satya N. Atluri. "Integrity of Aircraft Structural Elements with Multi-Site Fatigue Damage." Engineering Fracture Mechanics 51 (1995): 361-380.

[57] Kanninen, M. F., P. E. O'Donoghue, S. T. Green, C. P. Leung, S. Roy, and O. H. Burnside. "Application of Advanced Fracture Mechanics to Fuselage." ." Eds. S. N. Atluri, S. G. Sampath, and P. Tong. Structural Integrity of Aging Airplanes, Springer Series in Computational Mechanics. Berlin: Springer Verlag, 1991.

[58] Johansson, Thomas and Hans Ansell. Structural Reliability in Fatigue Design - Application to Riveted joints and the Presence of Multiple Site Cracking. Proc. of International Workshop on Structural Integrity of Aging Airplanes, 31 Mar - 2 Apr 1992, Atlanta, GA, Atlanta: Atlanta Technical Publications.

[59] Tong, P., R. Greif, and L. Chen, "Residual Strength of Aircraft Panels with Multiple Site Damage," Computational Mechanics 13 (1994): 285-294.

[60] Wang, G. S. "A Statistical Multi-Site Fatigue Damage Analysis Model," Fatigue and Fracture of Engineering Materials and Structures 18 (1995): 257-272.

[61] Sampath, S., and D. Broek. "Estimation of Requirements of Inspection Intervals for Panels Susceptible to Multiple Site Damage." Eds. S. N. Atluri, S. G. Sampath, and P. Tong. Structural Integrity of Aging Airplanes, Springer Series in Computational Mechanics. Berlin: Springer Verlag, 1991.

[62] Beuth, J. L. and J.W. Hutchinson. "Fracture Analysis of Multi-Site Cracking in Fuselage Lap Joints," Computational Mechanics 13 (1994): 315-331.

[63] Ingraffea, A. R. and M. D. Grigoriu. "Representation and Probability Issues in the Simulation of Multi-Site Damage." ." Eds. S. N. Atluri, S. G. Sampath, and P. Tong. Structural Integrity of Aging Airplanes, Springer Series in Computational Mechanics. Berlin: Springer Verlag, 1991.

[64] Broek, D. The Effects of Multiple-Site-Damage on the Arrest Capability of Aircraft Fuselage Structures, FractuREsearch-TR-9302, 1993.

[65] Horst, Peter. "A Simple Numerical Model for the Calculation of Multiple Cracks in Thin Skin Structures." Hamburg: Daimler-Benz Aerospace Airbus GmbH, 1995.

---

[66] Homan, J. J. and A. A. Jongebreur. "Calculation Method for Predicting the Fatigue Life of Riveted Joints." Proc. of the 17th Symposium of the International Committee on Aeronautical Fatigue, 9-11 Jun 1993, Stockholm, Swed. West Midlands, UK: EMAS, 1993.

[67] Nilsson, K.-F. and J. W. Hutchinson, "Interaction Between a Major Crack and Small Crack Damage in Aircraft Sheet Material." International Journal of Solids and Structures 31 (1994): 2331-2346.

This page intentionally left blank

3.

Experimental Investigations

3.1 Introduction

In the MSD scenario, the existence of many small, usually undetectable cracks reduces the structural integrity of the fuselage. Before any attempts can be made to address the MSD situation, the physics of the problem must be understood. Thus, the experimental investigation is focused on gaining insight in the crack growth behavior of small cracks when subject to the complex load conditions present in fuselage longitudinal lap-splice joints. The small cracks are of prime importance since during this stage of the crack growth life, the complications created by the rivet are of greatest influence. Those parameters associated with riveted connections that are the most difficult to quantify are:

(a.) Radial, circumferential, and axial (clamp-up) residual stresses due to rivet squeezing
(b.) Fretting corrosion as a result of contact not only between the faying surfaces but also between the rivet and sheets
(c.) Load transfer by friction between the faying surfaces
(d.) Secondary bending from the eccentric load application

The influence of these factors is so pronounced on the small crack growth behavior that if they are not isolated, the character of each cannot be understood. Toward this end, a typical lap-splice joint must be decomposed, shown in Figure 2.8, into it's simpler component parts in order to study each effect in turn. The simplest specimen to be fatigue tested is the open hole specimen and since the response to remote tension loading is well understood, this effort concentrates on loading by remote tension and secondary bending. The second type of specimens tested are simple lap splice joints. Two different joint designs are explored, the 1½ dog bone and asymmetric lap splice joints. Due to the complicated interaction between the adjoining sheets and rivets, the asymmetric lap splice joint, again refer to Figure 2.8, is the most evolved specimen design tested in this study.

In the most simple interpretation, a fatigue test yields the fatigue life for that particular specimen, and if multiple specimens are tested, an indication of the scatter in the fatigue life is also obtained. However, the fatigue life and scatter do not in themselves offer much insight into the crack growth behavior. Toward this end, in situ crack growth monitoring is employed to record the crack growth at various periods during the test which can then be used to construct crack history curves and further crack growth rate curves.

In the present chapter, methods for crack detection and crack size observations are discussed first (section 3.2), followed by a survey of experimental results of various crack growth test series (section 3.3). The purpose is to collect relevant empirical evidence for modeling prediction procedures for fatigue crack growth in riveted lap joints.

3.2 Crack Detection and Measurement

Crack detection and measurement in any lap splice joint is an arduous task due to the location of the crack nucleation site. Because of the secondary bending and load transfer in the joint, the outer rivet rows are most critical, see Figure 2.11. If the countersunk rivets are installed with low to moderate squeeze force, crack nucleation is most always in the upper rivet row as seen in Figure 3.1. Through extensive experimental investigations, Müller found the rivet squeeze force influences the crack nucleation location whereby increasing the rivet squeeze force shifts the crack nucleation site from the hole edge in the

Figure 3.1 Upper Rivet Row Cracking

net section toward the top of the hole as shown in Figure 3.2.[1] At relatively high squeeze forces resulting in a D/D_0 of 1.75 or greater, the nucleation site is no longer at the hole boundary, but in front of the hole at the location of maximum secondary bending. For a more thorough discussion on the effect of the rivet squeeze force on the crack nucleation site, see reference [1].

Unfortunately, crack growth in the critical upper rivet row is hidden by the head of the countersunk rivet for a large part of the fatigue life. To further complicate crack detection and measurement, the crack in the outer visible surface of the sheet may not be open due to compressive secondary bending stress. Two methods, traveling optical microscope and gel electrode, are used. The results were disappointing using the gel electrode method for crack detection and measurement due to limitations of this technique in highly stressed components. For a detailed discussion of the gel electrode method see reference [2]. The traveling optical microscope is described in the following section.

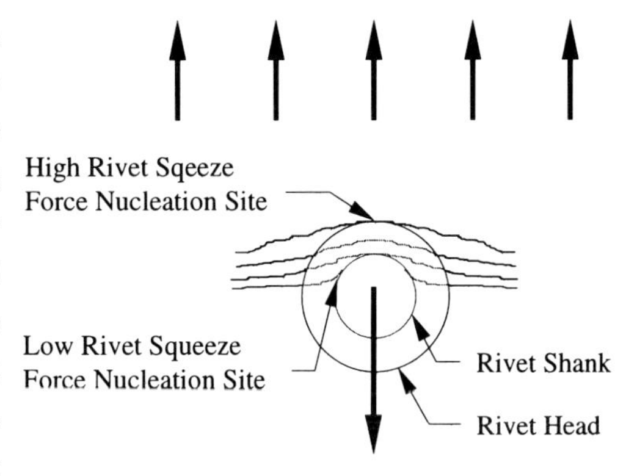

Figure 3.2 Shift of Crack Nucleation Site Due to Change in Rivet Squeeze Force

3.2.1 Traveling Optical Microscope

The forte of this method is the ease of setup and use. Microscopes of varying magnification, 12 – 120x, are used. With the help of millimeter paper attached to the sheet in close proximity to the crack nucleation site or a ruler with ½ mm tic marks, the crack is measured after growing out from under the rivet head. With the zero or compressive stress at the free surface, the crack is closed making determination of the crack tip difficult. For small crack lengths, cracks with a visible length less than 1.0 mm, the error in crack tip location alone can be as high as ±0.5 mm; thus, crack growth in the tested lap splice joints were not measured by this method. For the sheet specimens tested, discussed in section 3.2.2, the traveling optical microscope was sufficient with an accuracy of ±0.25 mm.

3.2.2 Marker Loads

No accurate and efficient method of in situ crack detection and measurement of lap joint specimens is currently available. An obvious approach then is to adopt marker load cycles in addition to the base line Constant-Amplitude (CA) cycles. Two marker load cycle histories have been adopted in the present investigation, see Figure 3.5 and Figure 3.6. In the first figure, one marker load cycle is introduced after 50 cycles of the baseline CA loading, which is the load cycle with the same σ_{max} as in the CA cycles, but with a lower σ_{min}. In the second figure, blocks of 100 marker load cycles are applied in between the baseline CA cycles. In this case, the marker load cycles have the same σ_{min} as the baseline cycles, but a lower σ_{max}.

The application of the marker loads presumes two requirements. First, the marker load cycles should leave striations or bands on the fracture surface that can be detected under the microscope. The markers then indicate where the crack front was at the moment that the marker load cycles were applied. Moreover, the crack front shape can also be determined, which for small and large cracks in riveted lap joints is essential information. Secondly, crack growth during the marker load cycles should have a negligible effect on crack growth during the baseline cycles in order not to disturb the crack growth phenomenon to be studied.

The basic principle of the marker load method is that the crack growth history can be reconstructed from the markers after final failure of a specimen. It presumes that markers can be associated with known numbers of the CA baseline cycles.

The use of marker loads, typically a instantaneous or short duration variation in the CA maximum stress or stress ratio, can perturb the striation spacing created by the CA loading. If the perturbations can be reliably detected in the electron microscope, the crack growth history can then be determined if the number of cycles to failure, also known as the fatigue life, is known. If the entire crack history is not of interest for a particular test series, the number of cycles when the test is halted must be recorded since this serves as the starting point for reconstructing the crack history. Most likely, the number of cycles between the last marker load and specimen failure is not known; therefore, the accuracy of the number of cycles at each marker load is plus or minus the number of cycles between marker loads. For clarity in describing how the crack history is determined, assume the specimen fails at a (last) marker load. By viewing the fracture surface with the unaided eye, optical microscope, or scanning electron microscope, the previous marker band is found. The number of cycles to obtain the crack length corresponding to this marker band is simply the fatigue life less the number of cycles between marker loads. This procedure is repeated until the marker bands can no longer be detected. The success of marking the fracture surface with marker loads hinges on detection of the marker bands throughout the fatigue life. Furthermore, the marker load is generally not the load cycle of interest; thus, the marker loads should not significantly effect the fatigue crack growth.

Extensive testing of center cracked tension and lap splice joint specimens of aluminum 2024-T3 subject to CA with periodic marker block loading has been completed by The National Aerospace Laboratory of The Netherlands.[3,4] Two separate spectra were tried, one with periodic overloads and the other with periodic blocks of high stress ratio with a constant σ_{max} but with an increase in σ_{min}. Overall, neither spectrum effectively marked the fracture surface for small crack lengths, less than 0.5 mm, through failure. The peak load spectrum created traceable marker bands only when the peak load was 140% of the maximum CA load which results in unacceptable crack retardation after application of the peak load. The high stress ratio spectrum created marker bands, but only for crack lengths greater than 0.5 mm.

A unique method for marking the fracture surface of 7075 aluminum of various heat treatments using cyclical condensation was developed by Johansson.[5] What Johansson calls "beach marks" are created in the initiation zone by periodic variations of the humidity in the test chamber. Although the mechanism by which the beach marks are created was not reported, he was able to follow the crack propagation for cracks less than 0.1 mm in length. The use of this method for lap splice joints is questionable since in order to create the beach marks, the fracture surface must be exposed to the cyclical humidity. With the cracks nucleating at the faying surface of the joint that is somewhat isolated from the environment, it is not known a priori whether the humidity can penetrate the joint. Marking the fracture surface by means of cyclical condensation is not pursued as a result of the anticipated experimental difficulties in this method.

3.2.2.1 Determination of Stress Levels and Sequencing

With 2024-T3 aluminum being a rather ductile material, care must be taken in choosing the fatigue stress level. If the stress level is too high, the fracture surface is quite torturous and dominated by micro-void coalescence shown by the white arrows in Figure 3.3, thus, the marker bands are discontinuous and difficult to see.

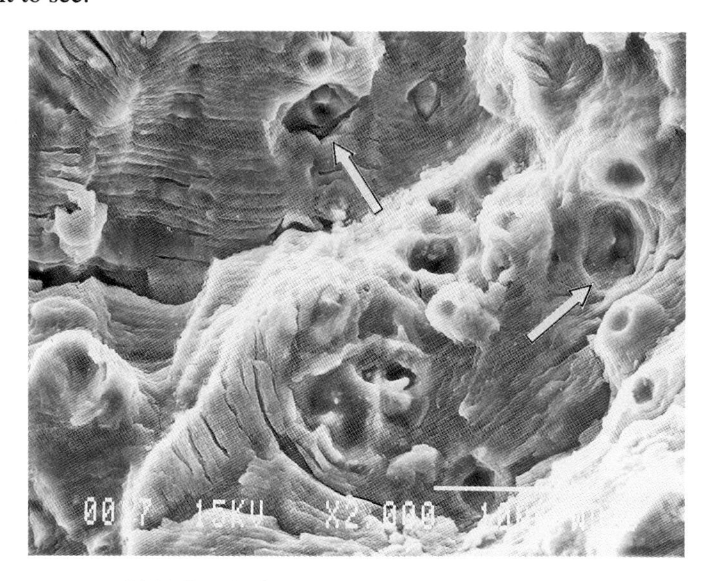

$\sigma_{tension}$ = 200 MPa, Fatigue Block R = 0.0918, Marker Load R ≈ 0
Half Crack Length, a = 4 mm

Figure 3.3 Fracture Surface Showing Micro-Void Coalescence

If the stress level is too low, crack nucleation and growth has a tendency to begin at the top of the rivet hole were fretting corrosion is maximum. Not only is this location not of interest from a fatigue viewpoint since in service cracking is not frequently found here; but also, the fracture surface is covered with fretting debris, shown in Figure 3.4, created as the sheet and rivet contact one another. In addition, the two opposite sides of the fracture surface can also come into contact destroying the marker bands. Even if the marker bands are not damaged by incidental contact, removing the fretting debris without harming the fracture surfaces is difficult and quite time consuming.

Three different applied remote stress levels were tried, 100, 70, and 50 MPa on the lap joint specimens. Recall that secondary bending is also created in the lap joints, thus the specimens were designed such that the stress in the net section was comprised of equal components of tension and bending to coincide with the findings of Müller. Another variable affecting the creation of marker bands is in what sequence the marker loads are applied. Two different marker load sequences were used schematically shown in Figure 3.5 and Figure 3.6. The most noticeable aspect of each spectrum are the lack of peak overloads;

Figure 3.4 Crack Nucleation and Growth at Top of Rivet Hole with Fretting Debris on Fracture Surface

thus, the fatigue life should not be affected by the period disruption of the CA loading. For the single under-load ($\sigma_{min} = 0$) spectrum, the fatigue stress ratio of the baseline cycles was varied from the R = 0.5 shown in Figure 3.5 down to R = 0.05 to determine the lowest R ratio at which marker bands could reliably be detected.

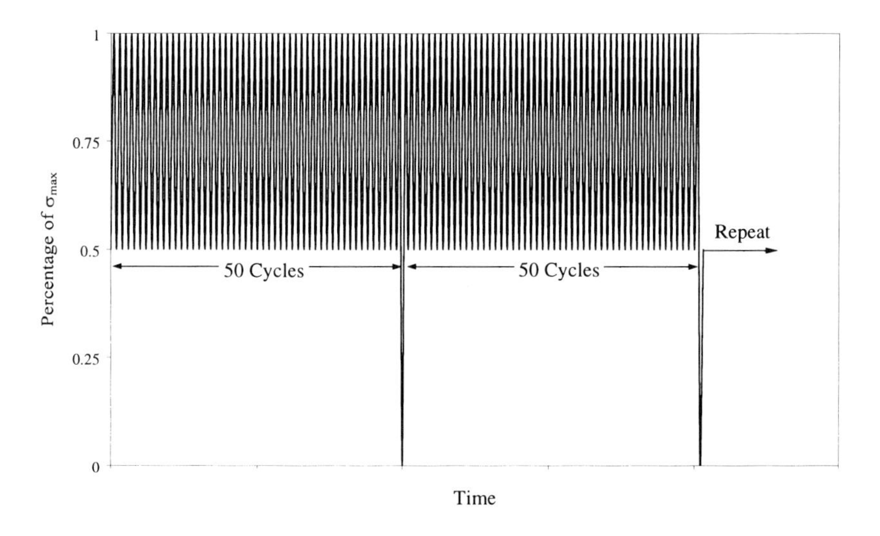

Figure 3.5 Single Under-Load Spectrum

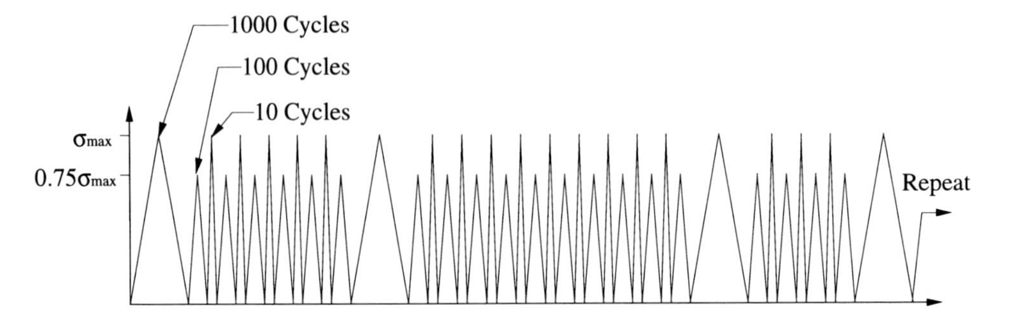

Figure 3.6 Program Spectrum

3.3 Fatigue Crack Growth

The three fatigue specimens tested follow the increase in structural complexity as described previously in Chapter 2. The most basic fatigue specimen is the center cracked tension (CCT) specimen, followed by a centrally located hole in a plate, and lastly, the asymmetric lap splice joint. Each test series has the same set of goals:

a. Collect crack growth history
b. Determine crack growth rate
c. Gather crack front shape data
d. Evaluate effectiveness of marker spectra

The crack growth history is used primarily to verify that the marker spectra does not affect the crack growth rate and to validate the prediction algorithm, discussed in Chapter 5. It represents the entire crack growth history and not just the fatigue life. The crack shape data may be the most interesting by providing insight into the relationship between the crack shape and loading conditions.

3.3.1 Center Crack Tension (CCT)

The CCT specimen serves as a baseline test bed meaning that if a new test procedure does not work on the CCT specimen, it will most likely not work for the more complex specimens. The specimen geometry is shown in Figure 3.7 and conforms to the ASTM E647-93 standard.[6] All specimen dimensions in this report are in millimeters unless otherwise specified.

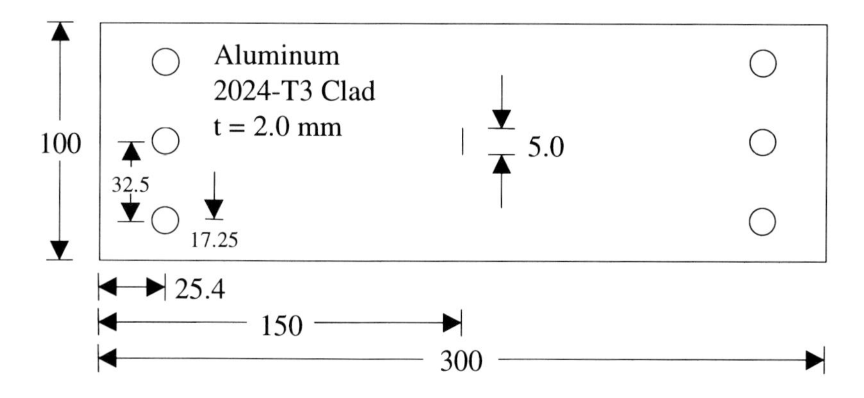

Figure 3.7 Center Crack Tension Fatigue Test Specimen

Introducing a bending stress while maintaining a relatively simple test specimen design was accomplished by Nam, Ando, Ogura, and Matui by using a relatively thick 3%NiCrMo steel plate and asymmetrically milling the length between the grips as shown in Figure 3.8.[7] The specimen of Nam et al. produced a predictable stress field at the crack location that is of prime importance in combined loading fatigue testing. This specimen is not used in this study since fuselage skin is not made from milled aluminum plates but from rolled sheets.

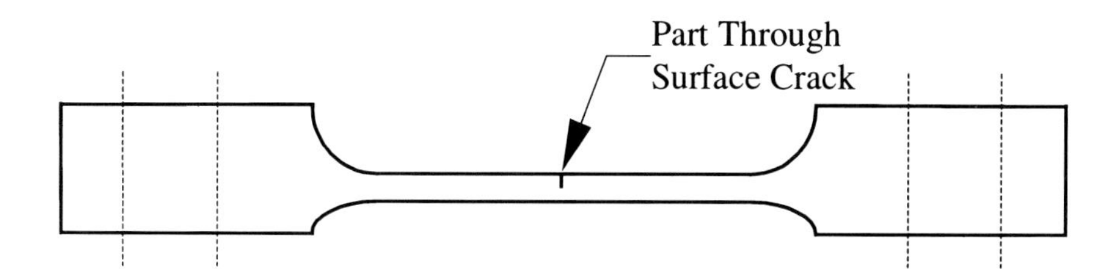

Figure 3.8 Nam et al. Tension and Bending Fatigue Specimen

A new tension and bending specimen is designed using commercially available aluminum sheets and an elevated temperature curing thermoset adhesive, see Figure 3.9. In addition to the baseline remote tension applied load, with a slight modification of the CCT specimen, cyclical remote tension and bending can be applied in a controllable manner. The modification induces the bending by the eccentricity of the load path through the specimen. The three sheets are hot bonded in a heat press using AF-163X at a temperature of 120°C for 130 minutes with an applied pressure of 3 bar.

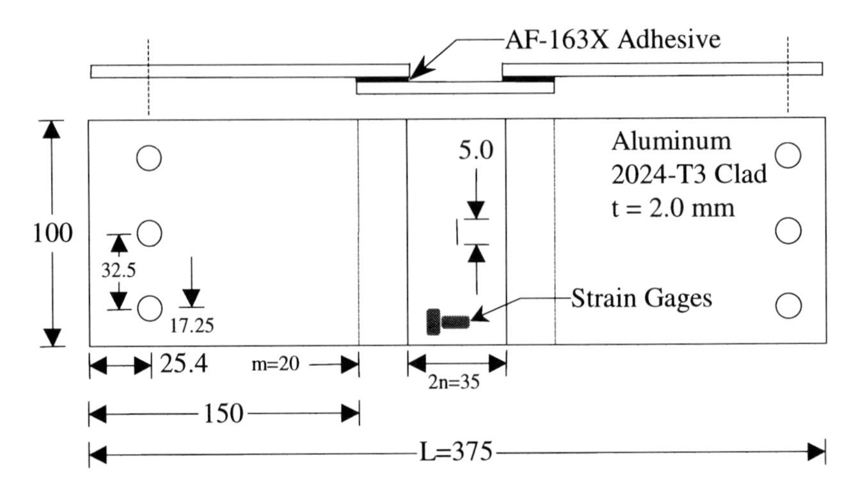

Figure 3.9 Center Cracked Combined Tension and Bending (CCTB) Fatigue Specimen

A remote tensile load is applied at the plate ends and due to the eccentricity of the joint, a secondary bending moments is created in the area of interest. To design the joint, a simple line model, one dimensional, of a joint is used as developed by Schijve.[8] The validity of this model for the joint can be verified by strain gage date gathered during preliminary static tests. Again referring to Figure 3.9, the dominant parameters for controlling the degree of bending created are m and n, the overlap and net section lengths, respectively. Using different thicknesses of the AF-163 adhesive could also be used to control the degree of bending, but for economic reasons, only the AF-163X with cured thickness of 0.1 mm is used. From a physical point of view, it is easy to surmise that as m and n increase, the bending moment decreases. As n is increased, the moment shifts from the center of the net section toward the ends of the overlap where the joint is stiffer. This latter effect is shown in Figure 3.10. The line model produces these trends with the derivation and a sample calculation given in Appendix A.

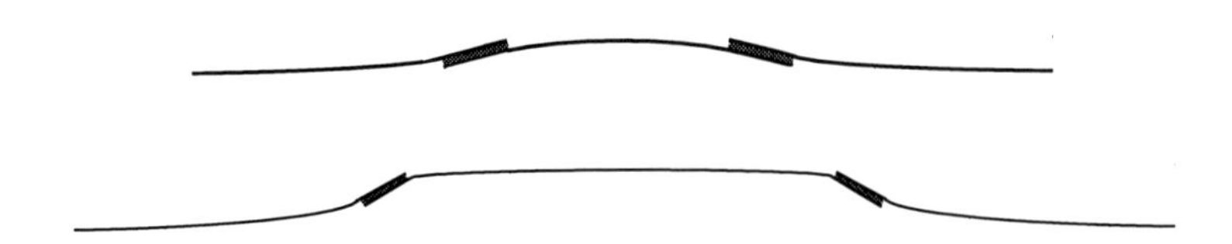

Figure 3.10 Shift of Bending Location by Increasing Net Section Length

To verify the stresses, two static tests were conducted with specimens having 12 strain gages in the test area where the center crack, and later, open holes are located. Four pairs of gages, one on each side of the sheet, are evenly distributed through the width at the center of the specimen, L = 375/2, to determine the stresses acting along the longitudinal axis of the specimen. The remaining 4 gages are adjacent and transverse to the "longitudinal" gages to determine the lateral stresses. For all of the fatigue test specimens, two pairs of strain gages were mounted, as shown in Figure 3.9, on both sides of the sheet. From the strain gage date collected during the static tests, the membrane and bending behavior is quantified as shown in Figure 3.11. The error between the test results and line model is less than 5% for both tension and bending at the desired maximum fatigue load of 13 kN where the tension and bending stresses are approximately the same at 90 MPa. The accuracy of the model is sufficient and can thus be used to design specimens with varying m and n in order to change the amount of bending generated.

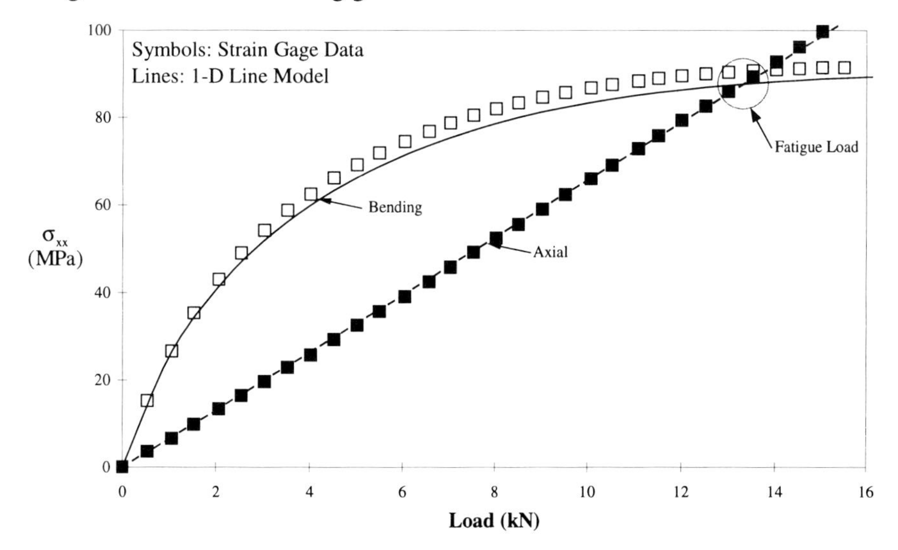

Figure 3.11 Experimental and Calculated Membrane and Bending Stresses in CCTB Specimen

The application of occasional under-loads has been shown to not affect the fatigue life.[9,10] However, for the program loading shown in Figure 3.6, a comparison of the crack growth rates between constant amplitude and program loading is made. Comparative crack propagation tests were carried out on sheet specimens with a central hole (width 100 mm, hole diameter 4 mm, thickness 2 mm), provided with two edge notches made by saw cutting. The specimens were loaded in cyclic tension only, because the bending component was not considered to be essential for the validation of the marker load application. Two tests were run without marker loads and two tests with the program loading of Figure 3.6, all tests with $\sigma_{max} = 100$ MPa. In the scanning electron microscope marker bands were well recognized. Striations of the intermediate blocks of 10 cycles at $\sigma_{max} = 100$ MPa could be observed as a separate batch of striations, occurring between adjacent featureless bands of the blocks of 100 smaller cycles at $\sigma_{max} = 75$ MPa, see Figure 3.12. The crack extensions during the blocks of 100 smaller cycles were approximately equal to the adjacent crack extensions of the blocks with 10 larger cycles. The overall contribution to crack growth by the small marker load cycles is thus very small. However, since crack growth apparently does occur during the small marker load cycles, some influence on crack growth during the full cycles (100 MPa) could be questioned. A fracture mechanics analysis was carried out based on the crack closure concept. The analysis of the results of tests with and without the program marker load cycles indicated that such an interference could well be ignored. The analysis is presented in Appendix B.

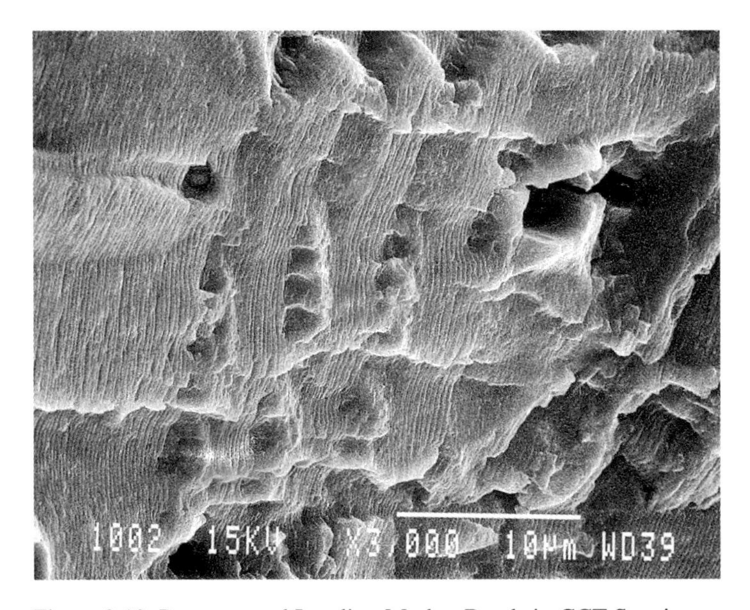

Figure 3.12 Programmed Loading Marker Bands in CCT Specimen

Two conclusions are drawn from the CCT and CCTB tests. One, from the stress-strain behavior measured during the static tension tests, the line model used to design the combined tension and bending specimen yields a design stress state in the area of interest of the specimen. Two, from the crack growth histories, the single under-load and program spectra do not noticeably affect the crack growth rate and is thus a viable candidate for fracture surface marking.

3.3.2 Open Hole Tests

In reviewing the literature on open hole tests for existing SIF solutions and their applicability to a joint, the part through crack growth portion of the fatigue life appears to be well quantified by the Newman/Raju solutions for one or two cracks growing from a hole in a plate subjected to tension, bending, and pin loading.[11,12] The same is not true for the through crack portion of the life where the crack front is oblique. The latter is necessary in order to model the entire crack growth life from crack nucleation as a part through crack at the faying surface to failure or link-up of adjacent through cracks. In addition, the through crack solutions can also be used for residual strength calculations although this is not investigated here.

The second set of experiments is twofold in purpose. First, the accuracy of the Newman/Raju solutions under combined loading, remote tension plus bending, is assessed. Second, fractographic data is generated to see the nature of the crack shape development under the combined loading. A tension and bending specimen similar to that shown in Figure 3.7 is developed for this task with the central crack replaced by an open hole. With this configuration, the effects of cyclic tension and bending on the crack growth of a crack located at the hole edge, which is also the net section of the joint, are isolated. Thus, the complexity induced by the load transfer in a typical joint is avoided.

Just as with the CCTB specimens, static tension tests are performed to determine the stress state and verify the validity of the line model. The presence of a single hole in the center of the plate does not significantly reduce the stiffness of the specimen; therefore, no notable difference should be seen between the CCTB and single hole specimens. However, since the line model makes no account for the width of the plate, specimens of a varying width, as the width increases the number of holes in the net section are also increased, are tested to explore the applicability of the line model to wide joints. As seen from what will be referred to as the 1, 5, 7-open hole joints, with widths of 100, 160, and 220 mm, the line model is applicable at least within this range of widths as seen in Figure 3.13 - Figure 3.15. The differences between the measured and calculated stresses amongst the three test series does not appear to be systematic; in other words, increasing the width does not adversely affect the accuracy in the line model in predicting the measured stress state. The accuracy for each of the three test series is within 5% that serves to validate the line model; however, for fatigue life predictions discussed in Chapter 5, the measured stresses are used.

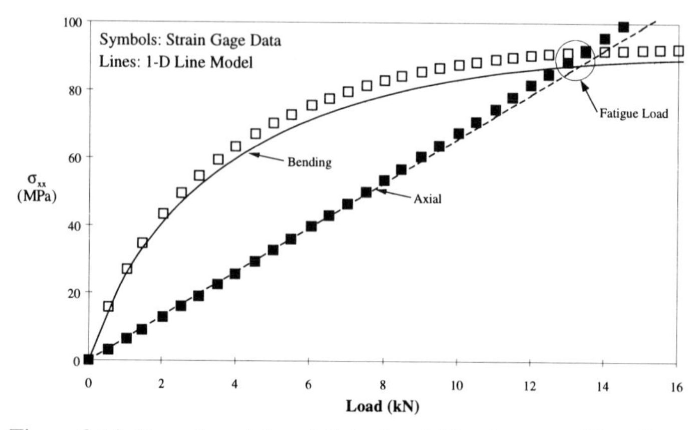

Figure 3.13 Experimental and Calculated Membrane and Bending Stresses in 1-Open Hole Specimen

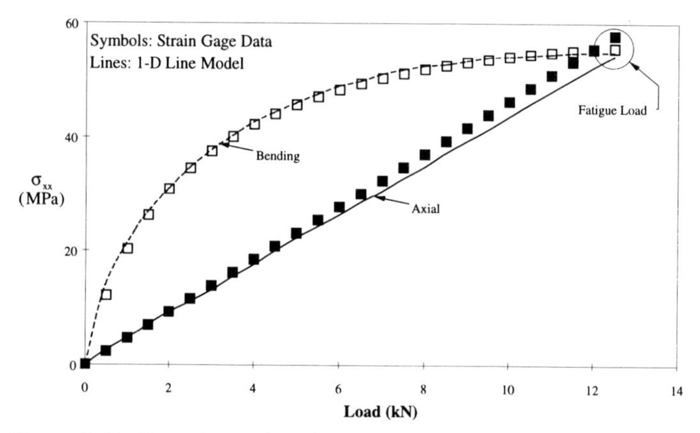

Figure 3.14 Experimental and Calculated Membrane and Bending Stresses in 5-Open Hole Specimen

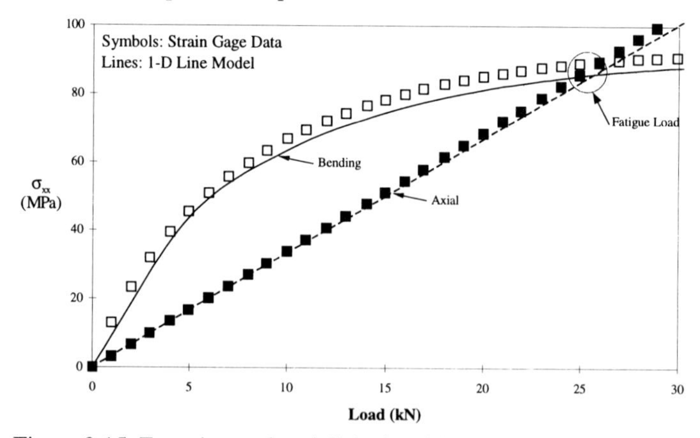

Figure 3.15 Experimental and Calculated Membrane and Bending Stresses in 7-Open Hole Specimen

The cyclical bending stress creates a nonuniform stress distribution through the thickness causing crack nucleation and growth from the side of the plate where the combined membrane and bending stresses are maximum, indicated here as thc front surface. As a result, the crack length at the front surface is larger than the penetrated crack length, the crack length on the side of the plate where the stress is a minimum, indicated here as the back surface. In a pure tension test, the penetrated crack experiences accelerated growth due to the high stresses in the cusp area of the plate, shown in Figure 3.16. However, this is not the case for combined tension and bending case since the penetrated crack experiences the maximum compressive stress from the bending load and when combined with the tensile load results in a lower stress level than at the front surface. Recall, the cracks nucleate and grow as part elliptical corner cracks until penetrating the back surface. The penetrated crack then continues to grow as a through crack maintaining a part elliptical, oblique, crack front; herein, this portion of the fatigue life is known as oblique crack growth.

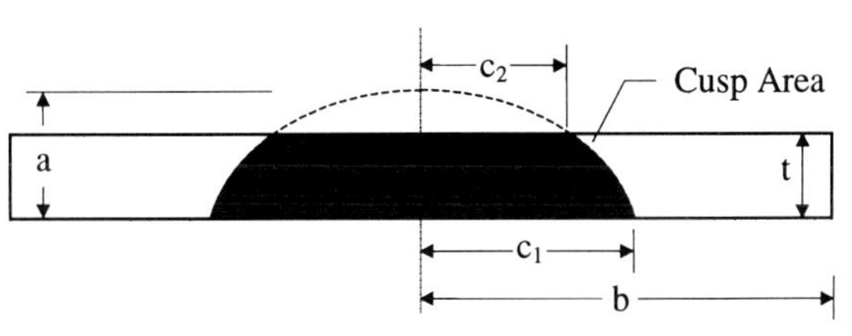

Figure 3.16 Oblique Crack Front Geometry

A series of 7-open hole specimens are tested to determine if specimens subject to combined tension and bending exhibit the catch-up behavior. The specimen dimensions, shown in Figure 3.17, are the same as that shown in Figure 3.9 except the central crack is replaced with seven 5 mm holes with a hole pitch of 25.2 mm resulting in a joint width of 220 mm. A useful parameter to describe the relationship between the tension and bending normal stress in the sheets of the joint is the bending factor, k defined as

$$k = \frac{\sigma_{\text{bending}}}{\sigma_{\text{tension}}} \tag{3.1}$$

The width of the net section and joint overlap dimensions are chosen such that a 100 MPa of tension and bending stress is produced creating a k = 1.0. The crack front shape cannot be monitored by nondestructive inspection; therefore, to obtain a qualitative understanding of the crack shape during the life the following procedure is adopted. Two specimens were tested until failure by fatigue that gives an estimate of the fatigue life for the given specimens and

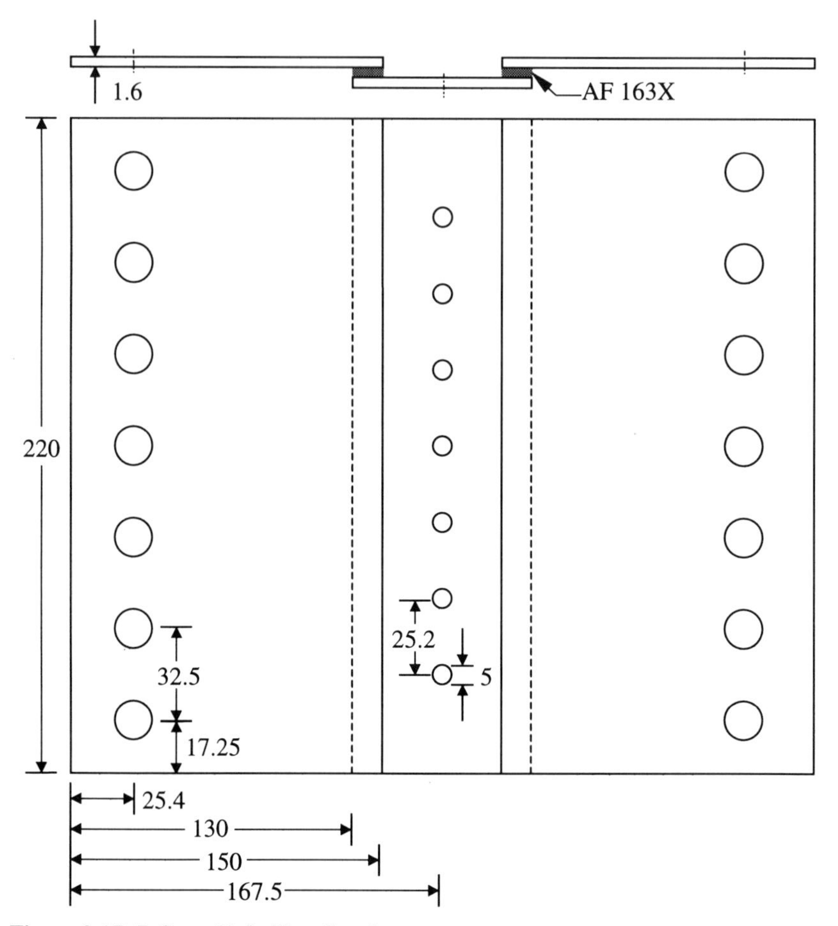

Figure 3.17 7-Open Hole Test Specimen

load conditions. Several additional specimens of the same geometry and load conditions were tested to a percentage of the fatigue life, then fractured by static overload. If the test was stopped at much less than 60% of the fatigue life, the cracks are quite small and difficult to accurately measure with an optical microscope. Cycling above 90% of the fatigue life is risky since this is possibly within the scatter band of the estimated fatigue life. Within the range of 60 - 90% of the fatigue life, eight percentages are chosen with two specimens for each percentage tested.

After static failure, two different methods are used to measure the crack. The crack can be measured directly with an optical microscope with an LVDT attached to the translatable microscope table or from a photograph of the crack, see Figure 3.18.

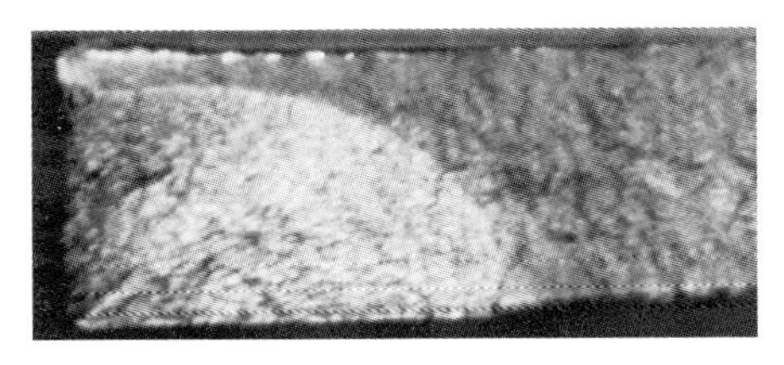

Figure 3.18 Crack Front Photograph from 7-Open Hole Fatigue Test Specimen

To measure the crack front with the optical microscope, a starting point at either the bore of the hole or at the front surface is identified. With the aid of the cross hairs etched in the microscope eyepiece lens, the microscope table is translated to the next point on the crack front to be measured. The crack front is curved; therefore, two translations of the table in orthogonal directions are required to measure the next crack front location. This procedure is repeated along the entire crack front. Unfortunately, this procedure is time consuming to complete, and the exact measurements are difficult to reproduce. Measuring the cracks from a photograph is far simpler since the measurements can be quickly taken using a ruler. For small cracks, the photograph can be enlarged as needed to increase the accuracy of the measurement. Since the photographs are to be made for fractographic examination, the cracks are measured from the photograph and not from the specimen. For ease in comparing crack fronts, an ellipse is fit by a least squares linear regression model through the crack front measurements of all cracks in the 7-open hole fatigue test specimens. The ellipse is assumed quarter elliptical with the minor and major axes coincident with the edge of the hole and front surface, respectively. The crack fronts shown in Figure 3.19 are different cracks from one specimen that clearly illustrates oblique crack growth throughout the fatigue life. The need for stress intensity solutions for part elliptical through cracks is paramount.

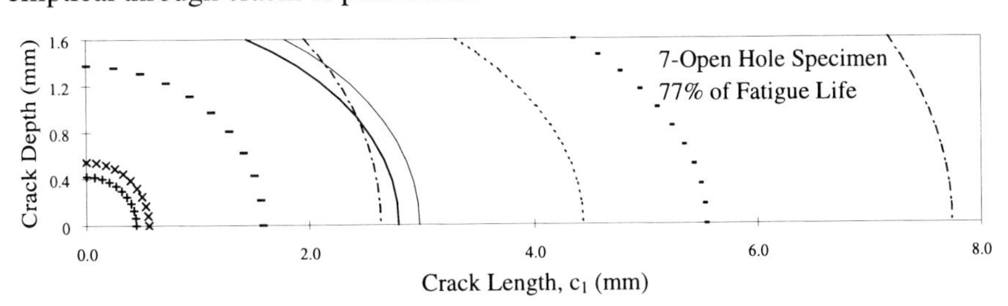

Figure 3.19 Crack Fronts at 77% of the Fatigue Life

More results of the 7-open hole specimens are given in Figure 3.20. At approximately 60% of the mean fatigue life, the cracks are still quite small, many measuring less than 1.0 mm as seen in Figure 3.20. The concave up line was drawn free hand to show what appears to be an upper limit.

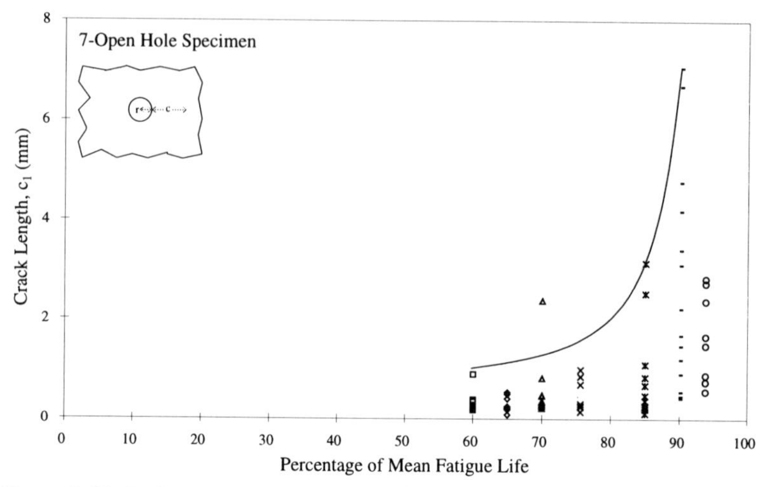

Figure 3.20 Fatigue Crack Length in the Open Hole Specimens after a Given Percentage of the Mean Fatigue Life

A similar test procedure was adopted by Ottens in the testing of lap joints with a staggered rivet pattern; seven rivets in the outer rows, and six rivets in the center row.[13] The role of the rivet in increasing the complexity of the crack growth mechanism is evident in comparing the trends of Figure 3.20 and Figure 3.21. The large range in fatigue crack initiation for the open hole specimens is not seen in the lap joints as a result of the load shedding that occurs in the joint. As the cracks nucleate and grow in the lap joint, more load

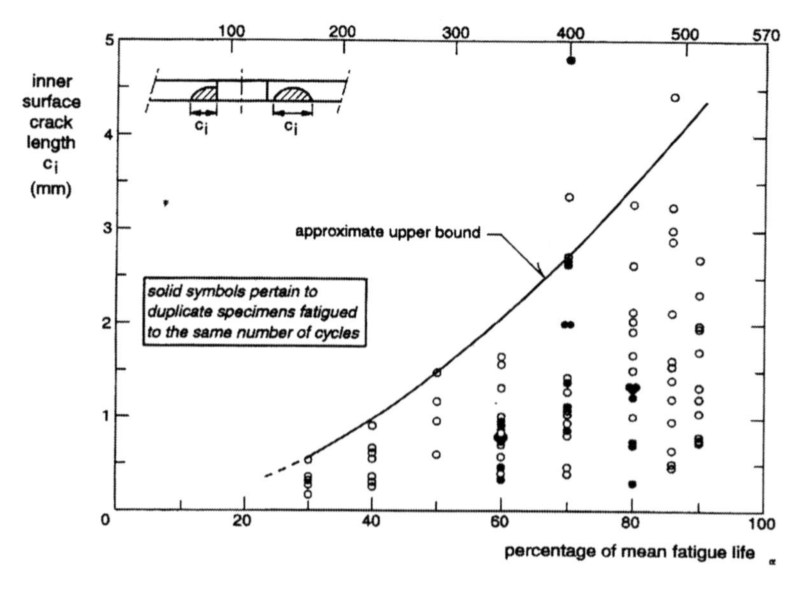

Figure 3.21 Fatigue Crack Length of a Riveted Lap-Splice Joint after a Certain Percentage of the Fatigue Life (NLR results)[13]

is shared by the adjacent rivets as the crack length at a given rivet increases. Now the adjacent rivets are transferring additional load thereby increasing the stresses near the hole. If no cracks are present at the adjacent holes, the higher stress level promotes crack nucleation and growth, or if cracks are present, they grow faster.

The stress amplitude can also play a role in the crack initiation period, the amount of time for the cracks to nucleate and grow to a visible size. The lap joints were loaded by remote tension to 79 MPa and for this geometry, the secondary bending stress can be estimated at 86 MPa using the line model discussed previously. A conservative estimate on the load transfer through the outer rows of rivets is 35% resulting in a bearing stress at each hole of 173 MPa; thus the combined stress is 338 MPa. The bearing stress is calculated using the following equation.

$$\sigma_{brg} = \frac{\gamma P}{nDt}$$

where

γ = Load transfer ratio, assumed to be 35%
P = Total applied load
n = Number of rivets per rivet row
D = Hole diameter
t = Sheet thickness

The open hole tests were subject to 100 MPa of both tension and bending. The higher stress level in the lap joint tests gives reason to the relative smaller crack initiation period.

The crack growth rate of the part through cracks is also of interest since this determines when the cracks grow through the thickness and penetrate the back surface of the sheet becoming detectable by visual inspection. Again for comparison, the same test specimens as in Figure 3.20 and Figure 3.21 are examined. For the open hole specimens shown in Figure 3.22 the cracks did not penetrate the back surface until approximately 80% of the mean fatigue life; while in the lap joints, shown in Figure 3.23 back surface penetration occurred at approximately 70%. The similar time periods for the crack to penetrate the back surface might be associated with the similar k factors for each, 1.0 and 1.08, for the open hole and lap joints, respectively.

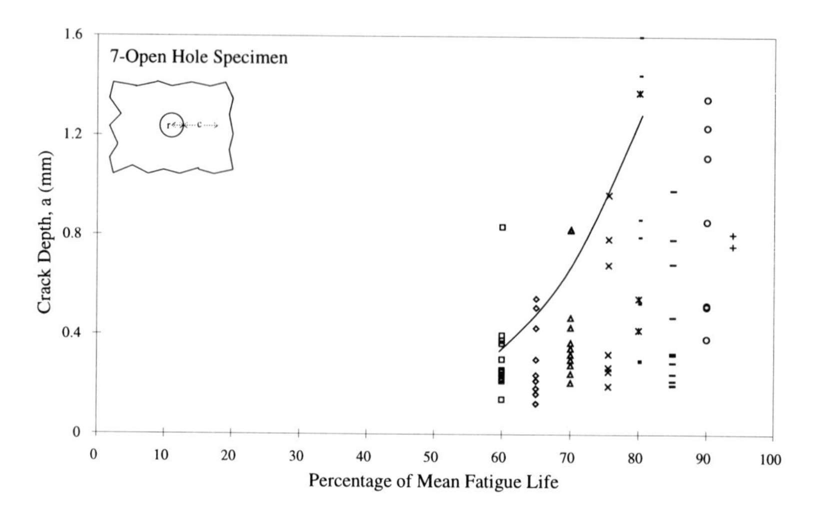

Figure 3.22 Fatigue Crack Depth in the Open Hole Specimens after a Given Percentage of the Mean Fatigue Life

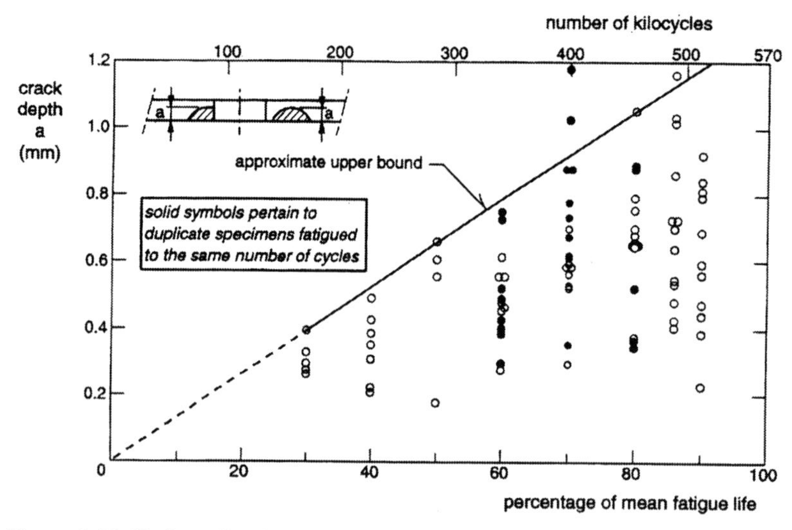

Figure 3.23 Fatigue Crack Depth of a Riveted Lap-Splice Joint after a Certain Percentage of the Fatigue Life (NLR results)[13]

3.3.3 Riveted Joints

In an attempt to reduce this inherent complexity two simplified lap joint test series are conducted. One of the simplest riveted connections is the 1½ dog-bone specimen shown in Figure 2.5. The primary advantage of this specimen is the reduced complexity of the load transfer since there is only one rivet row. Another specimen with a known load transfer is an asymmetric lap splice joint as shown in Figure 2.6. The latter specimen is the body of the experimental investigation of rivet lap joints. To avoid repetition in the proceeding sections, all joints are constructed of 2024-T3 clad aluminum with a dry installation of NAS 1097X-X AD rivets. The sheet thicknesses and rivet diameters for each series are varied to control the stress state in the net section of the joint.

3.3.3.1 1½ Dog bone single rivet joint

In the previous section, sheet specimens with open holes under a combined tension and bending fatigue load were considered. As pointed out before, this type of specimen represents still an essentially different notch effect as compared to sheet specimens with rivet loads on the holes under the same type of combined tension and bending.

It then was considered to be useful to start research on riveted joints with a 1½ dogbone specimen similar to that shown in Figure 2.5 except with only one rivet. Two arguments were that in a 1½ dogbone specimen with a single rivet, there is no interference between cracks from different rivets, and the load transmission by the single rivet was supposed to be well defined. This 1½ dogbone specimen was also extensively used for its simplicity in the AGARD round robin test program to evaluate the fatigue quality of a large variety of fasteners.[14-16] The specimen was designed with the goal to simulate the load transfer and secondary bending characteristics or runouts or stiffeners attached to the outer skin. It was developed by the Laboratorium für Betriebsfestigkeit (LBF) in West Germany. It was found in the AGARD program that the amount of secondary bending in the test section is influenced by the fastener fit.

As part of the same AGARD research program another still relatively simple specimen, the Q-joint, was designed in the UK as an alternative to the 1½ dogbone specimen, see Figure 3.24. The alternative design was based on a single fuselage lap joint, however, another load carrying sheet was added, which gives some more bending stiffness. It was assumed that the extra bending stiffness in a fuselage is provided by longitudinal stringers attached to the lap joint. Moreover, the double sheer connection at the

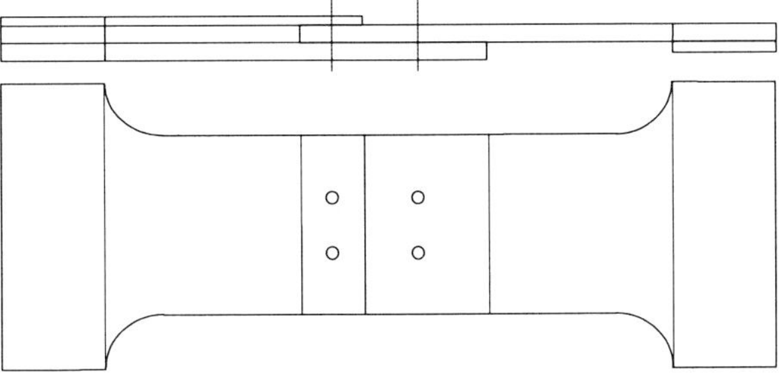

Figure 3.24 Q Joint

left fastener row ensures that fatigue failures do occur at the single shear connection in the right rivet row. Unfortunately, secondary bending and load transmission in the Q-joint still deviate considerably form those in a fuselage lap joint.

It was thought that a more realistic load transmission might be approached by a modified 1½ dogbone specimen with the dimensions shown in Figure 3.25.

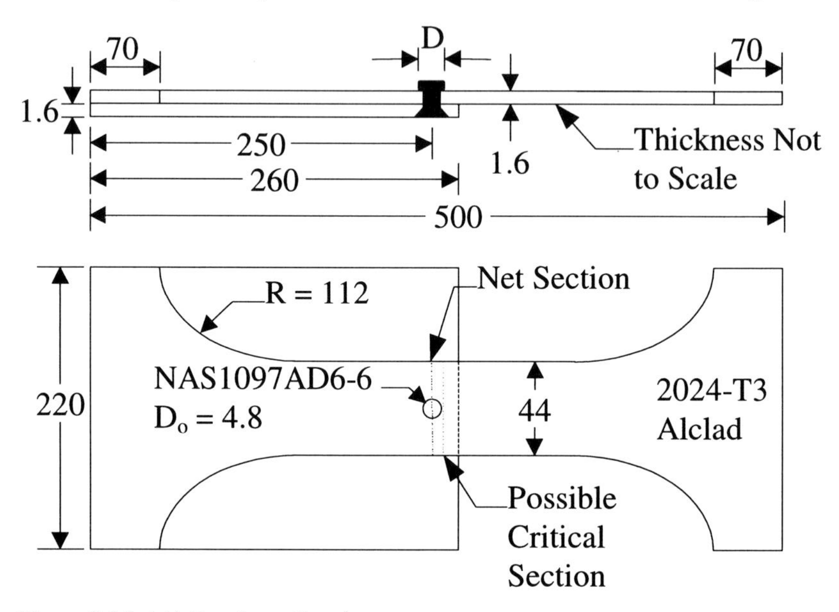

Figure 3.25 1 ½ Dog bone Specimen

Five fatigue tests were carried out, see Table 3.1, with the rivet squeeze force as the main variable. As shown by Müller, the squeeze force has a predominant influence on the fatigue behavior of a riveted joint.[1] Squeezing is done with a force controlled riveting machine developed by Müller. Some strain gaged specimens were used to determine the load transmission by the rivet, bypass load, and secondary bending. The fatigue tests were carried out at σ_{max} = 100 MPa, R = 0.03, and a 15 Hz frequency. During the fatigue tests, crack length measurements were carried out using an optical microscope, 40x and 60x. Fatigue loading was often interrupted trying to detect cracks before they became longer than one millimeter. Once a crack was detected, measurements were taken every 1000 to 5000 cycles using a millimeter scale. An aspect of special interest was the effect of the squeeze force on the location of crack initiation. The initiation should be expected at the faying surface of the base plate due to secondary bending, leading to corner cracks at the boundary of the hole at the faying surface. As pointed out before, high squeeze forces induce crack initiation away from those corners. The squeeze forces selected in Table 3.1 are relatively low.

The first specimen (specimen 1-1) was joined with a squeeze force of 18.3 kN and a rivet tail length, the length of rivet protruding above the base plate before squeezing, of 6.8 mm. This specimen presents two rather symmetrical cracks starting from two points close to the center of the top part of the hole, see Figure 3.26. Thus with a desire to shift the crack nucleation point towards the centerline of the hole, the subsequent specimens were joined with lower squeeze forces and shorter rivets. A shorter rivet was intended to yield a lower secondary bending ratio. In addition, the shorter rivet results in a smaller driven head D, which in turn offers less bending restraint thereby lessening the bending.

Table 3.1 1 ½ Dog Bone Fatigue Test Parameters

Specimen Name	Squeeze Force (kN)	Rivet Tail Length (mm)	D/D_o	σ_{max} (MPa)	Freq. (Hz)	kcycles to Failure
1-1	18.3	6.8	1.4	100	15	185.662
1-2	14	6.8	1.3	100	15	*
1-3	9.5	6.8	1.1	100	15	147.941
1-4	9.5	1.8	1.03	100	15	181.658
2-1**	9.5	1.8	1.03	100	15	62.201

* Statically failed by overload after 120 kcycles

** The rivet was rotated to have countersink in base plate

Figure 3.26 Cracking at the Top of the Rivet Hole

These trials gave less improvement then expected in the shift of the crack nucleation point. For normal lap joints used for laboratory tests, a nucleation point exactly on the centerline of the hole can be achieved with a ratio of the driven head to the nominal diameter less then 1.3.[1] Only test 1-3 and 1-4 present a shift of the crack nucleation site closer to the minimum net section, but still not at the net section of the base plate. Specimen 2-1 is joined with the manufactured head of the rivet on the base plate. Thus, the base plate is countersunk. This specimen resulted in two symmetrical cracks exactly on the centerline of the hole.

The 1½ dog bone was also used in the investigation of the fatigue behavior of mechanical joints by Palmberg, Segerfröijd, Wang, and Blom.[17] The geometric proportions of their specimen were completely different. The ratio between the width of the base and splice plate was 0.77 yielding vastly different values for the secondary bending ratio and load transfer. In order to force the crack to start at the net section, artificial defects at the fastener hole and at the faying surface of the base plate were manufactured by electro-discharge machining. In spite of this, the fatigue crack leading to failure of the joints started from the artificial defects only in 10 specimens out of 24. A majority of cracks started from points close to the center of the unloaded cylindrical part of the fastener hole. A nonuniform distribution of the bending stress at the hole cross section in normal lap joints has been found by other researchers.[1,17-20] A shift of the maximum bending stress from the net section to under the rivet head contour, shown in Figure 3.2, was first observed by Müller.[1] This effect is amplified in the single fastened base plate because the edges of the plate itself are not constrained by adjacent sheet material with rivets. Moreover, this effect is mitigated when the rivet is assembled with the manufactured head on the critical section due to different transmission of the bending moment between the countersunk edges of the rivet and hole. Indeed, in specimen 2-1 the two symmetrical cracks started exactly at the net section. To circumvent the limitations of the 1 ½ dog bone specimen, a more traditional lap splice joint test specimen is employed and discussed in the following section.

3.3.3.2 Asymmetric Lap Splice Joint

All joint configurations use 2024-T3 clad aluminum and are similar in that only the sheet thickness (t), rivet diameter (d), and rivet row pitch (p) are varied in order to control the degree of secondary bending and the magnitude of the load transferred by the rivets. A schematic representative of all joints with pertinent variables is shown in Figure 3.27. Again, note the rivets imply an antisymmetric configuration that implies that the noncountersunk part of the rivet holes is in the fatigue critical section of both sheets.

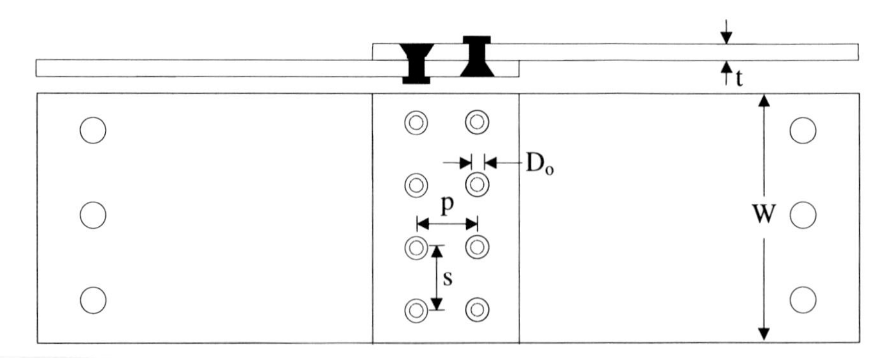

Figure 3.27 Schematic of Asymmetric Lap Splice Joint

The three primary goals with the asymmetric lap joint test series is to verify the sheet thickness effect, (section 3.3.3.2.1), to determine the relationship between the bending factor k and crack shape, (section 3.3.3.2.2), and to validate the marker load spectra, (section 3.4). The first two goals are investigated by changing the joint geometry to isolate the particular parameter of interest. Evaluation of the marker load spectra relies heavily on fractographic analysis of the fracture surface using a scanning electron microscope. Furthermore, the rivets are installed such that crack nucleation and growth occurs first in the noncountersunk hole; thus, the crack growth data can then be used to verify the stress intensity factor solutions presented in Chapter 4.

3.3.3.2.1 Sheet Thickness Effect

Joints to determine the thickness effect are designed with a varying sheet thickness, rivet diameter, and rivet pitch to isolate the sheet thickness as the only independent variable. The preceding static analysis revealed the impossibility of isolating a changing sheet thickness as the only independent variable. This dilemma is caused by the multiple dependencies of the geometric parameters on the local stresses; e.g. changing the thickness of the sheet not only affects the tensile stress, but also the bending and bearing stresses. In addition, the joint must be designed with commonly available aluminum sheets and rivets. At best, the number of independent variables can be reduced to two. Two test series were completed, one with three independent variables, sheet thickness, rivet load, and total stress; the other with two independent variables, sheet thickness and rivet load. The joint geometries, following the definitions shown in Figure 3.27, and resulting stresses are summarized in Table 3.2 - Table 3.5. In test series 1, the combinations of the rivet diameter, D_0, and the sheet thickness, t, agree with usual combinations in aircraft structures. It implies smaller D_0/t values for smaller thickness, which is significant for the bearing stress in the rivet holes. The purpose of the second test series is to remove the dependence of the sheet bearing stress on the sheet thickness as seen in the constant D/t ratio with increasing thickness (Table 3.3). A single, slightly deviating D/t ratio for test 96W2.5t is due to the available sheet thicknesses and rivet diameters.

The load levels refer to the maximum applied tensile load and are chosen to obtain the desired k factor. Except for the bending stress that is calculated using the line model (sample calculations in Appendix A), the stresses are calculated with the following set of equations.

Table 3.2 Thickness Effect Test Series 1 Joint Geometry

Test Name	Width W (mm)	Thickness t (mm)	Rivet Diameter D (mm)	Rivet Pitch s (mm)	Row Pitch p (mm)	s/D	p/D	D/t
64W1.2t	64	1.2	3.2	16	16	5	5	2.7
80W1.6t	80	1.6	4.0	20	20	5	5	2.5
96W2.5t	96	2.5	4.8	24	24	5	5	1.9

Table 3.3 Thickness Effect Test Series 2 Joint Geometry

Test Name	Width W (mm)	Thickness t (mm)	Rivet Diameter D (mm)	Rivet Pitch s (mm)	Row Pitch p (mm)	s/D	p/D	D/t
64W1.6t	64	1.6	3.2	16	16	5	5	2.0
80W2.0t	80	2.0	4.0	20	20	5	5	2.0
96W2.5t	96	2.5	4.8	24	24	5	5	1.9
128W3.2t	128	3.2	6.4	32	32	5	5	2.0

$$\sigma_{tension} = \frac{P}{Wt} \tag{3.2}$$

$$\sigma_{tension\ net} = \frac{P}{(W - nD)t} \tag{3.3}$$

$$\sigma_{bearing} = \frac{\gamma P}{nDt} \tag{3.4}$$

$$\sigma_{total} = \sigma_{tension} + \sigma_{bending} + \sigma_{bearing} \tag{3.5}$$

$$k = \frac{\sigma_{bending}}{\sigma_{tension}} \tag{3.1}$$

where γ is the load transfer ratio through one rivet row (50% for asymmetric lap splice joints since there are only two rivet rows) and n is the number of rivets per row, for all the asymmetric lap splice joints. The k factors used, 1.8 and 2.0 for test series 1 and 2, are toward the upper bound of k factors seen in the operational environment where $1.0 \leq k \leq 2.0$. Since use of commercially available aluminum sheets and rivets is required from an economic standpoint, using higher k factors is one of the design decisions made in trying to isolate the sheet thickness and rivet load as the only independent variables. Undoubtedly the k factors used can be reduced somewhat; however in order to do so, many joints would have to be designed, fabricated, and fatigue tested. In trying to achieve the design goals for the joint care must be exercised when choosing the remote tensile applied load. If the load level is too low, the crack nucleation site may shift from the location seen in service at the hole edge to the top of the hole due to fretting corrosion. The high k's used in these test series are not anticipated to diminish or mask the effect of the changing sheet thickness. Possibly, the effect is magnified since the degree of secondary bending is greatly affected by changes in the sheet thickness.

Table 3.4 Thickness Effect Test Series 1 Loads and Stresses

Test Name	Load (N)	$\sigma_{tension}$ (MPa)	$\sigma_{tension\ net}$ (MPa)	$\sigma_{bending}$ (MPa)	k	$\sigma_{bearing}$ (MPa)	σ_{total} (MPa)	Rivet Load (N)
64W1.2t	4454	58	73	106	1.8	145	309	557
80W1.6t	8448	66	83	121	1.8	165	352	1056
96W2.5t	26640	111	139	203	1.8	278	592	3330

Table 3.5 Thickness Effect Test Series 2 Loads and Stresses

Test Name	Load (N)	$\sigma_{tension}$ (MPa)	$\sigma_{tension\ net}$ (MPa)	$\sigma_{bending}$ (MPa)	k	$\sigma_{bearing}$ (MPa)	σ_{total} (MPa)	Rivet Load (N)
64W1.6t	6246	61	76	122	2.0	153	336	781
80W2.0t	9760	61	76	122	2.0	153	336	1220
96W2.5t	16080	67	84	134	2.0	168	369	2010
128W3.2t	24986	61	76	122	2.0	153	336	3123

The measured crack shapes for series 1 are shown below in Figure 3.29 with tabular data listed in Appendix C. Note, since the asymmetric lap joint has only two rivet rows and the rivets in each row or installed opposite of one another, there are two critical rows for fatigue cracking. Thus, the fracture surfaces of the critical row of each sheet are shown in Figure 3.29.

The most noticeable feature of all the cracks is the part elliptical shape throughout a majority of the crack fronts. Where the crack intersects a free surface, the shape appears to deviate from part elliptical, shown in Figure 3.28, due to the change in the stress field. The area near the free surface is commonly known as the boundary layer, and the stresses here are affected by the loss of constraint at the free surface and possibly residual stresses from manufacturing the sheet or joint. The holes for all the joints are drilled undersized and then reamed to the desired size with a tolerance of +12 μm (H7 tolerance).[19] Other possible factors promoting deviation from the ¼ ellipse are the residual stress in the plane of and perpendicular to the sheet due to squeezing the rivet. Since the exact cause of the deviations from the ¼ elliptical crack front shape is uncertain, a ¼ elliptical

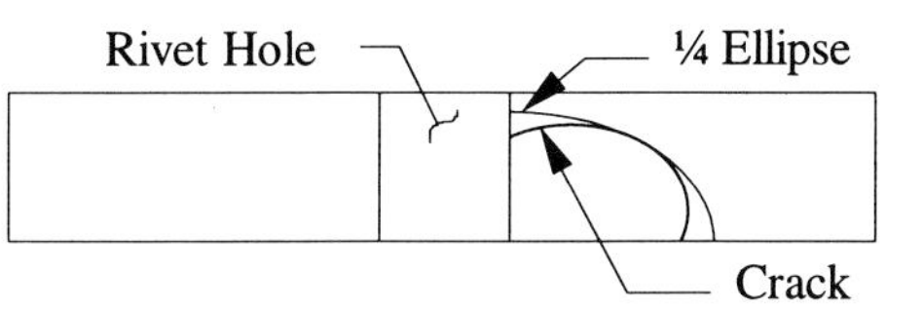

Figure 3.28 Boundary Layer Effect on Crack Front Shape

shape is assumed for all comparisons discussed below and analytical work presented in Chapters 4 and 5. Multiple site cracking occurs in all specimens with no preference to the rivets at the sheet edges; thus, the edge effect mentioned in Chapter 2 does not develop. Visual inspection of the fracture surface can be very informative but in this instance, it is not sufficient for determining the effect of the changing sheet thickness. Toward this end, for each crack an ellipse is forced through the measured data points by linear regression, and it is this fit data, listed in the tables in Appendix C, which is used to assess the thickness and k factor effects.

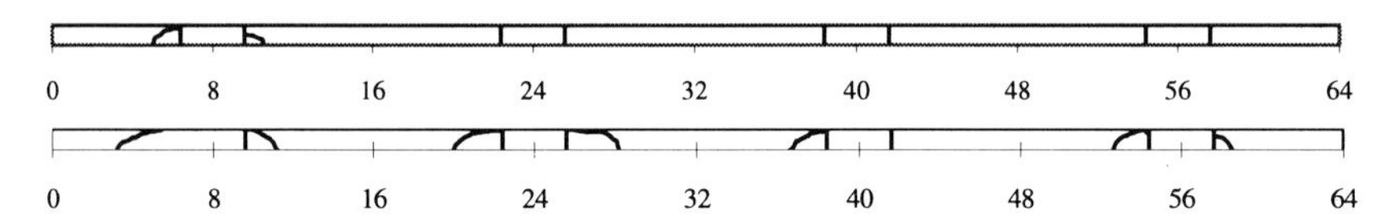

80W1.6t (Scale 1:2.22): t = 1.6 mm

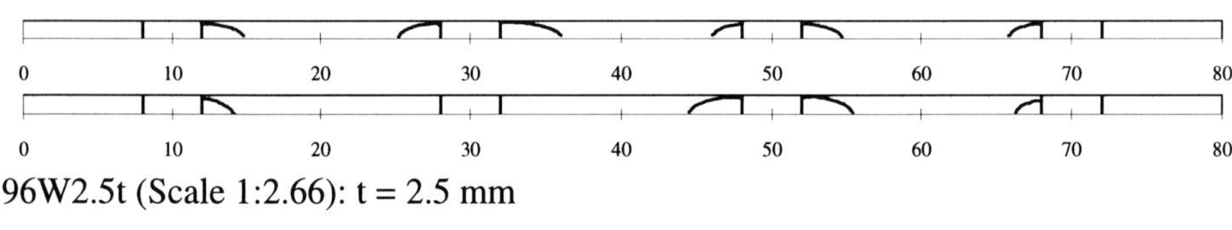

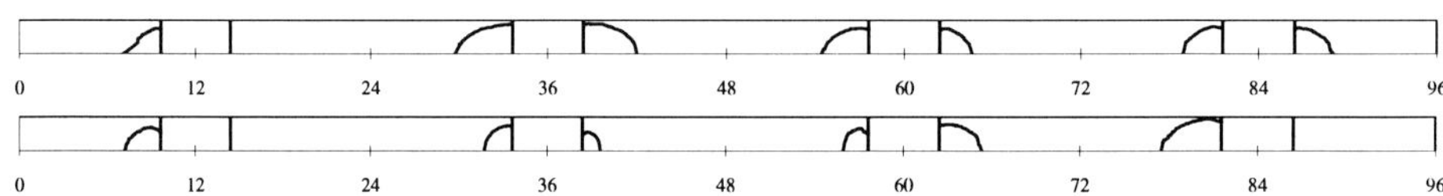

Figure 3.29 Fracture Surfaces from Asymmetric Lap Joint Test Series 1

The a/c and a/t ratios are measured using the convention shown in Figure 3.16. In each test, some cracks nucleate and grow very late in the life of the joint as a result of the severe cracking in various other locations of the joint and the ensuing load redistribution, load shedding, to the stiffer uncracked areas. These cracks are referred to as late initiators that experience a stress distribution much different from that, which caused the cracks at the other locations. The stress redistribution after cracking is difficult to quantify and more so to experimentally verify, neither of which is attempted here. For a discussion on the stress redistribution, see reference [20] by Wit who assumed the stress redistribution could be accounted for by the reduction of the load carrying area in a given rivet pitch. Thus, the late initiators are not included in the sample for calculating the mean and standard deviation of the a/c and a/t ratios.

After reviewing the experimental crack shape data and the elliptical curve fits of this data, a/t is not depending on the sheet thickness. The a/c ratios are most sensitive to the bending stress distribution. The maximum tensile component of the bending stress is at the faying surface and the maximum compressive component is at the free surface. The distribution of the bending stress through the thickness of the sheet is assumed to be linear although this may not be entirely accurate due to the contact behavior between the rivet and sheet along the longitudinal axis of the rivet. If the bending stress is low, the overall stress in the net section can still be tensile, shown in Figure 3.30; however, if the relative bending stress is high, shown in Figure 3.31, the stress at the faying surface is tensile but compressive at the free surface.

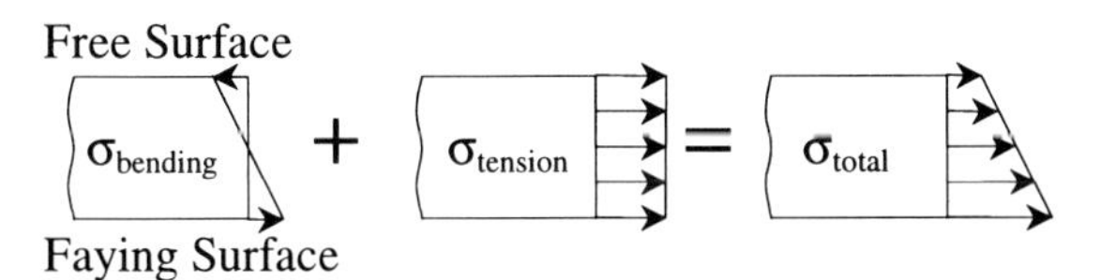

Figure 3.30 Stress Distribution through the Thickness of the Sheet with a Low Bending Stress

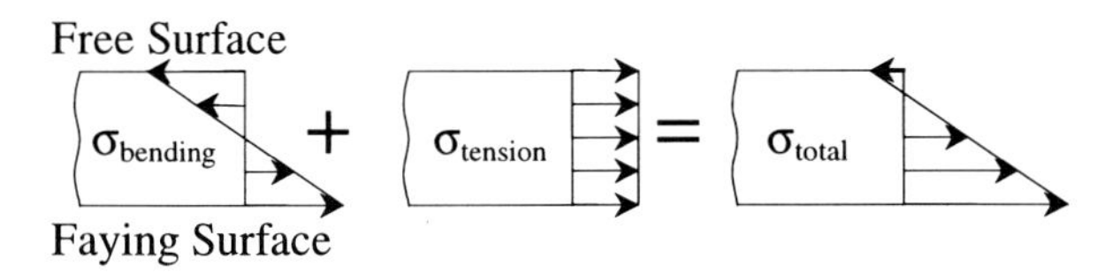

Figure 3.31 Stress Distribution through the Thickness of the Sheet with a High Bending Stress

By increasing the relative bending stress, the crack should grow slower through the thickness due to the decreasing tensile stress. In test series one, the a/c ratio is increasing as the sheet thickness increases; however, although the bending stress is increasing, the k factor, the ratio between the bending and tension stress, is held constant at 1.83. If the a/c ratio is changing only due to the change in sheet thickness, then a thickness effect is established. Upon further examination of the stresses affecting the stress state at the hole edge, the sheet bearing stress must be considered. As the sheet thickness and rivet diameter are increased, the bearing stress should decrease, but to maintain the desired k factor the load must also be increased which ultimately causes the increase in bearing stress. Due to the contact being maximum at the free surface of the plate and minimum at the faying surface, the bearing stresses will also follow the same behavior. Figure 3.32 schematically shows the stress state at the hole edge.

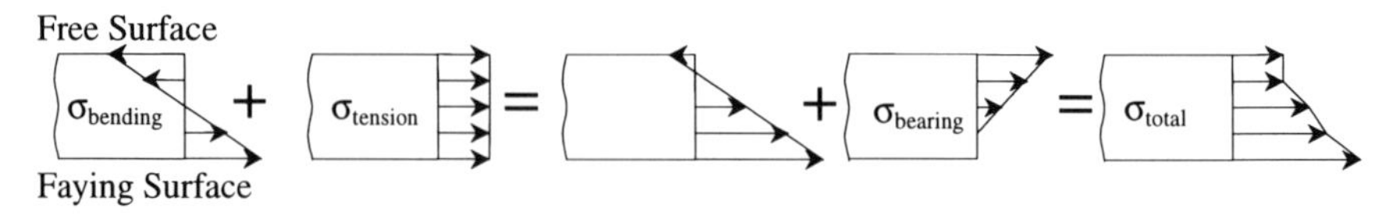

Figure 3.32 Stress Distribution through the Thickness of the Sheet at the Rivet Hole Edge

The maximum bearing stress being at the free surface mitigates the effect of the compressive bending stress; thus, the stress distribution through the thickness becomes more tensile. The part through and oblique through cracks are experiencing a dominant tensile field causing the a/c to increase. The increase in a/c in a dominant tensile stress field was also reported by Grandt, Harter, and Heath, which they called "catch-up".[21] Although the increase in a/c ratio with increasing thickness for test series 1, the change in crack shape cannot be solely attributed to the changing thickness since the state of stress is also changing. In order to remove the effect of the increasing bearing stress with increasing thickness, the lap joints were redesigned for the second test series whereby the tensile, bending, and bearing stresses remain constant for the four different joints.

Just as in test series 1, a/t is not depending on the sheet thickness. More specifically, increasing the amount of load transferred by the rivet has no effect on the a/t ratio. The sheet thickness effect on the a/t ratio might be masked by errors in the least squares linear regression of the penetrated crack front data. The linear regression is strongly affected by the crack front measurements taken from near the penetrated side of the crack; unfortunately, the crack front is the least clear in this area. As the crack is growing through the thickness, the growth accelerates as the cusp area becomes smaller. Eventually the crack is large enough to cause static overload in the cusp area leaving a shear lip at the free surface of the sheet as seen in Figure 3.33. The accelerated growth and shear lip, not to mention possible contact between the opposing sides of the fracture surface, cause the crack front to be less defined which eventually may induce errors in the linear regression. However, with one exception, the a/c ratios, listed in the tables in Appendix C,

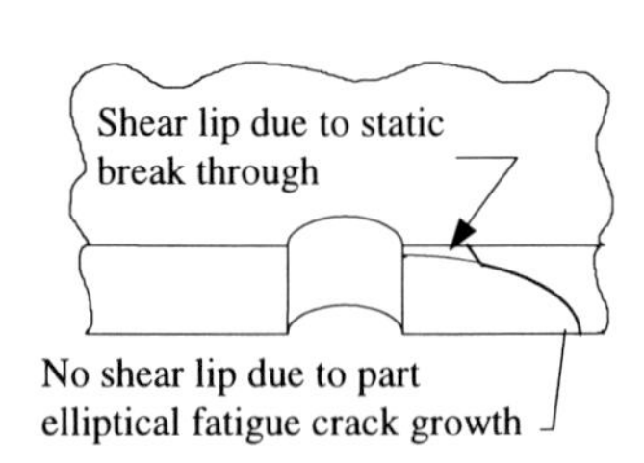

Figure 3.33 Crack Front Morphology

64W1.6t (Scale 1:2): t = 1.6 mm

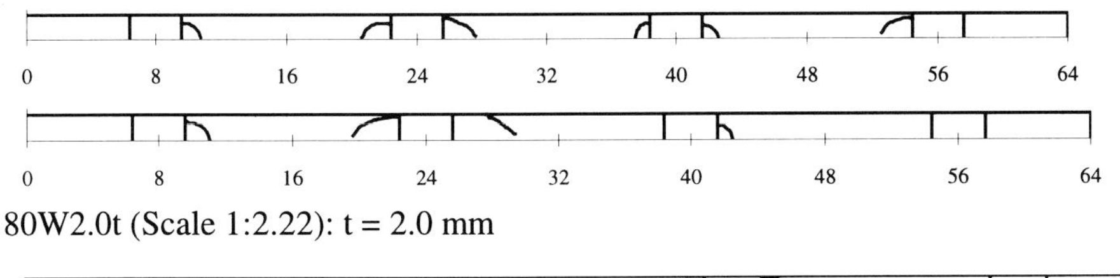

80W2.0t (Scale 1:2.22): t = 2.0 mm

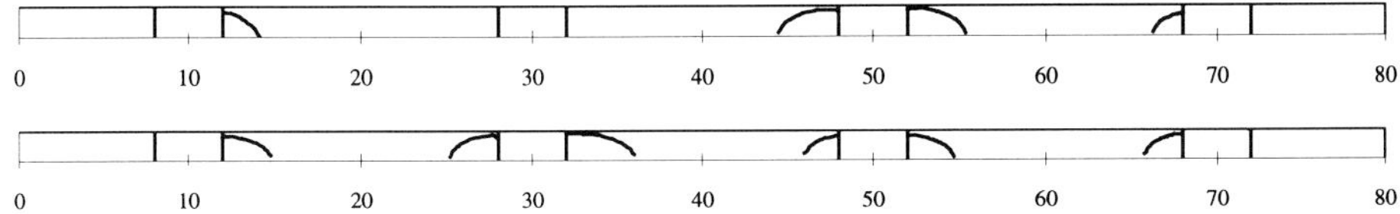

96W2.5t (Scale 1:2.66): t = 2.5 mm

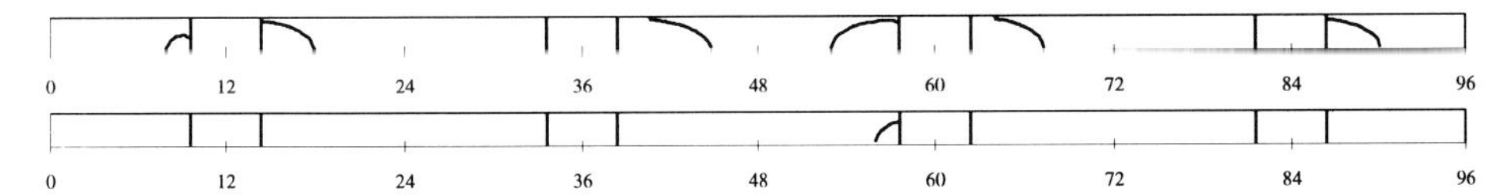

128W3.2t (Scale 1:3.33): t = 3.2 mm

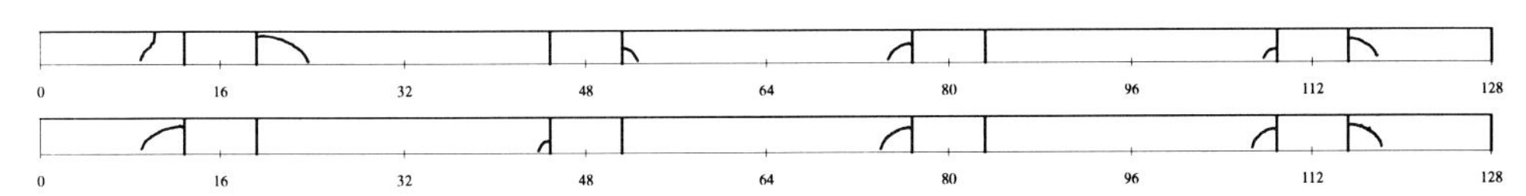

Figure 3.34 Fracture Surfaces from Asymmetric Lap Joint Test Series 2

do decrease with increasing sheet thickness which is indicative of a compressive or smaller tensile stress field toward the free surface of the sheet. As seen in Figure 3.34, the crack fronts become shallower with increasing thickness. Just as in test series 1, each asymmetric lap joint has two critical rows, and the fracture surface of each critical row is shown in Figure 3.34.

The results from joint 128W3.2t, which had the thickest sheets (t=3.2 mm) of all joints tested, have not been considered due to edge cracking. One possible explanation for this behavior is the width effect, discussed in Chapter 2, common to wide panel MSD testing where the edge rivets are transferring more load than the interior rivets. The crack sizes illustrated in Figure 3.34 show the largest cracks at the edge rivets. Although only data for one of the 128W3.2t joints is presented, two tests were completed but the edge cracking was so severe in the first specimen, the testing was aborted. The edge cracking in the thicker specimens is due to the high stress in the sheets resulting in increased lateral contraction. The outside rivets react a majority of this load ultimately causing crack nucleation before the two interior rivets of the same row. This same trend was seen by Müller in his finite element calculations.[1] For the commercially available sheet thicknesses and rivet diameters, the sheet thickness cannot be isolated as the only independent variable. In this series the rivet load, the amount of load transferred by a rivet, increased along with the thickness while maintaining a constant bearing stress. The higher rivet loading increases the amount the rivets rotate which in turn reduce the amount of clamping of the sheets created during installation of the rivets. Reduction in the clamping forces in a joint shifts the point of maximum bending from the top of the hole back to the net section.[1] Furthermore, the rivet head is less likely to bend as the rivet tilts which causes the moment to be reacted by the sheet. Finally, the increased rivet tilting causes the point of maximum bending to shift from the top of the rivet hole back toward the net section of the joint; thus the bending stress at the net section increases with increasing rivet load (the rivet load is the load transferred by the rivet and not the force used to squeeze the rivet during installation). Müller found a similar result investigating the effect of the rivet squeeze force on the secondary bending in a joint.[1] He found that as the rivet squeeze force is decreased, rivet tilting increases resulting in the shift of the maximum bending from the top of the hole back to the net section.

3.3.3.2.2 Effect of k Factor on Crack Shape

The secondary bending stress is most often ignored in MSD prediction models due to its nonlinear relationship with the membrane loading caused by cabin pressurization. Intuitively, as the bending stress increases the a/c ratio decreases resulting in shallow cracks for high bending stress and deep cracks for low bending stress as alluded to in the previous section. A third series of asymmetric lap splice joints are designed and fatigue tested in an attempt to experimentally verify the relationship between the secondary bending stress and crack shape. The same basic joint geometry is used as in the previous asymmetric lap joint tests, shown in Figure 3.27, using a sheet thickness of 2.0 mm and a rivet diameter of 4.8 mm. The rivet row pitch, p, is the only independent variable whereby increasing p decreases the bending stress and vice versa. Five different joints are designed to produce a k factor ranging from 1.0 - 2.0 in response to a remote tensile stress of 100 MPa. In addition to determining the relation between the bending stress and crack shape, the development of the crack shape is also of interest. Specifically, the question is whether the shape changes through the life of the crack. From a prediction point of view, a constant crack shape throughout the life is simpler since K solutions for only one crack shape are required.

Using the same protocol as in the 7-open hole fatigue testing discussed earlier, five joints for each joint design are tested. The first test is run to fatigue failure with the resulting number of cycles used as an estimate of the joint fatigue life. The remaining four specimens for each joint geometry are tested to 60, 70, 80, and 90% of the fatigue life. Crack shapes comparisons between the different joints and thus bending stress levels are made at 60% and 90% of the fatigue life, N, as listed in Table 3.6 (except for joint 96W2T24N where the crack shapes at 90%N are not available). Diagrams of the fracture surfaces for the joints at 60% and 90% are in Appendix C along with tables of the corresponding crack shape data. Any discussion on the trends which may exist in this data must be predicated by the fact that mean values of a/c and a/t for a given joint at a given %N are used; thus, the trends can only be qualitative.

Table 3.6 Effect of $\sigma_{bending}$ on Crack Shape in the Asymmetric Lap-Splice Joint Test Series 3

Test Name	Row Pitch (mm)	k	60% N		90%N	
			Mean a/t	Mean a/c_1	Mean a/t	Mean a/c_1
96W2T8N	16	1.988	0.509	1.443	0.771	0.787
96W2T12N	24	1.682	0.398	1.472	0.829	0.918
96W2T16N	32	1.510	0.508	1.473	0.837	0.992
96W2T24N	48	1.246	0.592	1.442	0.692*	1.261*

* Tested to 80%N

In addition, when comparing the crack shapes at different %N, recall these are not the same cracks at different stages of their lives, but different cracks in a joint of the same design. The %N for comparison are chosen to see if the crack is changing shape throughout its life. Comparing the crack shapes at 90%N, the mean a/c and a/t ratios increase with decreasing bending stress. The increase in the a/c ratio is expected since the stress state becomes more tensile with decreasing bending stress resulting in catch-up of the c_2 crack. Also, as the number and lengths of cracks in a joint increase, the bending stiffness decreases with the remaining uncracked ligaments between the rivets being loaded as though they were subjected to uniaxial tension. At 60%N, the cracks are relatively small, less than 2.0 mm, and no trend is apparent in either a/c or a/t. For cracks of this size and smaller, the complexities introduced by the hole and rivet are maximum; for example, the residual stress system due to rivet squeezing, pin loading caused by the rivet, and shifting of the line of maximum bending from the top of the rivet to the net section. It is difficult to separate the contributions of these behaviors; however, the last effect may lead to the unsystematic response of the a/c and a/t ratios with increasing bending stress. Müller showed that as the line of maximum bending shifts from the top of the hole to the net section, the area close to the hole edge experiences a relatively lower bending stress. If the hole edge is shielded from changes in relative bending stress, changes in the k factor won't be manifest by changes in a/c and a/t at 60%N.

Another interesting discovery is the appreciable difference in the a/c and a/t ratios at 60% and 90% as shown in Figure 3.35 and Figure 3.36. The a/c_1 and a/t values are near those listed in Table 3.6 and the boundary layer effect is quite evident at the bore of the hole, crack depth axis. In the fracture surfaces of those specimens that are tested to failure (100%N), the crack shape is often no longer part elliptical, but more resembles a slant crack, see Figure 3.37. Either the crack shape change during the life can be inherent in the crack growth behavior of lap joints or the estimate of 90% of the fatigue life is near

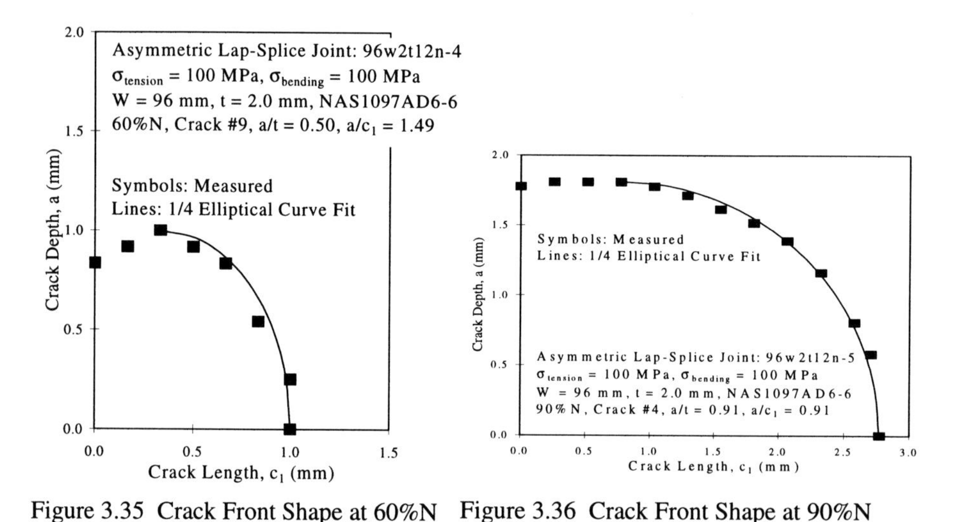

Figure 3.35 Crack Front Shape at 60%N Figure 3.36 Crack Front Shape at 90%N

the actual fatigue life of the particular joint. The latter explanation is not so likely in view of the results for the 96W2T24N joint whose crack shapes show a similar difference at 60% and 80% of the estimated fatigue life. Due to the random nature of fatigue crack nucleation, some rivet holes crack earlier than others. Those cracks that nucleate late in the fatigue life of the joint are not only smaller, but also have larger a/t and a/c ratios.

Figure 3.37 Transition from Part Elliptical to Slant Crack Front Shape

With a decrease in the rivet row pitch and ensuing increase in the bending stress obviously increases the total stress in the joint. This has a pronounced effect on the fatigue life of the joint as shown in Figure 3.38.

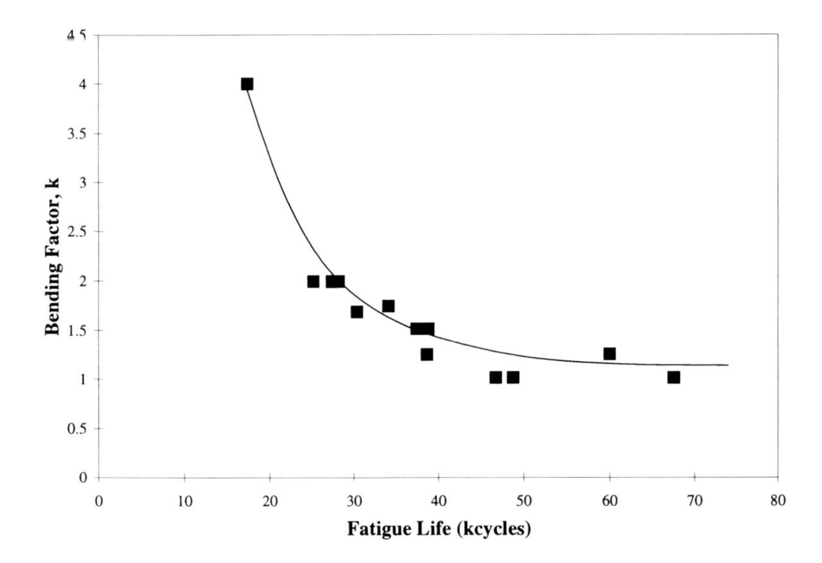

Figure 3.38 Effect of Increasing Secondary Bending Stress on Fatigue Life

Commonly, scatter is assessed by calculating the standard deviation of the logarithm of the fatigue life, $\sigma_{\log N}$ for a series of similar fatigue tests. Table 3.7 also indicates the scatter in the fatigue life of joints tested at low, intermediate, and high stress levels corresponding to a range in k factors between 1.0 and 2.0. Values of $\sigma_{\log N} < 0.1$ in general indicate low scatter.[22] Also evident from this data is the larger scatter band for the lower total stress amplitude tests, which was also reported by Schijve in reference [22]. The scatter in fatigue lives of test series 1 - 3 is acceptable; therefore, the testing protocol is not in question. Thus, the inconclusive results only indicate the difficulty in determining relationship between crack shape and secondary bending stress level and further emphasizes the need for in situ crack shape measurements. In consideration of a prediction model, the crack shape may be changing throughout the life that must be accounted for in the model.

Table 3.7 Scatter vs. Bending Factor

k	$\sigma_{\log N}$
1	0.088
1.5	0.011
2	0.021

The three asymmetric lap splice joint test series provided insight into the complicated nature of fatigue crack growth in riveted connections. Test series 1 and 2 gave credence to the thickness effect whereby an increase in the sheet thickness leads to the crack shape becoming shallower. The only conclusive result from test series 3 is the larger a/t ratio for the cracks that initiate late in the fatigue life.

3.4 Fractographic Observations

Marker load spectra were used on several CCT, CCTB, 1-open hole, and asymmetric lap splice joints to evaluate the utility of each spectrum on marking the fracture surface without affecting the macro crack growth rate. The fracture surfaces of all the specimens used in the marker load investigation were examined visually and with a scanning electron microscope. The stress-time histories of the single under-load and program loading spectra are shown in Figure 3.5 and Figure 3.6, respectively. The purpose of creating the marker bands on the fracture surface is to reconstruct the crack history after the specimen has failed with no in situ crack growth monitoring required. For cracks that are difficult or impossible to measure during the test, this method provides a means to recreate the crack history, thus providing information on the crack growth rates. Furthermore, the crack history can then be used to validate prediction models for the given crack configuration and loading. Without the crack history, the validity of the prediction model cannot be determined.

3.4.1 Single Under-load Spectrum

As evident from the stress-time history, the single under-load spectrum is quite benign in terms of deviations from the constant amplitude fatigue loading. The marker band is created by the increase in ΔK for the marker cycle in relation to the fatigue cycle. This one cycle increase in ΔK is manifest by a larger striation spacing on the fracture surface as seen in Figure 3.39 for a CCTB specimen. A more magnified view of the same marker bands is shown in Figure 3.40 where the large striations created by the marker load bound the small striations created by the CA block.

$\sigma_{tension}$ = 122 MPa, $\sigma_{bending}$ = 90 MPa, Fatigue Block R = 0.656,
Marker Load R ≈ 0.0, Half Crack Length, a = 10 mm

Figure 3.39 Marker Bands at Large Crack Lengths Created by Single Under-Load Spectrum

Figure 3.40 Magnified View of Figure 3.39

The clarity of the marker band is higher for the larger cracks than smaller ones due to the relatively larger ΔK values for the former; however, the marker bands are detectable in the small crack regime as shown in Figure 3.41.

$\sigma_{tension}$ = 106 MPa, $\sigma_{bending}$ = 0 MPa, Fatigue Block R = 0.129
Marker Load R ≈ 0.0, Half Crack Length, a = 0.12 mm

Figure 3.41 Marker Bands at Small Crack Lengths, Single Under-Load Spectrum

Looking solely at Figure 3.41 to determine whether the visible striations are marker bands or striations from the fatigue cycle is impossible. By comparing the striation spacing from this crack length, 0.12 mm, to that from a fractograph of a larger crack length, the striation type is ascertained. For example, the average striation spacing from Figure 3.41 is 0.198 µm which is greater than the 0.110 µm spacing from a fractograph, not shown here for brevity, of a crack that is 2.0 mm. Since the striation spacing increases with increasing crack length, the striation spacing of the smaller crack cannot be greater than that of the larger crack unless the striations for the smaller crack are actually marker bands representing 50 cycles per visible striation.

For these constant K_{max} spectra, the marker bands are the most distinct for a fatigue block with a high stress ratio since the increase in ΔK between the fatigue block and marker cycle (R = 0) is large. As this difference declines, so does the clarity of the marker bands. Although marker bands are detectable in both the micro and macro crack regimes, marker bands could not be detected throughout the crack history. In order to reconstruct the crack history, marker bands must be distinguishable from the final crack length at specimen failure through the macro crack regime and hopefully well into the micro crack regime. Segerfröjd and Bogren were able to reconstruct a continuous crack history by coupling the marker bands and in situ crack growth measurements.[23] A similar technique could have been used here; however, since in situ crack growth measurements are difficult and unreliable for lap joints, this method was not used.

This preliminary study of the single-under-load spectrum on CCT and CCTB specimens gave insight as to what spectrum parameters dictate effective, continuous fracture surface marking. Marker bands are distinguishable in both the CCT and CCTB specimens indicating no dependence on the type of loading, remote tension versus remote tension and secondary bending. Most importantly is the effect of the stress ratio difference between the fatigue block and marker cycle where the distinctness of the marker bands increases with increasing fatigue block stress ratio. To further investigate the stress ratio dependence in addition to exploring the effect of the local stress field on marker band formation, a series of eight 1-open hole specimens were tested, six subject to remote tension and two subject to remote tension and bending, with R ranging from 0.0918 - 0.656. The first open hole specimen is subjected to 200 MPa of remote tension with R = 0.0918 which results in a very rough fracture surface with barely detectable marker bands, see Figure 3.39. All the visible striations in Figure 3.42 are marker bands. However, the fractograph shows the marker bands can still be detected on a rough fracture surface if a sufficiently high R is used. The maximum stress plays a secondary role in marker band formation. Marker bands are not only visible but also measurable for cracks as small as 30 μm. At this scale, the fatigue block striations are not visible, only the marker bands. Moreover, no general trend in the effect of R on marker band delectability or measurability is evident. Although some marker bands at small crack lengths are found, whether or not the first visible marker band is the first one created is impossible to determine. Thus, no attempt is made at finding the first marker band, but only the smallest marker band detectable within the resolution limits of the given SEM. Thus, only groups of measurable marker bands could be photographed in the SEM. From the fractographs taken in the small crack regime, $c \leq 0.5$ mm, the marker bands occur in discrete groups with no rational method of determining the number of marker bands between groupings. It implies, unfortunately, that reconstruction of the crack history at this scale is not possible with the single under-load spectrum.

$\sigma_{tension}$ = 101 MPa, Fatigue Block R = 0.309, Marker Load R ≈ 0.0, Half Crack Length, a = 0.029 mm

Figure 3.42 Marker Bands on 1-Open Hole Specimen Fracture Surfaces in Small Crack Regime (Near the Hole Edge)

In the intermediate crack regime, $0.5 < c \leq 10$ mm, the marker bands are quite distinct over a larger range of the crack length as seen in Figure 3.43. The good definition of the marker bands allows for reliable measurement, and when converted to crack growth rates, correlates well with the crack growth rates derived from the measurements gathered during the fatigue test. The crack growth rate data is presented in Appendix D. Also in Appendix D are fractographs of relatively small and large cracks of all the single under-load spectrum marker load tests. By comparing all the images in Appendix D, the dependence of marker band width on R is evident. The marker band is simply a fatigue striation created by a higher ΔK than that used in the fatigue block; thus, the width of the striation caused by the marker cycle must be larger than that of the fatigue block since the ΔK is higher for the marker cycle. The larger crack growth increment appears as a dark band in comparison to the fatigue block striations. Although the marker bands are quite distinct, they are not continuous either in the crack growth direction or along the crack front, but appear in discrete groups making crack shape determination and reconstruction of the crack history impossible. Fracture surface roughness, inclusion particles, microvoid coalescence, and fretting damage are the most prevalent causes for disturbing the continuous formation of marker bands. The first two causes are

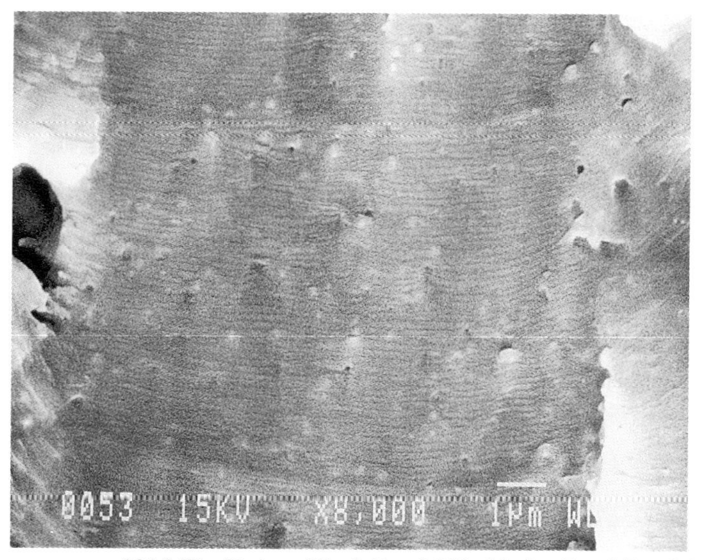

$\sigma_{tension}$ = 101 MPa, Fatigue Block R = 0.309, Marker Load R ≈ 0.0
Half Crack Length, a = 2.0 mm

Figure 3.43 Marker Bands on 1-Open Hole Specimen Fracture Surfaces in Large Crack Regime

inherent in the material; however, microvoid coalescence can be reduced by reducing the stress level. The fretting damage is most likely caused when the opposing fracture surfaces make contact when the applied stress is less than the crack opening stress. The latter stress is the stress required to open the crack which can be a stress level distinctly above σ_{min} of the fatigue cycle.[24] If the crack opening stress is relatively high, the crack may be closed and the fracture surfaces are in contact with one another for a significant portion of the load cycle. Thus, fretting damage is possible while the crack is closed.

The ductility of 2024-T3 is manifest in a very rough fracture surface even at low stress levels that can make the marker bands difficult to detect especially with a small difference between fatigue block and marker cycle R. This behavior is exacerbated at higher stress levels as seen in Figure 3.44 which shows disruption of the marker bands as a result of microvoid coalescence. Recall microvoids are created as the higher stress levels separate the metal at the interface between the base material and inclusion particles. The newly created voids act as a stress concentration that promotes further void growth which ultimately join forming large cavities. The inclusion particles not only disrupt the formation of striations, but also locally affect the growth rate whereby the crack decelerates as it approaches the inclusion particle and accelerates after it passes the particle.

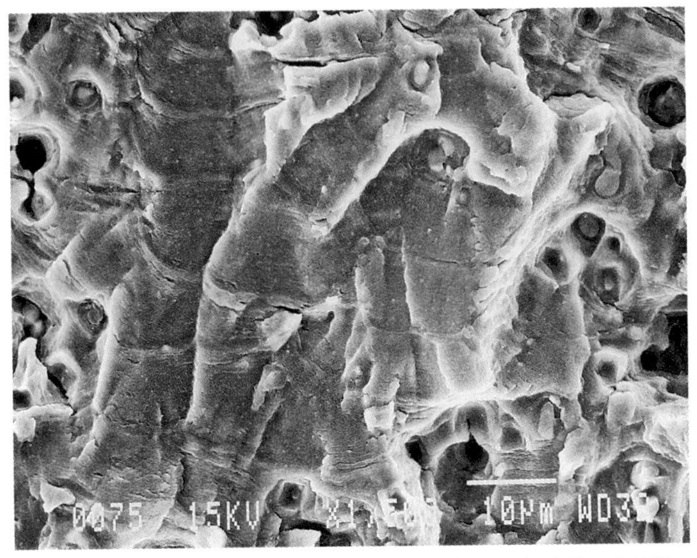

$\sigma_{tension}$ = 122 MPa, $\sigma_{bending}$ = 90 MPa, Fatigue Block R = 0.656
Marker Load R ≈ 0.0, Half Crack Length, a = 10.0 mm

Figure 3.44 Fracture Surface Roughness in High Stress Level Fatigue Test

One final comment on the marker bands in the macro crack regime. As the crack nears the point of unstable cracking, the crack growth rate of the fatigue block and marker cycle is high resulting in an indistinguishable difference in striation spacing. As expected, this behavior is more pronounced in the tests of low fatigue block R. Therefore, the number of cycles between failure and the last detectable marker band is indeterminable.

The single under-load spectrum generated marker bands which were detected in tests using fatigue block R ranging from 0.092 - 0.656 and at crack lengths as small as 30 μm. Discrete groups of marker bands are created for a majority of the crack history without any affect on the crack growth rate after applying the marker cycle. However, the discontinuous formation of marker bands results in an incomplete crack history; therefore, the single under-load spectrum cannot be used for generating crack growth data ultimately to be used for verifying the analytical crack growth prediction model.

Although the single under-load spectrum did not sufficiently mark the fracture surface for the entire crack history, this same spectrum was used on asymmetric lap splice joints. From the open hole series, the dependence of marker band formation on the local stress field and R is established. The local stress field in the open hole specimens was comprised of tension and bending, but the asymmetric lap splice joint also has a bearing stress component caused by the load transfer through the rivet. The most significant result of using the single under-load spectrum with the lap joints was easy detection of marker bands at small crack lengths, $c \leq 20$ µm. The clearly defined marker bands shown in Figure 3.45 correspond to a crack length of 15 µm. Again, the visible striations in Figure 3.45 are marker bands since they are larger than fatigue cycle striations measured from a fractograph of the same crack later in its life. The marker bands found in the open hole specimens at crack lengths of this scale did not provide the same definition.

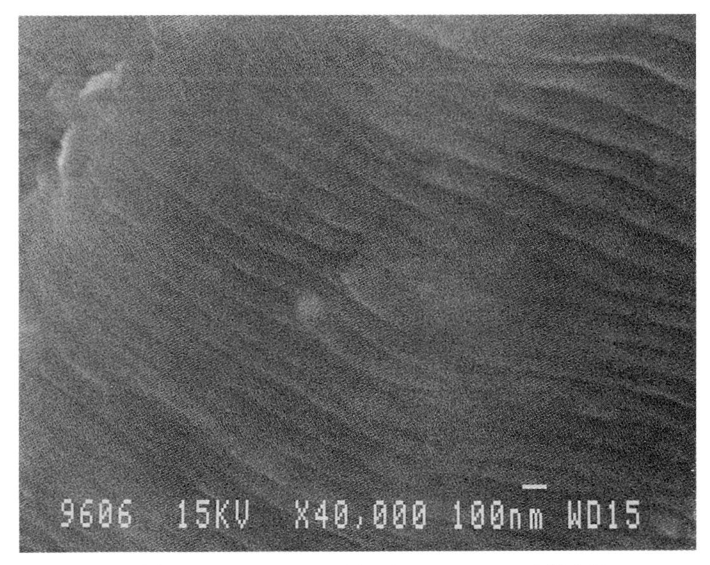

$\sigma_{tension}$ = 100 MPa, $\sigma_{bending}$ = 100 MPa, $\sigma_{bearing}$ = 260 MPa
Fatigue Block R = 0.115, Marker Load R ≈ 0.0, Half Crack Length, a = 0.015 mm

Figure 3.45 Marker Bands on the Fracture Surface of an Asymmetric Lap Splice Joint Crack

Although the marker bands for the lap joint specimens are easier to detect at small crack lengths, the discontinuous occurrence of the bands prevails through the crack from nucleation to failure just as in the open hole and CCT specimens. The formation of marker bands without affecting the crack growth rate is possible as seen in the previous experimental results. The deficiency of the single under-load spectrum is in sequencing; in other words, since the sequence of fatigue blocks and marker cycles is constant if one or a series of marker cycles are undetected there is no way to discern at what point in the spectrum the last marker band was detected. The program loading spectrum discussed in the next section addresses the poor sequencing used in the single under-load spectrum.

3.4.2 Program Loading Spectrum

The program loading spectrum as shown in Figure 3.6 uses a varying number of marker blocks occurring in a repetitive sequence separated by blocks of fatigue cycles. This spectrum has been used successfully by Piascik in marking countersunk single edge notch specimens subject to remote tension.[25] Marking the fracture surface is based on the same premise as the single under-load spectrum by changing the ΔK between the fatigue and marker cycle. Note the fatigue and marker cycles are at an $R = 0.0$. Just as with the single under-load spectrum, the program loading spectrum does not affect the crack growth rate as discussed previously in section 3.3.1. Reconstructing the crack history from marker bands is quite time consuming; therefore, since the preliminary results of the program loading spectrum for the 1-open hole specimen were promising, all further marker band investigations were conducted on the asymmetric lap splice joints. The mark is created after the 100 cycles at the lower stress level when the $\Delta\sigma$ (which is σ_{max} since σ_{min} is zero) increases from 75% of σ_{max} to 100%. For the 100 cycles at 75% of σ_{max}, the striations are difficult, if not impossible to detect; whereas the marker cycles are quite distinct as seen in Figure 3.46 for a crack length of 2.5 mm. Even at nearly three times the magnification used in Figure 3.46 the fatigue striations cannot be clearly seen in Figure 3.47.

Referring to the program loading spectrum in Figure 3.6, each marker band is composed of a group of ten cycles at 100% σ_{max} separated by 100 cycles of 75% σ_{max}. The sequencing of the fatigue and marker cycles creates a distinct pattern of marker bands in groups of six, four, and ten (6-4-10). If a group of marker bands is not detected, the position in the spectrum can be deduced based on the progression of marker band groupings. For example, if the marker groups are detected in a 6-4-10-4-10 pattern, then based on the 6-4-10 sequence the second group of 6 is undetected. The ordered sequence of the spectrum is recognizable throughout the crack history from a crack size of approximately 75 μm to fracture. At the smaller crack lengths and growth rates the 6-4-10 pattern is captured in one photograph as seen in Figure 3.48.

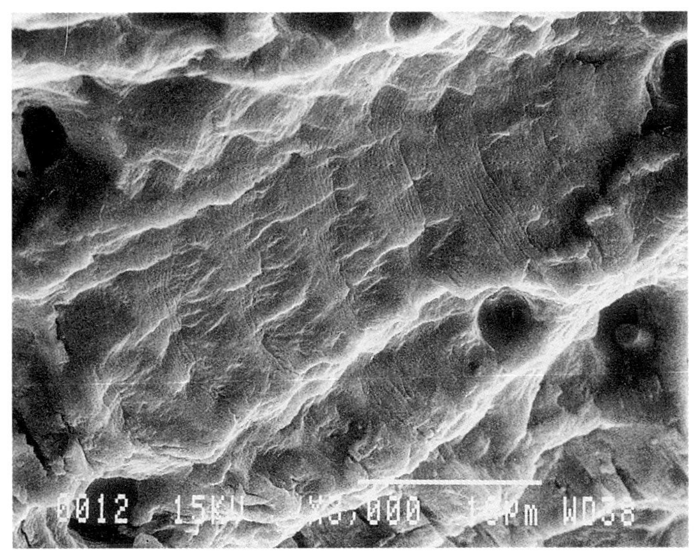

$\sigma_{tension}$ = 75 MPa, $\sigma_{bending}$ = 135 MPa, $\sigma_{bearing}$ = 187 MPa
Fatigue Block R ≈ 0.0, Marker Load R ≈ 0.0, Half Crack Length, a = 2.5 mm

Figure 3.46 Group of Six Marker Bands

$\sigma_{tension}$ = 75 MPa, $\sigma_{bending}$ = 135 MPa, $\sigma_{bearing}$ = 187 MPa
Fatigue Block R ≈ 0.0, Marker Load R ≈ 0.0, Half Crack Length, a = 2.5 mm

Figure 3.47 10,000X View of Figure 3.46

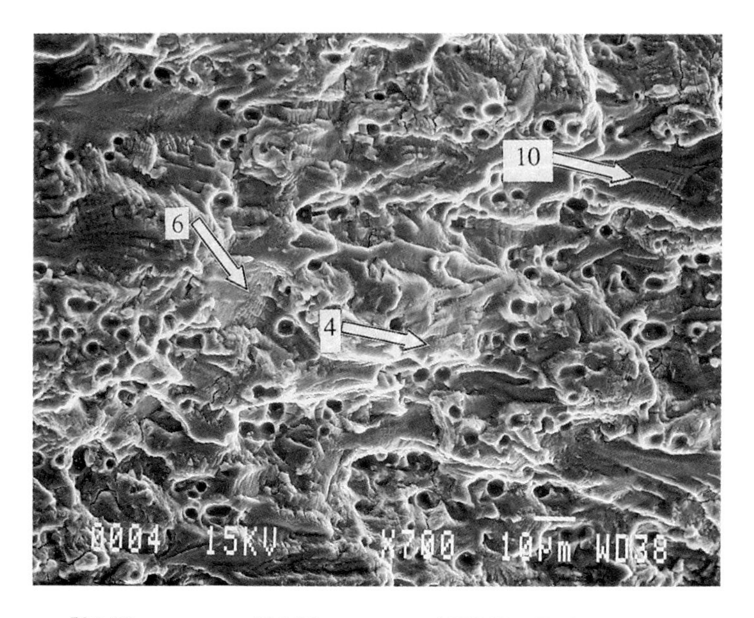

$\sigma_{tension} = 50$ MPa, $\sigma_{bending} = 59$ MPa, $\sigma_{bearing} = 125$ MPa, Fatigue Block $R \approx 0.0$, Marker Load $R \approx 0.0$, Half Crack Length, $a = 4.875$ mm

Figure 3.48 6-4-10 Marker Bands in Asymmetric Lap Splice Joint

The most difficult part of the crack history in which to see the marker bands is just prior to failure of the joint or link-up with the crack growing from the adjacent fastener hole. During this stage of crack growth, the striations from both the fatigue and marker cycles are large and the relative difference in width between the two is difficult to detect. Reconstruction of the crack history begins by finding the last marker bands prior to failure then moving toward the nucleation site taking measurements for each set of marker bands. The coordinates of the marker bands are measured from the hole edge as close to the faying surface as possible. The clad layer in 2024-T3 is approximately 5% of the thickness of the sheet; thus for the 2.0 mm thick sheet, the clad layer is 0.1 mm. Striations and marker bands were not regularly found in the clad layer most likely due to fretting damage caused by contact of the opposing sides of the fracture surface; therefore, all crack length measurements are taken outside the clad layer. Reconstructed crack histories are shown in Figure 3.49 and Figure 3.50. Two separate crack length measurements at the same number of cycles indicate the oblique nature of the crack front. The first three data points in Figure 3.49 are not near the faying surface as are the remaining points. The fracture surface close to the hole is damaged by fretting near the faying surface; thus, these three points are measured at a different depth.

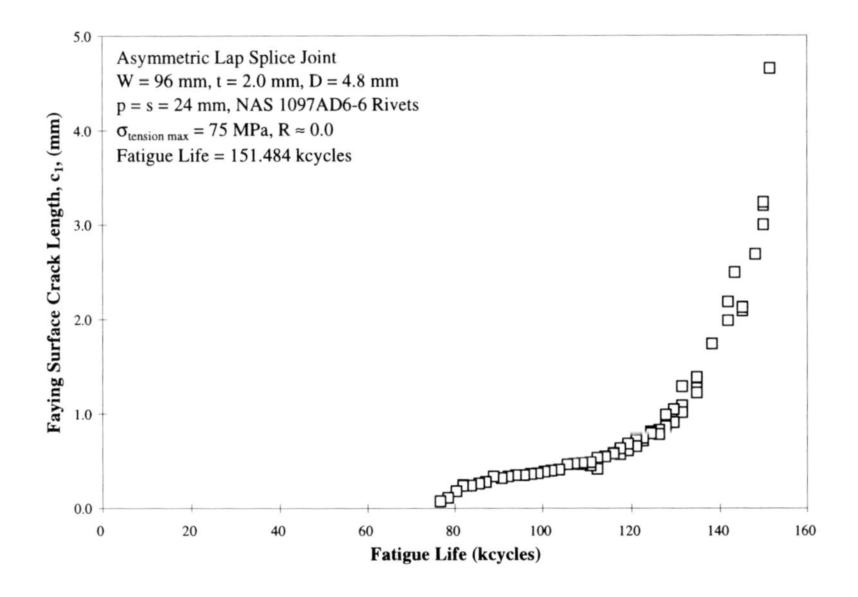

Figure 3.49 Reconstructed Faying Surface Crack History from Asymmetric Lap Splice Joint

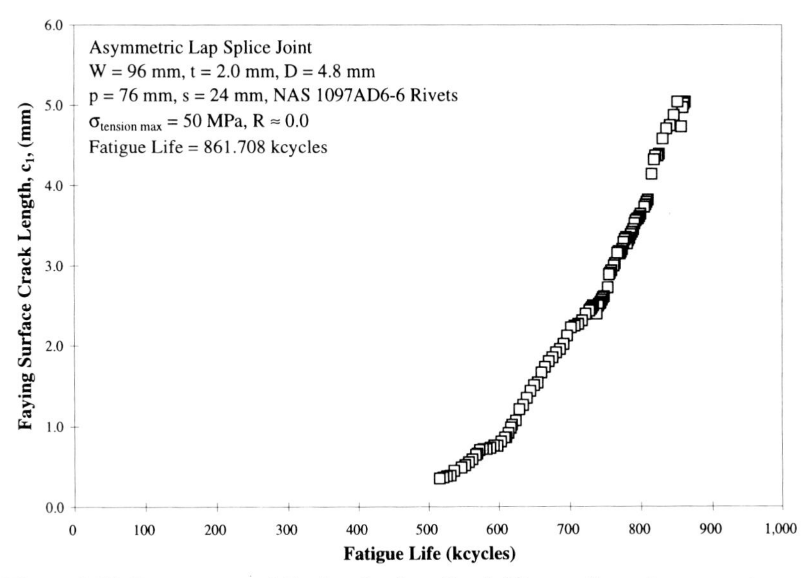

Figure 3.50 Reconstructed Faying Surface Crack History from Asymmetric Lap Splice Joint

The accuracy of this crack history reconstruction method is based on the number of fatigue cycles between the 6-4-10 groupings of marker bands which was chosen to be 1000 cycles. Since the number of cycles between the last marker band and failure is difficult to discern, the error can then be as large as 1000 cycles which is only 0.6% of the fatigue life of this joint.

Using the program loading spectrum and the scanning electron microscope, the crack history of a lap splice joint was reconstructed. The only drawback of this method is the time required finding and measuring the marker bands using the SEM. Approximately 50 hours is required to obtain the measurements shown in Figure 3.49; however, with the experience in detecting the marker bands, only 30 hours is needed to obtain the data shown in Figure 3.50. The reconstructed crack history can now be used to verify the prediction algorithm discussed in Chapter 5.

3.5 Conclusions

The purpose of the present test series is to observe fatigue crack growth under combined cyclic tension and bending loads. The observation should include the development of size and shape of cracks starting as very small part through cracks up to oblique through cracks. Fractography has been essential for the observations. Tests were carried out on 3 types of specimens: (i) a newly developed combined tension and bending sheet specimen (Figure 3.9), (ii) a 1½ dogbone specimen (Figure 3.25), and (iii) an asymmetric riveted lap-splice joint (Figure 3.27). The major findings are summarized below.

Results of the combined tension and bending specimens

- The specimen yields a controllable stress field in the test area with a satisfactory agreement between secondary bending calculated with the Schijve line model and measured by strain gages.
- Tests were carried out on center cracked and 1, 5, and 7-open hole specimens. In the open hole specimens all cracks nucleated in the minimum net section, initially growing as corner cracks until they penetrate the back surface of the sheet to continue propagation as through cracks with oblique crack fronts.

Results of the 1½ dogbone specimens

- A modified 1½ dogbone specimens was adopted to obtain a bending factor k (= $\sigma_{bending}/\sigma_{tension}$) representative for fuselage lap joints. Strain gage measurements indicated an inhomogeneous secondary bending distribution in the width direction, not representative for fuselage lap joints.
- Specimens riveted with different squeeze forces confirmed that a higher squeeze force promotes crack initiation away from the minimum net section towards the top of the rivet hole.

Results of the riveted lap joints

- Asymmetric lap joints with two rivet rows were designed to different geometry's. All specimens revealed an MSD behavior (Figure 3.29 and Figure 3.34) similar to that found in the more structurally complex fuselage longitudinal lap-splice joints. Also secondary bending and crack growth appears to be similar.
- For an increasing sheet thickness the ratio a/t of the crack shape is not affected, but the a/c ratio decreases.
- The bending ratio k was varied in the range 1.2 to 2.0 by using different distances between the rivet rows. Increasing secondary bending reduced the fatigue life (Figure 3.38). It also led to lower a/c values for larger cracks.

Fractographic techniques

- Two types of marker loads were added to the constant-amplitude (CA) loading. The first type is using an under-load cycle (UL) to $\sigma_{min} = 0$ after every 50 cycles of the basic load cycles, which should have $\sigma_{min} > 0$ (Figure 3.5). Distinct striations of the UL cycles could be observed in the electron microscope for small cracks, $c_1 \leq 1.0$ mm. For larger cracks, such observations became difficult and marker bands could remain undetected.
- In the second approach a program loading spectrum was adopted by introducing blocks of 100 smaller cycles with maximum stress of 75 % of s_{max} of the basic CA loading, but the same $\sigma_{min} = 0$ stress level (Figure 3.6). Clear marker bands were produced from a crack size of about 75 μm to specimen failure. The crack history can thus be reconstructed. This method is particular useful for small inaccessible cracks occurring in riveted lap joints.

- Observations on crack size and shape were also made by interrupting fatigue tests at certain percentages of the average fatigue life, followed by pulling the specimen to failure. The size and the shape of the crack nuclei could then be measured. It showed that crack nuclei in the open hole specimens developed relatively later than in riveted lap joints.

Crack front shapes

- The shapes of a large number of fatigue crack fronts obtained under combined tension and bending were examined. In general, the crack front could well be approximated as a part elliptical shape with the two axes along the sheet surface and the edge of the hole. Values of a, c_1 and c_2 were then obtained by regression analysis.

---

[1] Müller, Richard Paul Gerhard. An Experimental and Analytical Investigation on the Fatigue Behaviour of Fuselage Riveted Lap Joints, The Significance of the Rivet Squeeze Force, and a Comparison of 2024-T3 and Glare 3. Diss. Delft University of Technology, 1995. Delft:NL, 1995. ISBN 90-9008777-X, NUGI 834.

[2] Fawaz, S. A. Application of the Gel Electrode Method in Thin Sheet Fatigue Specimens, B2-97-04. Delft, NL: Faculty of Aerospace Engineering, Delft University of Technology, 1997.

[3] Wanhill, R. J. H. Damage Tolerance Engineering Property Evaluations of Aerospace Aluminum Alloys with Emphasis on Fatigue Crack Growth, NLR-TR-94177L. Amsterdam: National Laboratory of The Netherlands, 1994.

[4] Mussert, K. M. Fracture Surface Marking of Alclad 2024-T3 Sheet, NLR-CR-94456C, Amsterdam: National Laboratory of The Netherlands, 1994.

[5] Johansson, Sten. Corrosion Fatigue and Micro-Structure Studies of the Age Hardening Al-Alloy AA-7075. Diss. Linköping Studies in Science and Technology, No. 110, Linköping, Swed.: VTT-Grafiska, 1984. ISBN 91-7372-810-1.

[6] "Standard Test Method for Constant-Load-Amplitude Fatigue Crack Growth Rates Above 10^{-8} m/Cycle." American Society for Testing and Materials Specification E647-91, 1991.

[7] Nam, K. W., K. Ando, N. Ogura, and K. Matte, "Fatigue Life and Penetration Behaviour of a Surface-Cracked Plate Under Combined Tension and Bending." Fatigue and Fracture of Engineering Materials and Structures. 17 (1994): 873-882.

[8] Schijve, J., Some Elementary Calculations on Secondary Bending in Simple Lap Joints, NLR-TR-72036. Amsterdam, NL: National Aerospace Laboratory, 1972.

[9] Schijve, J. and De Rijk, P. The Effect of Ground-to-Air Cycles on the Fatigue Crack Propagation in 2024-T3 Alclad Sheet Material. NLR-TR-M-2148. Amsterdam, NL: National Aerospace Laboratory, 1965.

[10] Misawa, Hiroshi and J. Schijve. Fatigue Crack Growth in Aluminium Alloy Sheet Material Under Constant-Amplitude and Simplified Flight-Simulation Loading. Report LR-381. Delft, NL: Delft University of Technology UP, 1983.

[11] Newman Jr., J. C. and I. S. Raju. Analysis of Surface Cracks in Finite Plates Under Tension or Bending Loads, NASA-TP-1578. 1979.

[12] NASGRO Fatigue Crack Growth Computer Program, Version 2.01, NASA JSC-22267A, 1994.

[13] Ottens, H. H. Multiple Crack Initiation and Crack Growth in Riveted Lap Joint Specimens, Proc. of the 18th Symposium of the International Committee on Aeronautical Fatigue, 3- 5 May 1995, Melbourne, Austral. West Midlands, UK: EMAS, 1995.

[14] Cook, Robin. Standard Fatigue Test Specimens for Fastener Evaluation, AGARD-AG-304. Neuilly-Sur-Seine, Fr. 1987.

[15] Van der Linden, H. H. Fatigue Rated Fastener Systems - An AGARD Coordinated Testing Programme, AGARD Report 721. Neuilly-Sur-Seine, Fr. 1985.

[16] Van der Linden, H. H., L. Lazzeri, and A. Lanciotti. Fatigue Rated Fastener System in 1½ Dogbone Specimens, NLR-TR-86082. National Aerospace Laboratory of The Netherlands, NL, 1986.

[17] Palmberg, B., G. Segerfrojd, G.-S. Wang, and A. Blom. Fatigue Behaviour of Mechanical Joints: Critical Experiments and Numerical Modeling, Proc. of 18th Symposium of the International Committee on Aeronautical Fatigue, 3-5 May 1995, Melbourne, Austral.

[18] Jarfall, Lars. Shear Loaded Fastener Installations, SAAB-SCANIA Rapport KH-R-3360. 1983.

[19] Creemers, M. R., P. H. H. Leijendeckers, M. C. M. van Maarschalkerwaart, J. E. Rijnsdorp, and Sj. Tysma. Polytechnisch Zakboekje. Arnhem, NL: Koninklijke PBNA, 1984.

[20] Wit, G. P. MSD in Fuselage Lap Joints - Requirements for Inspection Intervals for Typical Fuselage Lap Joint Panels with Multiple Site Damage, LR-697. Delft, NL: Delft University of Technology UP, 1992.

[21] Grandt, Jr., A. F., J. A. Harter, and B. J. Heath, "Transition of Part-Through Cracks at Holes into Through-the -Thickness Flaws," Fracture Mechanics: Fifteenth Symposium, ASTM STP 833, R. J. Sanford, Ed., American Society for Testing and Materials, Philadelphia, 1984, pp. 7-23.

[22] Schijve, J. "Fatigue Predictions and Scatter." Fatigue and Fracture of Engineering Materials and Structures. 17 (1994): 381-396.

[23] Segerfröjd, Gabriel, and Jonas Bogren. Three Different Marker Load Techniques , An Evaluation, FFA-TN-1995-20. Bromma, Swed: The Aeronautical Research Institute of Sweden, 1995.

[24] Elber, Wolf. "Fatigue Crack Closure Under Cyclic Tension." Engineering Fracture Mechanics. 2 (1970): 37-45.

[25] Piascik, Robert S. The Growth of Small Fatigue Cracks in Corroded Al Alloys. Proc. of Air Force 4th Aging Aircraft Conference, 9-11 Jul 1996, USAF Academy, CO.

This page intentionally left blank

4.

Analytical Investigations

4.1 Introduction

As a result of recent catastrophic failures of transport aircraft riveted lap-splice joints due to fatigue crack growth and unstable fracture, many attempts have been made to better understand this phenomena. Toward this end and as mentioned in the previous chapter, a fatigue test program of thin sheet material typically found in aircraft fuselages, 2024-T3 clad aluminum alloy, with a centrally located hole subject to remote tension and bending is completed in addition to tests on asymmetric lap-splice joints. The intent of these investigations is to gain insight on the effect of the bending and bearing stress on fatigue life and crack-front shape development. Specifically for the bending component, as an aircraft is pressurized the in-plane forces resolved from the hoop stress apply an eccentric load in the lap joint. This eccentricity creates a bending moment, typically called secondary bending, and therefore normal stresses due to bending in the joint. The secondary bending has two components, a linear component due to the eccentric loading, and a nonlinear component due to the relatively large out of plane displacements as a result of the eccentric load application and joint geometric configuration. Müller found that the bending stresses could be as large or larger than the normal stress due to the in-plane loading.[1] For brevity, herein the normal stress due to bending and tension are referred to as bending and tensile stress respectively. The primary focus in the present chapter is to calculate the influence of crack front shape on stress intensity factors for a variety of loadings, typical of those that occur in lap-splice joints.

Cracks usually nucleate as surface or corner cracks in close proximity to the rivet hole on the faying surface of the joint. As crack growth continues, the cracks penetrate the back surface, the free surface on the inside or outside of the fuselage that is the surface opposite the nucleation site. If the loading is pure tension, the penetrated crack length experiences high growth rates due to the large stress gradient in the small ligament of material between the crack front and the back surface. Grandt et al. using the finite element alternating method (FEAM) verified the higher stress intensity factors, K's, at the back surface leading to back surface crack "catch up."[2] The FEAM solutions agreed well with fatigue crack growth rate changes obtained during fatigue tests of polymethyl methacrylate (PMMA). In lap joints, the large secondary bending stress prohibits the back surface catch-up; therefore, a penetrated crack, as shown in Figure 4.1, maintains a part-elliptical, oblique, shape until rapid crack growth just prior to the onset of unstable fracture.

In service, fatigue cracks have been found which indicate a large degree of secondary bending.[3,4] The fracture surfaces of cracks of various lengths exhibited oblique shapes indicative of penetrated surface or corner cracks subject to bending. Similar shapes are found in fatigue test specimens conducted in this study.

The three dimensional virtual crack closure technique, 3D VCCT, is used to obtain stress intensity factors, SIFs, for crack geometries and load conditions for which no published solutions are available. The structure of interest is a longitudinal riveted lap-splice joint found in the skin of a transport aircraft fuselage. This structure is modeled by the finite element method as a finite width plate with a centrally located hole. The crack geometry of interest is a through crack which has nucleated as part through crack and grown sufficiently to penetrate the surface opposite the nucleation site. The load conditions in the structure of interest are quite complex; however, for the verification studies completed here only biaxial tension, remote bending, and rivet loading are applied.

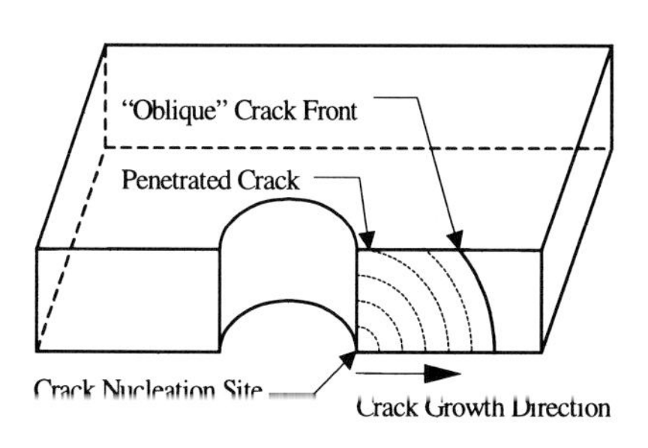

Figure 4.1 Diagram of an Oblique Quarter Elliptical Crack Front

Substantial emphasis is directed toward validating the 3D VCCT when used in conjunction with a finite element model that has a non-orthogonal crack plane mesh. A non orthogonal mesh has elements that are not normal to adjacent elements. To appreciate the flexibility and utility of the 3D VCCT, a brief discussion of the other finite element based methods for determining K's is found in section 4.2. Theoretical and application considerations of the 3D VCCT are given in section 4.3. Section 4.4 contains the results of the ten verification studies completed. Application of the 3D VCCT to obtain K's for two new crack geometries is presented in section 4.5. New stress intensity factor solutions for part elliptical through cracks with an oblique crack front are calculated and discussed in section 4.6, with the conclusions found in the final chapter, section 4.7.

4.2 Background

To determine the closed form stress intensity factor solutions for a cracked three-dimensional finite body is a difficult, and most often an intractable task; therefore alternate methods have been developed. The most prevalent for engineering applications is the finite element method (FEM). In the FEM, the K's are determined either by direct or indirect methods. In the direct method, the stress intensity is calculated directly from the finite element solution. Three classes of elements are available in the direct approach; conventional, singularity, and hybrid. Due to the difficulties in incorporating hybrid elements in a general purpose finite element program, these elements have not been widely used; therefore, only the conventional and singularity elements are discussed here.[5] For a thorough examination on the use of hybrid elements in obtaining fracture parameters via FEA, see reference [1]. If conventional elements are used, the stress intensity factor is estimated by evaluating the behavior of the stresses, forces, or displacements in the vicinity of the crack tip, which are known from the finite element solution, and then extrapolating back to the crack tip. The force and crack opening displacement methods are the most prevalent implementations of this extrapolation technique. When singularity elements are used, the stress intensity factor can be calculated directly. The indirect methods use nodal information, displacements and forces, to obtain the energy release rate that is then used to calculate K. Examples of indirect methods are the virtual crack closure technique, equivalent domain integral (three-dimensional extension of the J-integral), and the stiffness derivative method. In addition, de Koning and Lof extended the stiffness derivative approach by making use of the stress intensity rates for a given crack extension.

Note, in a finite element model the non-singular strain terms can only approximately describe the singularity at the crack tip resulting in an underestimation of the stress intensity factor. Further, it should be noted that K values calculated from the singular part of the interpolation functions of such elements are also inaccurate because of the presence of non zero nonsingular terms in their interpolation functions.[6] In all cases, the use of singularity elements in the direct or indirect methods is beneficial.[7] The method used in this investigation is the three dimensional virtual crack closure technique which is discussed in detail in Chapter 3, however, the various other methods mentioned above are briefly discussed in the following sections.

4.2.1 Direct Methods

In a displacement finite element formulation, the fundamental result of solving the system of equations is the nodal displacements from which the stresses and strains are calculated. The stresses and strains are calculated at either the nodal coordinates or integration points depending on whether exact or numerical integration is used. The direct method of obtaining the stress intensity factor via FEA is attractive by making use of the standard output from a general purpose finite element program. However, in commercial finite element packages, evaluation of fracture parameters, K, may not be implemented;

therefore, significant post processing of the standard output must be completed in order to extract K. The two direct methods reviewed in the proceeding sections are the crack opening displacement and force methods with a more comprehensive discussion of direct methods available in reference [5].

4.2.1.1 Crack Opening Displacement Method

Using the crack opening displacements (COD), as seen in Figure 4.2, from finite element analyses (FEA), the stress intensity factor is derived from the following plane strain relation.[8]

$$\mathrm{COD} = 2\mathrm{v} = \frac{8\mathrm{K_I}}{\mathrm{E}}\sqrt{\frac{\mathrm{r}}{2\pi}}\left(1-\nu^2\right) + \mathrm{O}\left(\mathrm{r}^{\frac{3}{2}}\right) + \cdots \tag{4.1}$$

where

K_I ≡ Mode I Stress Intensity Factor

E ≡ Modulus of Elasticity

r ≡ Normal distance from crack front to displacement node

ν ≡ Poisson's ratio

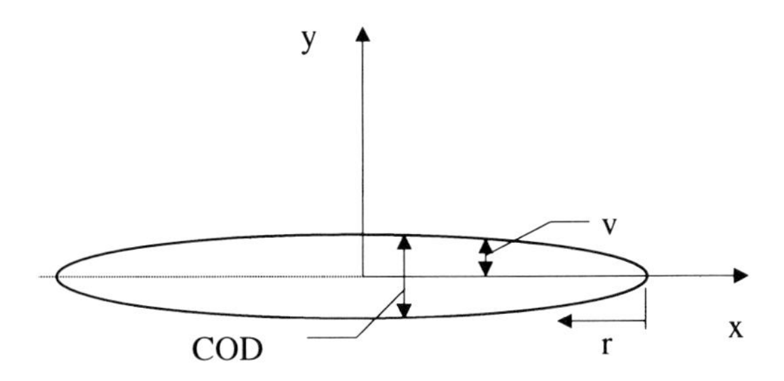

Figure 4.2 Crack Opening Displacement

Upon solution of the finite element model (FEM), all integration point values are known, e.g.; displacements, forces, strains, and stresses. Then Eqn. (4.1) is rearranged in the form,

$$\frac{\mathrm{Ev}}{\sqrt{\frac{\mathrm{r}}{2\pi}}4\left(1-\nu^2\right)} = \mathrm{K_I} + \mathrm{A_I r} + \cdots \cong \mathrm{K_I} + \mathrm{A_I r} \tag{4.2}$$

where K_I is as defined previously and A_I is an unknown constant. By substituting the COD at a given r location for several locations in the crack wake normal to the crack front as shown in Figure 4.3, K_I and A_I can be determined by a least squares linear regression.[6] The singular portion of the crack opening stresses, σ_{yy}, in the crack plane vary with $1/\sqrt{r}$ for small values of r; therefore, care must be taken in choosing the maximum r. To better represent this variation, quadratic and not linear elements should be used near the crack tip. In addition, since the constants are evaluated at nodes j, K_I is not explicitly calculated, but must be extrapolated back to the crack tip from the values at nodes j. By definition of distance from the crack front, r, the COD method is not easily implemented with non-orthogonal meshes.

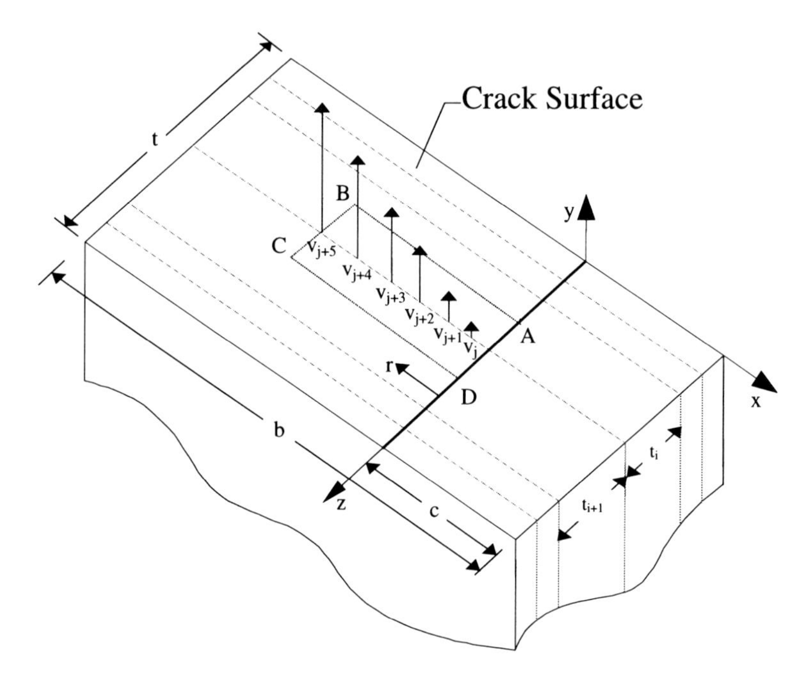

Figure 4.3 Nodal Displacements on y = 0 Plane at the Interface Between Layers i and i+1

Classification of a mesh as being orthogonal or non-orthogonal depends on the orientation of the elements adjacent to the crack front on the crack plane. The difference between the two mesh types is quite evident as seen in Figure 4.4 where the elements surrounding the crack front in Figure 4.4A have sides normal and parallel to the crack front; conversely, in Figure 4.4B, orientation of the element sides is arbitrary. At a given location along the crack front for which K is desired, the displacements at several distances, r, normal to the crack front on the crack surface are needed; therefore, if no nodes are located at r locations, the displacements must be interpolated from those displacements at neighboring nodes.

4.2.1.2 Force Method

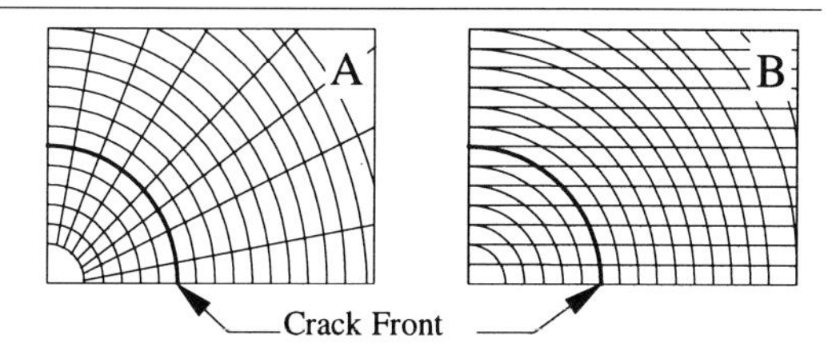

Figure 4.4 A. Orthogonal Mesh, B. Non Orthogonal Mesh

The force method offers more generality in that unlike the COD method it does not require an a priori assumption of plane stress or plane strain.[9] Similar to the COD method, if singularity elements are used the $1/\sqrt{r}$ stress singularity is obtained by placing the singularity elements on the crack front and K is then calculated directly. Even though K can be calculated directly from the singularity elements, not all singularity elements are formulated with this capability, the following outlines an extrapolation procedure to evaluate K when using singular or non-singular elements. Assuming a two dimensional stress state is valid along every infinitesimal portion of the crack front, the normal stress perpendicular to the crack front is written as

$$\sigma_y = \frac{K_I}{\sqrt{2\pi r}} + A_1 r^{\frac{1}{2}} + A_2 r^{\frac{3}{2}} + \cdots \tag{4.3}$$

where K_I and A_1 are as defined in the COD method.[8] The total force normal to the crack plane, over an area bounded by $z_1 \le z \le z_2$ and $0 \le r \le r_D$ can be represented as

$$F_y = \int_{z_1}^{z_2} \int_{0}^{r_D} \sigma_y \, dr dz \tag{4.4}$$

By substitution of Eqn. (4.3) in Eqn. (4.4), the total force is approximated as,

$$F_y = \frac{K_I}{\sqrt{2\pi}} 2\sqrt{r_D}(z_2 - z_1) + A_1(z_2 - z_1) + \cdots$$

$$F_y \cong \frac{K_I}{\sqrt{2\pi}} 2\sqrt{r_D}(z_2 - z_1) + A_1(z_2 - z_1) \tag{4.5}$$

The nodal forces in the region enclosed by ABCD, shown in Figure 4.5, are known from the FEA and are used with Eqn. (4.5) to solve for K_I and A_1 using a least squares linear regression. In other words, a node in the region ABCD is located a distance r from the crack front with a force normal to the crack plane, F_j, which are substituted into Eqn. (4.5) resulting in an equation in terms of K_I and A_I. Just as in the COD method, K is not explicitly calculated, but extrapolated from the values calculated at nodes j; therefore, Eqn. (4.5) is applied at several nodes in close proximity to the crack front to increase the accuracy of the linear regression and extrapolation.

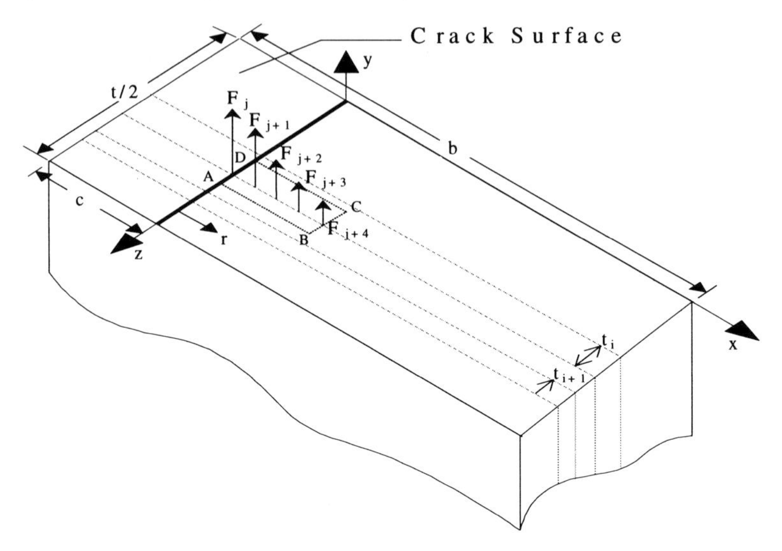

Figure 4.5 Nodal Forces on y = 0 Plane at the Interface Between Layers i and i+1

Through numerical experimentation, Raju and Newman have found consistent K values when Eqn. (4.5) is used for five forces and the maximum value of r_D is less than a/10 where a is the crack depth of a semi-elliptical crack or crack length for a straight crack.[9] Similar to the displacements used in the COD method, the nodal forces used in calculating K must come from nodes normal to the crack front. This condition restricts use of non-orthogonal meshes for K calculations with this method.

4.2.2 Indirect Methods

In general, the indirect methods determine the stress intensity factor from the elastic energy release rate. The elastic energy release rate can be obtained by determining the changes in compliance, stiffness, or energy available for crack growth during a given crack extension; in addition, the J-integral can also be used since J is equal to the elastic energy release rate for linear elastic behavior. Due to additional post processing of displacement data required by the compliance method, it has not received wide spread use and is not discussed here.[5,8,10] The stiffness derivative method, a variation of the stiffness derivative method developed by de Koning and Lof, and the J-integral are briefly discussed.

4.2.2.1 Stiffness Derivative Method

The stiffness derivative method independently developed by Parks and Hellen obtains K by calculating the change in the element stiffness matrices at the crack tip.[11,12] In a finite element solution, the potential energy is expressed as

$$P = \frac{1}{2}\{u\}^T[S]\{u\} - \{u\}^T\{f\} \tag{4.6}$$

where

$\{u\} \equiv$ Nodal displacement vector

$[S] \equiv$ Global stiffness matrix

$\{f\} \equiv$ Prescribed nodal forces

The strain energy release rate is obtained by differentiating the potential energy with respect to the crack length, which yields

$$\frac{\partial P}{\partial c} = \frac{\partial\{u\}^T}{\partial c}[[S]\{u\} - \{f\}] + \frac{1}{2}\{u\}^T\frac{\partial[S]}{\partial c}\{u\} - \{u\}^T\frac{\partial\{f\}}{\partial c} \tag{4.7}$$

where the first term on the right hand side of Eqn. (4.7) is exactly zero by the finite element solution.[11] The strain energy release rate is now represented by

$$\frac{\partial P}{\partial c} = \frac{1}{2}\{u\}^T\frac{\partial[S]}{\partial c}\{u\} - \{u\}^T\frac{\partial\{f\}}{\partial c} \tag{4.8}$$

The partial derivative of the global stiffness matrix with respect to the crack length represents the change in the former per unit crack advance. In application, the crack advance is provided by moving the crack tip nodes an amount specified by the analyst. By further ignoring body forces and crack face loadings resulting in loading only by remote surface tractions, Eqn. (4.8) reduces to

$$\frac{\partial P}{\partial c} = \frac{1}{2}\{u\}^T\frac{\partial[S]}{\partial c}\{u\} \tag{4.9}$$

Eqn. (4.9) is the fundamental relation used by Parks and Hellen to introduce their technique. To implement the differential stiffness method, the global stiffness matrix is written in terms of the element stiffness matrices. In their development, Parks and Hellen prove only the elements on the crack front contribute to the change in the global stiffness matrix thus an computationally efficient method is obtained to calculate K. To estimate the error in assuming K is linearly related to the virtual crack extension, the higher order terms, products of the incremental terms, in Eqn. (4.9) can be included. Several error estimates are given in reference [5]. Unfortunately, wide spread use of this technique remains limited due to the complexity of implementation into a general purpose finite element code. For further information on the development, application, and accuracy of the derivative stiffness method, see references [5], [11] - [13].

4.2.2.2 Extension of the Stiffness Derivative Method

Stress intensity variations along three-dimensional crack fronts have been calculated by de Koning and Lof using the stiffness derivative method to obtain stress intensity rates along the crack front.[14] Their formulation is driven by the desire to obtain stress intensity factors for numerous crack lengths from one finite element analysis. From known analytical solutions, the dependency of the stress intensity distribution on the crack size and shape parameters for a wide range of crack sizes can be approximated by the following linear relation.[14]

$$K_i = K_i^o + \frac{\partial K_i^o}{\partial a_j} \Delta a_j \tag{4.10}$$

where

K_i^o ≡ Stress intensity factor at a given position along the reference crack front

$\frac{\partial K_i^o}{\partial a_j}$ ≡ Stress intensity rates

Δa_j ≡ Crack size increment

The variation of the crack front in the finite element model is done by shifting the corner nodes of the crack front elements to obtain the new crack front shape and size. The direction of the shift is normal to the reference crack front.[14] The key to using this method is accurately calculating the stress intensity rates that can be derived from the displacement field in the vicinity of the crack front given by the following plane strain relation.

$$K_i = \frac{v_i E}{2(1-\upsilon^2)} \sqrt{\frac{2\pi}{r_i}} \tag{4.11}$$

where

$\nu_i \equiv$ One half the crack opening displacement

$r_i \equiv$ Distance to the crack front

Both ν_i and r_i are measured in an intersection perpendicular to the crack front at the corner node location, i. Then the stress intensity rate is obtained from Eqn. (4.11) by

$$\frac{\partial K_i}{\partial a_j} = \frac{E}{4(1-\upsilon^2)}\sqrt{\frac{2\pi}{r_i}}\frac{\partial \nu_i}{\partial a_j} \tag{4.12}$$

$$\frac{\partial K_i}{\partial a_j} = \frac{K_i}{\nu_i}\frac{\partial \nu_i}{\partial a_j} \tag{4.13}$$

As the crack front nodes shift, the distance r_i is not affected since all relevant nodes in an intersection normal to the crack front are assumed to shift the same amount. In Eqn. (4.13) at node i, the crack opening displacements, ν_i, are known from the finite element solution and the stress intensity factor, K_i, can be obtained using standard procedures like those mentioned previously. The second term in Eqn. (4.13) is generated using a stiffness derivative method similar to the one described previously. From the displacement equilibrium equation,

$$\frac{\partial \nu_i}{\partial a_j} = S_{ki}^{-1}\left[\frac{\partial S_{kl}}{\partial a_j}\nu_l\right] \tag{4.14}$$

where

$S_{ki}^{-1} \equiv$ Inverse of the stiffness matrix

the first term inside the brackets is available from the Choleski decomposition of the system matrix which is calculated during the normal finite element solution procedure.[14] From this straightforward procedure, the stress intensity rates can then be calculated without a considerable increase in computational effort.

4.2.2.3 The J-Integral

Traditionally, the J-integral is of interest for those crack configurations where the plasticity effects are not negligible. In this case, the elastic strain energy release rate, G, being based on the elastic stress field is inadequate in describing the energy release rate. The J-integral can then be used to determine the energy release rate. In trying to avoid solution of the complex, detailed boundary value problem associated with strain concentration fields near cracks, Rice identified the J-integral (a line integral) which has the same value for all paths surrounding the tip of a notch in a two-dimensional strain field.[15,16] As defined along an arbitrary contour, Γ, around the crack tip, see Figure 4.6, the J-integral is represented by,[16,17]

$$J = \int_{\Gamma} \left[W dy - \mathbf{T} \frac{\partial \mathbf{u}}{\partial x} \right] ds \tag{4.15}$$

where

$W \equiv$ Strain energy density per unit volume

$\mathbf{T} \equiv$ Traction vector acting along outward normal to Γ

$\mathbf{u} \equiv$ Displacement vector in the x-direction

$ds \equiv$ Arc length on Γ

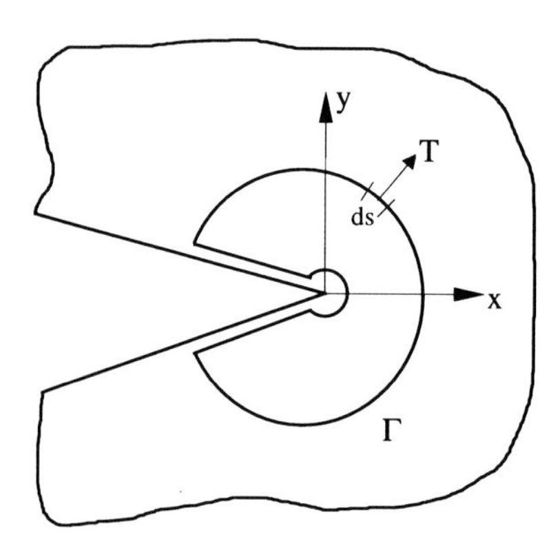

Figure 4.6 Arbitrary Contour around the Crack Tip Used in J-Integral

For a linear elastic material, the J-integral is equal to the elastic energy release rate,

$$J = G \tag{4.16}$$

Now the J-integral can be related to the stress intensity factor in view of Irwin's assertion that the energy lost in extending the crack some distance, Δc, is equal to the work required to close the crack to its original length. This relation in polar coordinates with the origin at the crack tip takes the following form.[18]

$$G_I = \lim_{\Delta c \to 0} \frac{1}{2\Delta c} \int_0^{\Delta c} \sigma_y(r,0)\,\bar{v}(\Delta c - r, \pi)\,dr \tag{4.17}$$

where

$\sigma_y \equiv$ Stress distribution ahead of crack front

$\bar{v} \equiv$ Displacement distribution behind crack front

$r \equiv$ Distance normal to crack front

For opening mode, mode I, substitution of Westergaard's solution for the stress and displacements in the vicinity of a crack into Eqn. (4.18), where E is the modulus of elasticity, yields

$$G_I = \frac{K_I^2}{E}\left(1-\nu^2\right) \Rightarrow K_I = \sqrt{\frac{G_I E}{1-\nu^2}} \quad \text{plane strain}$$

$$G_I = \frac{K_I^2}{E} \Rightarrow K_I = \sqrt{G_I E} \quad \text{plane stress} \tag{4.18}$$

The key to using the J-integral with the finite element method to obtain stress intensity factors is choosing a contour such that Eqn. (4.15) can be evaluated using the element stresses and displacements readily available from a standard finite element solution. As one might expect, choosing an appropriate contour, which is a surface for three-dimensional analyses, is paramount. Since the J-integral is based on the conservation of energy, the material volume enclosed by the integration surface must be in equilibrium, thus maintaining the energy balance. To avoid [19]extrapolation errors, the integration points of the elements and not the nodes should define the integration surface.

By substituting $\mathbf{T} = \sigma\mathbf{n}$ in Eqn. (4.15), the J-integral is now in terms of the element stresses and displacements.

$$J = \int_{\Gamma}\left(W\mathbf{n} - \sigma\frac{\partial \mathbf{u}}{\partial x}\mathbf{n}\right)ds \tag{4.19}$$

where σ is the element stress tensor with all other quantities as defined in Eqn. (4.15).

Using the stress and displacements at the integration points for a given element on the integration surface, the local J-integral is calculated using Eqn. (4.19). The strain energy, W, and displacement derivatives, $\partial\mathbf{u}/\partial\mathbf{x}$, use the FEA output; whereas, the normal vector, $\mathbf{n}$, uses the nodal point coordinates. For a numerical procedure to determine these three quantities see reference [19]. The local J-integral must now be calculated for those elements that are inside and adjacent to the integration surface. The final step is to define how the J-integral varies along the integration surface. Bakker found that a simple average of the J-integral values between adjacent elements is adequate and that when using quadratic elements with third order gaussian integration little improvement is seen when using a more elaborate interpolation method.[19] For further discussion on evaluating the J-integral in a finite element model see references [8, 20-22].

4.3 Methodology

The principal obstacle inhibiting use of the COD, force, and derivative stiffness methods is not related to accuracy but application. The COD and force methods require the elements to be normal to the crack front that restricts the types of crack geometries that can be solved. For example, creating elements normal to a curved crack front is time consuming and may not be possible at locations where the crack intersects a free surface. Although the stiffness derivative method does not require a specific element orientation with respect to the crack front, the solution algorithm to calculate the stress intensity factors is not easily incorporated into a general purpose finite element code. Therefore, the desire to have a general purpose finite element code that can also calculate stress intensity factors for complex crack shapes served as the impetus for the development of the three dimensional Virtual Crack Closure Technique (3D VCCT). Furthermore, the 3D VCCT can be easily used with any commercially available finite element analysis software.

4.3.1 Three Dimensional Virtual Crack Closure Technique

The 3D VCCT used for calculation of stress intensity factors is based on Irwin's crack closure integral.[18] The formulation for use with FEA is originally addressed by Rybicki and Kanninen[23] for two dimensions and extended to three dimensions by Shivakumar, Tan and Newman.[24] Extending Irwin's relation, Eqn. (4.20), to three-dimensional bodies with the intent of application to the finite element method, Shivakumar et al. proposed[24]

$$G_i = \frac{1}{2w_i\Delta}\int_{s_{i-1}}^{s_{i+1}}\int_{0}^{\Delta}\sigma_y(r,s)\bar{v}(\Delta - r, s)drds \tag{4.20}$$

where

w_i ≡ Element length along crack front

Δ ≡ Element length on each side and normal to crack front

σ_{yy} ≡ Stress distribution ahead of crack front

$\bar{v}$ ≡ Displacement distribution behind crack front

r $\equiv$ Distance normal to crack front

s $\equiv$ Distance along crack front

The limits of integration for s in Eqn. (4.20) are such that the force contributions of the elements adjacent to element i are included. Figure 4.7 illustrates those parameters used in the calculation of G_i. Application of Eqn. (4.20) presumes a continuous crack front can be approximated by discrete segments as typically found in FEMs. The right hand side of Eqn. (4.20) is equivalent to the product of the nodal forces ahead of the crack front and the nodal displacements behind the crack front for the ith segment with contributions from elements on each side of the crack front.[24] Eqn. (4.20) is in terms of the ith segment alone and a typical FEA solution gives nodal quantities, force and displacement contributions from all elements connected to a given node. As a consequence, a method of partitioning the nodal forces must be devised. Assuming the nodal forces are proportional to the element length normal to the crack front, the strain energy release rate for an eight node element is written as

$$G_i = \frac{1}{2w_i\Delta}\sum_{j=1}^{2} C_j F_j \bar{v}_j \tag{4.21}$$

where

$$C_1 = \frac{w_i}{w_{i-1} + w_i} \text{ and } C_2 = \frac{w_i}{w_i + w_{i+1}}$$

Note the forces come from nodes on the crack front, and the displacements from nodes behind the crack front.

For a more complete discussion and use of this method with higher order elements see reference [22]. Eqn. (4.21) is exact for a uniform stress field and approximate for a non-uniform stress field.[22] As described above, the 3D VCCT appears to require an orthogonal mesh neighboring the crack front; however, looking more closely at Eqn. (4.20) orthogonality is not essential. The local strain energy release rate is the virtual work required to close the crack over a surface area, $w_i\Delta$, and for application to FEA is the element area in the crack plane. The normality requirement in Eqn. (4.20) is only assumed to simplify the original derivation. The only information related to element shape is used to correctly partition the nodal forces, which again, have no normality

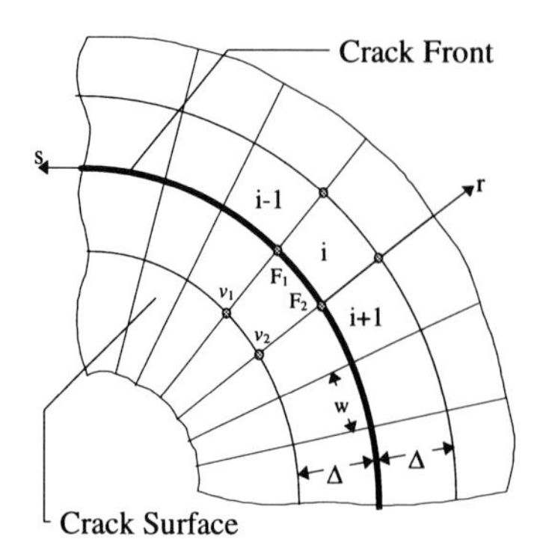

Figure 4.7 Diagram of Crack Plane Parameters Used in Calculation of the Local Strain Energy Release Rate

requirement. Eqns. (4.17), (4.20), and (4.21) are used for the full field strain energy release rate, which translates to one FEA to obtain G_I, G_{II}, and G_{III}

4.3.2 Finite Element Analysis Methodology

The finite element analysis codes used, ZIP3D, was developed by the NASA Langley Research Center specifically for obtaining fracture parameters, strain energy release rate, stress intensity factor, and J-integral, in three dimensional elastic and elastic-plastic bodies.[25] In addition, uncracked bodies may also be analyzed to obtain stress, strain, and displacement fields. Although incorporation of linear elastic fracture mechanics analysis capability in commercially available finite element packages is increasing, they are usually restricted to analysis of two dimensional bodies or for evaluating only the total strain energy release rates of a three dimensional body. The code has only one element type, eight noded isoparametric solid, and is capable of using a special reduced shear integration scheme for bending dominant problems. The isoparametric element formulation is used to define the shape functions, which are then used to create the stiffness matrix.[26-29] Linear finite elements are arbitrarily stiff in the transverse direction if all stresses are included in the integration, full integration, therefore, by reducing the number points for shear integration the flexibility of the model is increased. Several researchers have given more detailed explanations of the applicability and benefits of reduced integration.[26-30]

The models can be loaded at the nodes by applying displacements, concentrated loads, or surface tractions, which offers great flexibility in combined loading analyses. By superposition, the contributions of each load case to the three fracture modes, modes I, II, and III, is determined. The total mode I stress intensity factor is simply the addition of the individual stress intensity factors, which can be expressed in equation form as

$$K_{TOTAL} = \sum_{i=1}^{n} K_i \tag{4.22}$$

For example, the stress intensity for a plate subject to remote tension and bending is obtained by loading the model separately in tension and in bending then extracting the K's for each analysis individually. Expanding to the basic definition of stress intensity yields

$$K_{TOTAL} = \left(\sum_{i}^{n} \beta_i \sigma_i \right) \sqrt{\pi a} \tag{4.23}$$

where

$\beta_i \equiv$ Boundary correction factor for each load condition

$\sigma_i \equiv$ Remote stress for each load condition

$a \equiv$ Crack length

For additional discussion of the solution methods used in each code, see reference [25].

4.3.3 Finite Element Models

Many commercially available finite element preprocessors have the capability to write node and element information to simple text files which is the format requirement for ZIP3D. All models in this study were generated using McNeal-Schwendler Corporation MSC/PATRAN versions 1.4.2 – 6.0. As can be expected, much care must be exercised in generating the mesh pattern on and near the crack plane. As mentioned previously, the COD and force methods require orthogonal meshes around the crack front to obtain accurate K solutions. No such stipulation is known for the 3D VCCT; therefore both orthogonal and non-orthogonal meshes are generated. If a particular K solution is desired for one geometry, then maintaining an orthogonal mesh is quite simple. However, if multiple solutions are obtained from modifying one mesh, especially in the case of oblique crack fronts, orthogonality cannot be sustained due to the changing geometric requirements. The orthogonality requirement is discussed in further detail in the proceeding sections.

Finite element models were generated to develop K solutions for a circular internal crack embedded in an infinite solid subject to uniform tension, center crack tension (CCT), single edge crack tension (SECT), semi-elliptical surface crack subject to tension and bending, diametrically opposed through cracks at a hole subject to tension and bending, and semi-elliptical crack in post penetration subject to tension and bending. All models are constructed employing symmetry arguments where available. Except for the circular internal crack embedded in an infinite solid which is modeled with one eighth of the plate; in general, only one quarter of a plate, shown in Figure 4.8, is required to model the entire plate. Symmetry planes are located at $x = 0$ and $y = 0$; in addition, for plane strain analyses, $z = 0$ and $z = t$ are also fully constrained. Furthermore, sufficient global dimensions are used to avoid perturbations of the stress field due to mesh transitions, load application, and boundary conditions. Therefore, the following ratios are maintained for all models.

$$\frac{h}{b} \geq 1.5 \qquad \frac{t}{b} \leq 0.2 \qquad \frac{c}{b} \leq 0.5 \tag{4.24}$$

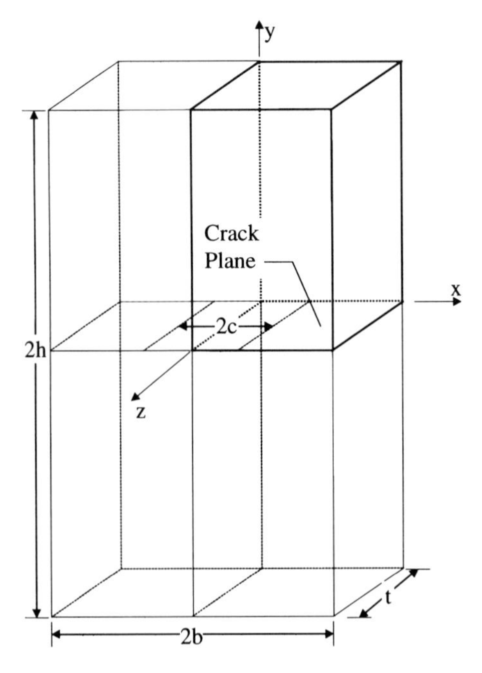

Figure 4.8 Baseline Finite Element Model

The computer requirements for generating the models and K solutions are quite modest. All preprocessing and model generation/manipulation is done on Sun Microsystems SPARC 5 and SPARC 20. Individual finite element solutions were completed on each of the following machines; two single processor workstations, Digital Equipment Corporation DEC 3000/900 and DEC Alpha 300; and two supercomputers, Convex 4640 with four processors, and Cray Research Corporation CRAY YMP with eight processor. Dr. James C. Newman, Jr. of the NASA Langley Research Center, provided use of the analysis code and CRAY YMP.

4.4 Validation Results

In order to evaluate the accuracy of a finite element model with regard to the degree of discretization error, convergence studies are usually employed. In view of the various crack geometries of interest, the preprocessing requirements for a convergence study are prohibitive; therefore, the model results are compared to known stress and stress intensity solutions. Since mode I stress intensity solutions are of prime interest, the normal stress in the y-direction, σ_{yy}, are examined thoroughly. For example, for a model to be used to generate K_I solutions for a CCT specimen, a stress analysis without a crack is performed. By constraining all of the nodes on the crack plane, y = 0, no perturbations of the stress field in the y-direction are acceptable when away from the point of load application. For models with a centrally located hole in a plate, the stress concentration factor at the edge of the hole is an additional parameter used for model verification. Furthermore, closed form stress intensity solutions when available are also used for confirmation of model adequacy.

4.4.1 Verification of Stress State

For all of the models used to generate K_I solutions for through the thickness cracks whether straight or oblique, the stress analysis with no crack present yielded no deviation in the stress field from the theoretical solution within the limits of computer precision.[31] For the models with centrally located holes, the stress distribution in the net section was also used. Figure 4.9 and Figure 4.10

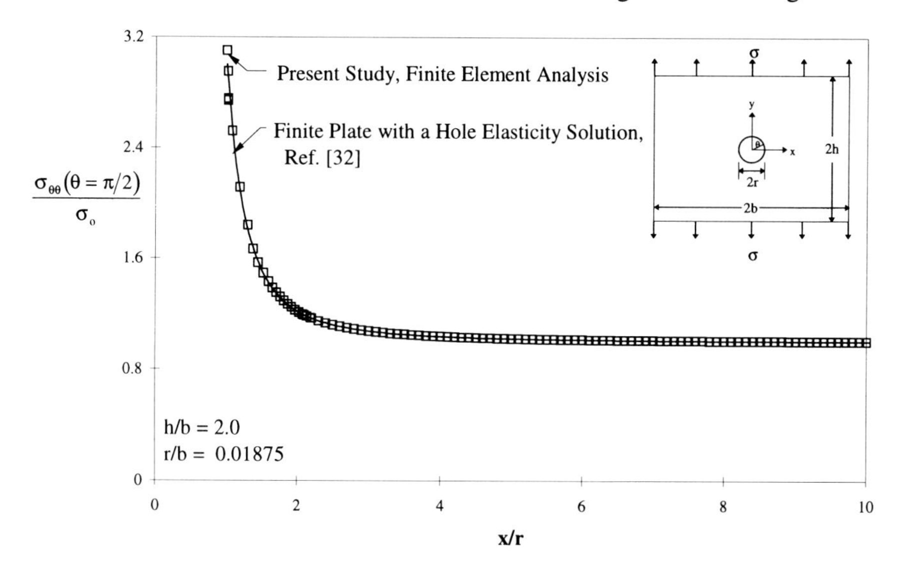

Figure 4.9 Comparison of Theoretical and FEA Normalized Stress in a Finite Width Plate with a Centrally Located Hole Subject to Uniform Remote Tension

illustrate the accuracy of the present solution. In a displacement formulation finite element analysis, solution of the system of equations yields the displacements of the unconstrained degrees of freedom from which the stresses are calculated. Integration of the stiffness matrix occurs by numerical integration at the Gauss points of the elements. By using the interpolation (shape) functions, the nodal quantities are then extrapolated. The small deviation from the theoretical value at the root of the notch, x/r = 1, plotted in Figure 4.9 and Figure 4.10 is due to the extrapolation of the stress from the Gauss points to the nodes. Recall the 8 noded isoparametric brick elements being used here are linear elements; thus extrapolation is also linear.

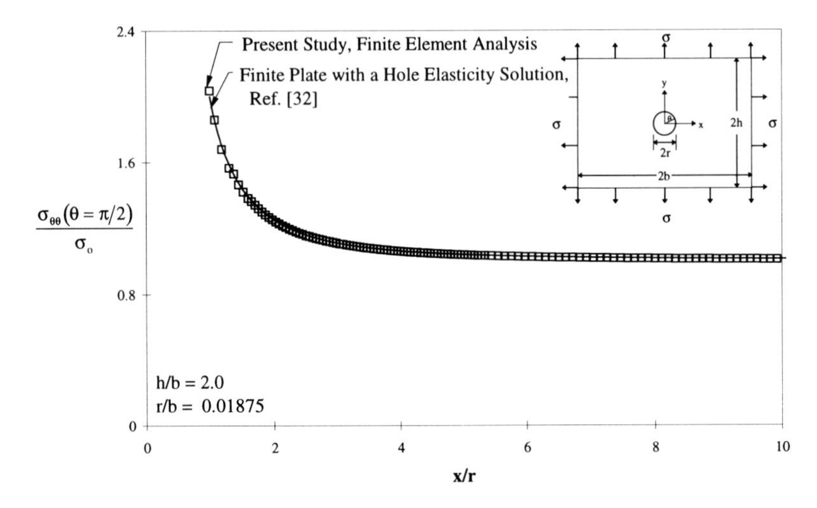

Figure 4.10 Comparison of Theoretical and FEA Normalized Stress in a Finite Width Plate with a Centrally Located Hole Subject to Uniform Biaxial Tension

4.4.2 Circular Internal Crack Embedded in an Infinite Solid Subject to Uniform Tension

This crack configuration was analyzed first since there is a closed form solution available for comparison. In addition, the crack does not intersect a free surface where the calculation of the strain energy release rate is questionable. The finite element model created to generate K solutions for this crack geometry has 6032 elements, 7182 nodes and 21,546 degrees of freedom. The crack plane mesh, y = 0, is circular at the crack location to better represent the circular front as shown in Figure 4.12. Due to the symmetry of the problem only one eighth of the plate is modeled and a unit stress is applied at the top of the plate. The mode I stress intensity factors have been normalized using the following relation.

$$\beta = \frac{K}{\sigma\sqrt{\pi a}} \tag{4.25}$$

Herein, all stress intensity solutions obtained are assumed to be mode I and plane strain is assumed when converting strain energy release rates to stress intensity factors. Note, commonly K's for semi-elliptical cracks are depicted as a function of the parametric angle, ϕ, of an ellipse as defined in Figure 4.11. Since a circle has the same functional form as an ellipse, the same convention is used.

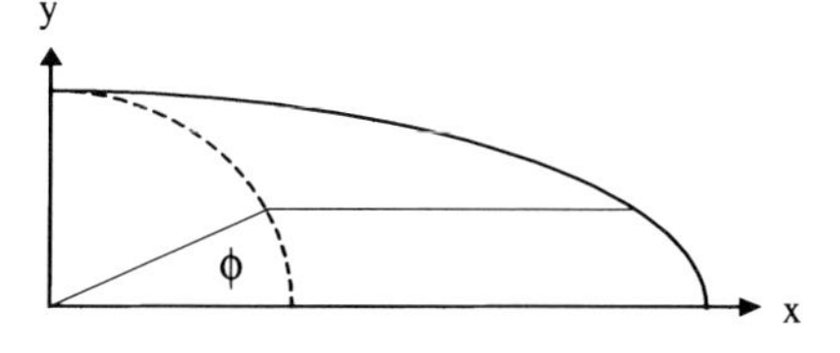

Figure 4.11 Parametric Angle Definition

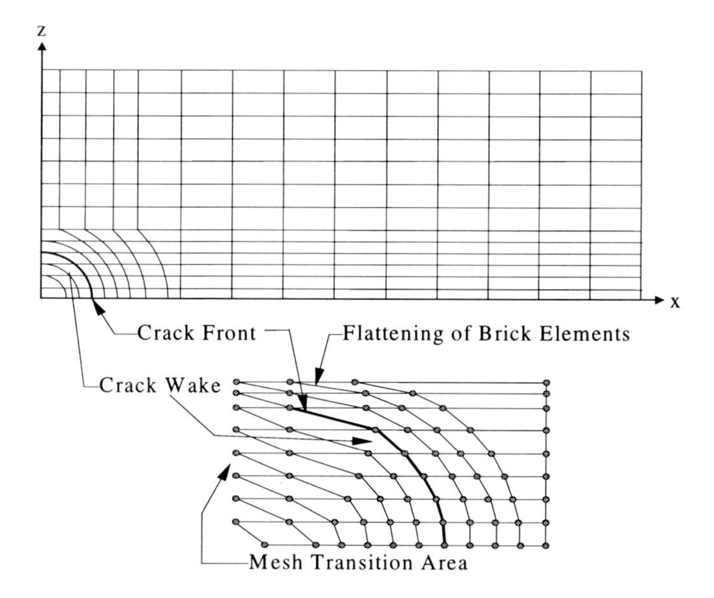

Figure 4.12 FE Mesh Pattern for Internal Circular Crack Embedded in an Infinite Solid

Sneddon derived the solution of a circular crack in an infinite solid.[32]

$$K = \frac{2}{\pi}\sigma\sqrt{\pi c} \tag{4.26}$$

A comparison of the FEA results and theoretical solution is made in Figure 4.13. At first glance, the variation between the FEA results and theoretical solution in the range $0 \le 2\phi/\pi \le 0.7$ appears extreme. However, the crack plane finite element mesh explains much of the variation seen in this range of ϕ. The somewhat regular variation indicates large mesh transitions where elements on one side of the crack front are of different size relative to the adjacent element.

Recall, G is calculated from nodal forces on the crack front and displacements one element behind the crack front. At the mesh transitions, the assumption is made that the nodal quantities are proportional to the element size; therefore in calculating G an average area is used in addition to scaling the forces as a function of the element size. As will be seen in results presented later, this variation disappears with a more uniform mesh where the element size transitions at the crack front are minimized.

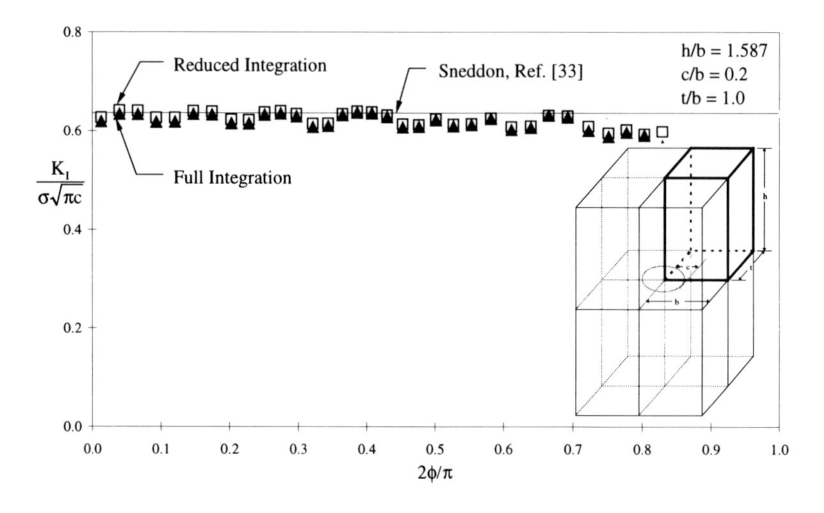

Figure 4.13 Comparison of Theoretical and FEA Solutions for Circular Internal Crack Embedded in an Infinite Solid Subject to Uniform Tension

Furthermore, at $\phi = 0$, the mesh is not skewed which indicates the element sides on the crack front are smaller than those at larger ϕ resulting in a better representation of the circular crack. Also at this location, the crack is furthest away from the boundaries of the plate mitigating any disruption of the stress field due to the boundary better representing an infinite solid. The deviation in the two solutions as ϕ approaches maximum is due to extremely skewed elements where the crack front is no longer circular since enforcing a circular boundary at this location would require collapsing the brick elements. Figure 4.12 illustrates the "flattening" of the circular crack at maximum ϕ. In addition, a slight difference in the analyses using full and reduced integration is also evident, which is further indication the mesh is coarse. Reduced integration eliminates the displacements ultimately used to obtain the transverse shear stress in the global stiffness matrix of a finite element model. In doing so, the arbitrary stiffness associated with this shear stress is eliminated making the model more flexible. For fine mesh finite element models subject to tension only, full and reduced integration solutions should be coincident. A coarse mesh is arbitrarily stiff due to a lack of degrees of freedom. By using reduced integration on a coarse mesh, the stiffness associated with the available degrees of freedom decreases making the model more flexible. As mentioned previously, reduced integration is only beneficial for bending dominant problems.

4.4.3 Cracks in Three Dimensional Finite Bodies

In attempts to better represent the physics of cracks occurring in operational structures, developing three dimensional stress intensity solutions has received much attention. In addition, the advances in computer technology have made possible solution of problems once too time consuming and cumbersome to generate. Recall the main goal of this effort is to generate three-dimensional solutions for a semi-elliptical crack that has penetrated the back surface. Since there are limited solutions in which to compare the present results, verification of the methodology is done by generating K's for known solutions. In addition, where the crack intersects the free surface, the K's are calculated using the plane stress relation of Eqn. (4.18). The plane stress assumption at the free surface is to accommodate for the changes in constraint and crack closure. The following five sections present comparisons to center crack tension (CCT), single edge crack tension (SECT), semi-elliptical surface crack subject to tension and bending, diametrically opposed quarter elliptical corner crack at a hole subject to tension and bending, and center crack tension with a skewed mesh at the crack front.

4.4.3.1 Center Crack Tension

Three dimensional solutions for this and the following crack geometry, in addition to several other commonly used fracture specimens, were first generated by Raju and Newman in 1977 using the finite element method with singularity elements and the finite element method employing the force method.[9] The model generated in this study contains 16112 elements, 18972 nodes, and 56,916 degrees of freedom; nearly 75 times more degrees of freedom than [9]. The increased refinement of the mesh is done to allow use of one mesh to generate multiple solutions, where in [9] each model was used for only one solution. Furthermore in [9], the mesh pattern surrounding the crack front is orthogonal to the crack front. A unit stress was applied at y = h and the boundary conditions are $u(0,y,z) = 0$ and $v(x,0,z) = 0$. Figure 4.14 presents the results from [9] and the present study for a central cracked specimen loaded in tension. Excluding the boundary layer, correlation is good throughout the thickness of the model. The slight variation at the mid-plane is due to differences in the model height to width ratio, h/b, which is known to effect the stress distribution in the model. This St. Venant's behavior is accentuated as h/b decreases because the crack plane becomes closer to the point of load application. This height effect typically causes variations of one percent or less for h/b ratios in this range.[33] The height effect is small in the CCT models and can be ignored for $h/b \geq 2.0$. Near the peak K value in the boundary layer, the 0.67% difference is attributed to the current model having more degrees of freedom; thereby being more flexible resulting in a slightly higher K.

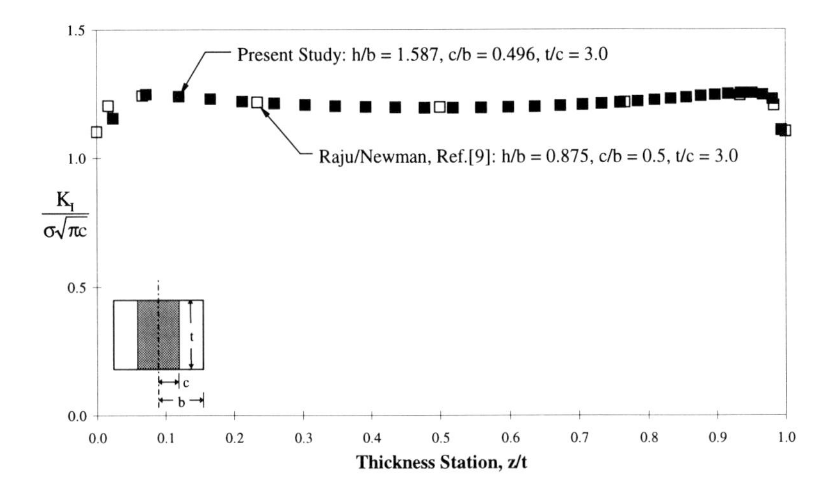

Figure 4.14 Comparison of CCT FEM Mode I Stress Intensity Factor Solutions

4.4.3.2 Single Edge Crack Tension

In reference [9] and the current study, the same model used for the CCT analysis is used for the Single Edge Crack Tension by simply removing the $u(0,y,z) = 0$ boundary condition which in the CCT analysis is used for creating the symmetry plane. A unit stress was applied at $y = h$ and no bending restrictions were enforced. Figure 4.15 shows the comparison between [9] and the present results. The 1.1% difference evident through most of the cross section can only be attributed to an increase in the accuracy of the solution due to the increased degrees of freedom in the present model. The height effect is

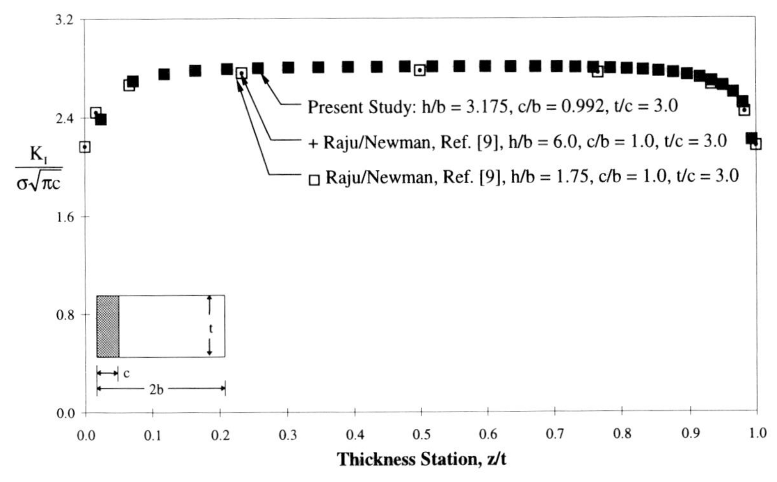

Figure 4.15 Comparison of SECT FEM Mode I Stress Intensity Factor Solutions

extremely small in the SECT models and can be ignored for h/b ≥ 1.5. The results of the two h/b values used in [9], shown in Figure 4.15, are coincident.

4.4.3.3 Semi-Elliptical Surface Crack Subject to Tension and Bending

The model used to generate the circular internal crack embedded in an infinite solid (penny crack) is also used for the semi-elliptical surface crack subject to tension and bending by changing the boundary conditions. For the penny crack, symmetry planes lie at (0,y,z), (x,0,z), and (x,y,0) with sufficient explicit modeling in each of the positive axis directions to represent an infinite body. By removing the $w(x,y,0) = 0$ constraint, the penny crack model now represents a surface crack. The Newman/Raju solutions have become a standard of comparison for new K solutions of semi-elliptical cracks.[34] The Newman/Raju results shown in Figure 4.16 are derived from the equations presented in [34] not directly from the FEA results. The equations were obtained by a regression analysis of the FEA results. As expected, the general trend of the present analysis correlates well to the referenced solution, since this model is a derivative of the penny crack model, see the penny crack discussion for the explanation of the variation in solutions. Also recall, the back surface of the penny crack, and thus for the surface crack also, had to be flattened in order to prevent the elements from being extremely skewed which can lead to a singular stiffness matrix in the FEA.

The mode I stress intensity factor for cracks of a semi-elliptical form is normalized in the following manner.

$$\beta = \frac{K}{\sigma\sqrt{\frac{\pi a}{Q}}} \tag{4.27}$$

$$\text{where } Q = 1 + 1.464\left(\frac{a}{c}\right)^{1.65} \quad \text{for} \quad \frac{a}{c} \le 1.0$$

$$Q = 1 + 1.464\left(\frac{c}{a}\right)^{1.65} \quad \text{for} \quad \frac{a}{c} > 1.0 \tag{4.28}$$

In Figure 4.16 and all combined loading cases, the subscript "t" and "b" refer to tension and bending, respectively; if no subscripts are used, assume tension. The flaw shape parameter, Q as given by Eqn. (4.28), was derived by Rawe as an approximation to the square of the complete elliptical integral of the second kind, which is required to represent the stress distribution around an elliptical crack in an infinite body.[35] Newman and Raju found the maximum error in the stress intensity factor by using these approximations is about 0.13% for all a/c values.[34] Subsequently, Schijve derived a more accurate solution to the

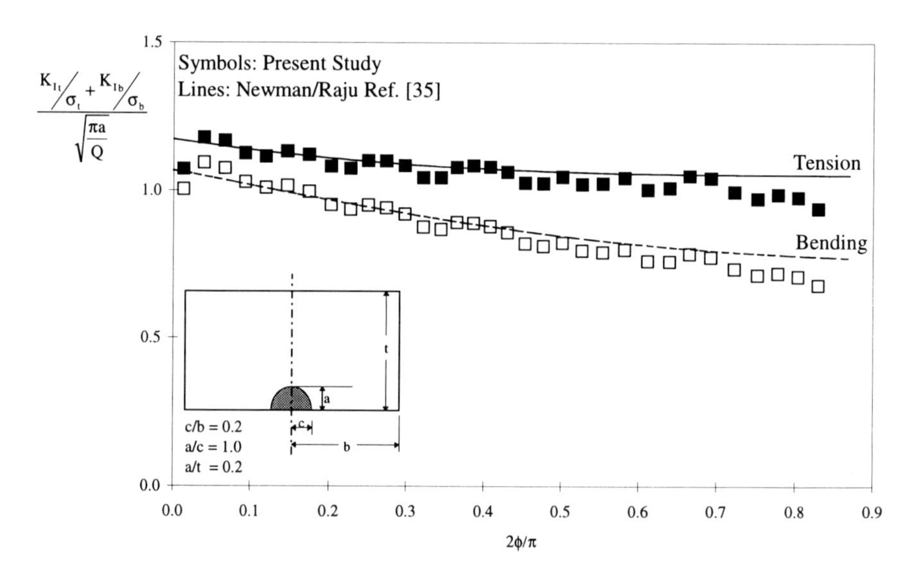

Figure 4.16 Comparison of Semi-Elliptical Surface Crack Subject to Tension and Bending FEM Mode I Stress Intensity Factor Solutions

complete elliptical integral, Φ:[36]

$$\Phi = \frac{\pi}{2(1+m)}\left[1 + \frac{m^2}{4} + \frac{m^4}{64}\right] \tag{4.29}$$

$$\text{where } m = \frac{1 - \frac{a}{c}}{1 + \frac{a}{c}}$$

With Eqn. (4.29), the normalized mode I stress intensity factor can accurately be represented by the following equation.

$$\beta = \frac{K(\varphi)}{\sigma\sqrt{\pi a}} \tag{4.30}$$

4.4.3.4 Diametrically Opposed Through Cracks at a Hole Subject to Remote Biaxial Tension, Remote Bending, and Uniform Internal Pressure

In searching the literature, few stress intensity solutions exist for this crack geometry and load condition. The solutions that are available, Newman[37], NASGRO TC09[38], Tweed/Rooke[39] are two-dimensional and for an infinite plate; thus the Irwin[18] finite width correction is applied. Furthermore, the Tweed/Rooke and NASGRO TC09 solutions are for a single crack; therefore, the Shah[40] correction is applied to account for two cracks. The difference in solutions is shown in Table 4.1 for the range of crack lengths shown in Figure

Table 4.1 Solution Comparison of Diametrically Opposed Through Cracks at a Hole Subject to Remote Tension

Solutions	Maximum Difference
Newman vs. Tweed/Rooke	4.3%
NASGRO TC09 vs. Newman	3.4%
Tweed/Rooke vs. NASGRO TC09	2.0%

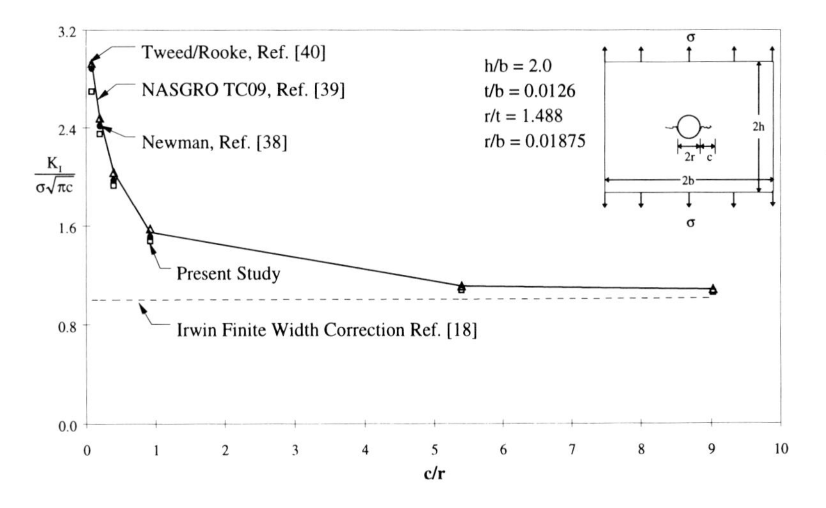

Figure 4.17 Comparison of Diametrically Opposed Through Cracks at a Hole Subject to Tension

4.17. The agreement is satisfactory with the Newman and NASGRO TC09 solutions varying approximately 5% on average with the current 3D FEA results except for the smallest crack length where the mesh refinement in the crack wake of the current study is insufficient.

The results for biaxial tension, uniform tension applied along all four plate edges, also correlates well with references [37] and [38] as shown in Figure 4.18. References [37] and [38] differ by less than 4% for all cracks lengths shown below. The maximum difference with the current 3D FEA results occurs at the smallest crack length which is again due to insufficient mesh refinement in the crack wake.

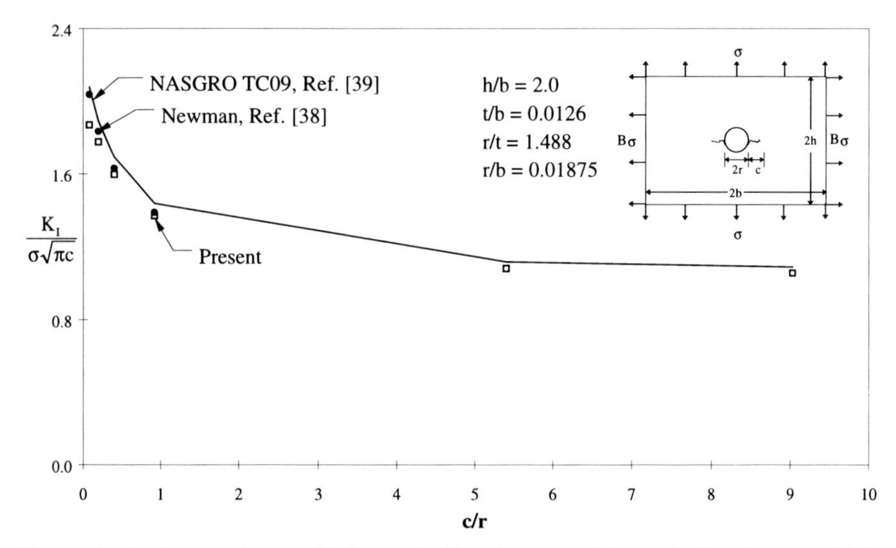

Figure 4.18 Comparison of Diametrically Opposed Through Cracks at a Hole Subject to Uniform Biaxial Tension

Only two solutions for comparison were available for this geometry subject to remote bending as shown in Figure 4.19. The linear distribution of K through the thickness for the 2D solutions is assumed for comparison with the present 3D solution. Good correlation is seen with the NASGRO TC09 solution which was derived using conformal mapping by several authors.[41-44] Other than the solution being two dimensional, it is difficult to discern why the Tweed and Cartwright[45] solution is 23% lower than the present results.

A uniform internal pressure on bore of the hole only, not on the crack faces, is used extensively for the rivet loading analyses that are done later; therefore, a validation analysis is completed. Again, no three dimensional solutions exist, thus the comparison is with a two dimensional boundary collocation solution shown in Figure 4.20. The difference between the present plane strain analysis and reference [37] varies with the maximum being 3.5%.

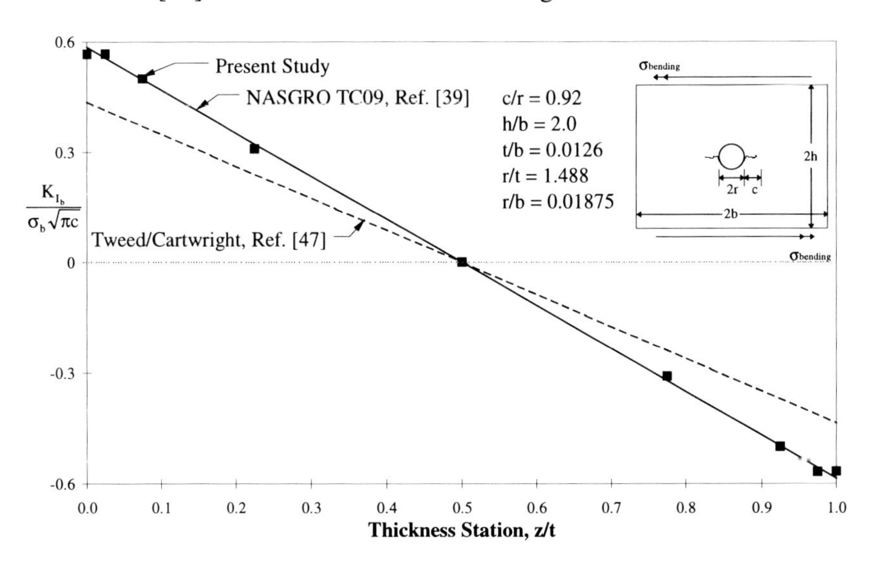

Figure 4.19 Comparison of Diametrically Opposed Through Cracks at a Hole Subject to Bending

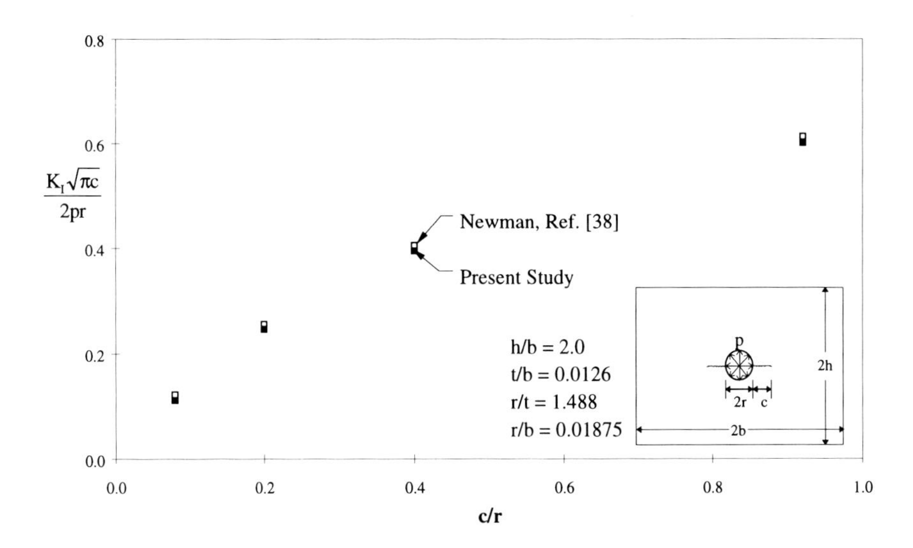

Figure 4.20 Comparison of Finite Width Plate with a Centrally Located Hole Subject to Uniform Internal Pressure

4.4.3.5 Through Cracks with an Oblique Elliptical Crack Front Subject to Remote Tension and Bending

The final verification models were designed to investigate the K variation along an oblique through crack front since this will be of prime interest in future studies. Miyoshi et al. are the only researchers who have developed a solution for this geometry and load condition.[46-48] They used the boundary element method to generate their solutions for varying a/t and a/c ratios. The accuracy of these solutions is "undefined."[46] The model previously used for the straight through crack analyses was also used for this configuration by transforming the mesh to a part elliptical pattern as seen in Figure 4.21. Figure 4.23 shows the good correlation for the tension case. The solutions of references [47] and [48] do not account for the boundary layer effect; whereas, the present results do show the K drop off at the free surface where $\phi = 0$. Little variation in magnitude of either solution is evident through most of the plate thickness. As the point of back surface penetration is reached, the ligament between the crack front and back free surface is at a minimum elevating the stress field resulting in a sharp increase in K. Two other comparisons of the Miyoshi et al. solution were made yielding similar results, but are not shown here for brevity.

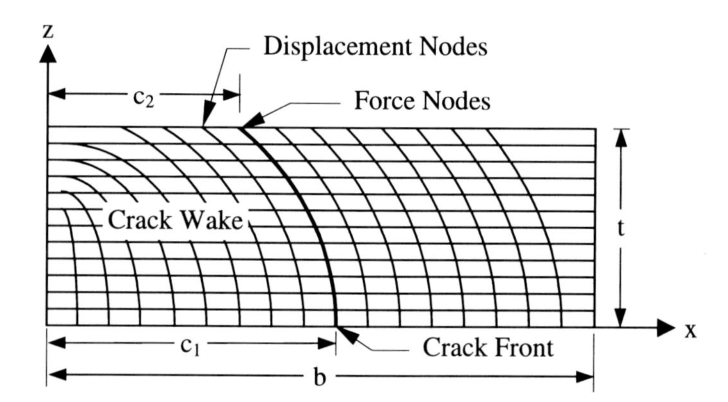

Figure 4.21 Diagram of Through Crack with Oblique Elliptical Crack Front

The bending solution comparison is shown in Figure 4.22 for the same crack configuration as Figure 4.23. The stress intensity factor is calculated from the strain energy release rate in accordance with Eqn. (4.25). To calculate negative K's, for example in the bending solution shown in Figure 4.22, G is calculated using the absolute value of displacements with the sign of K being determined based on the original sign of the displacements.

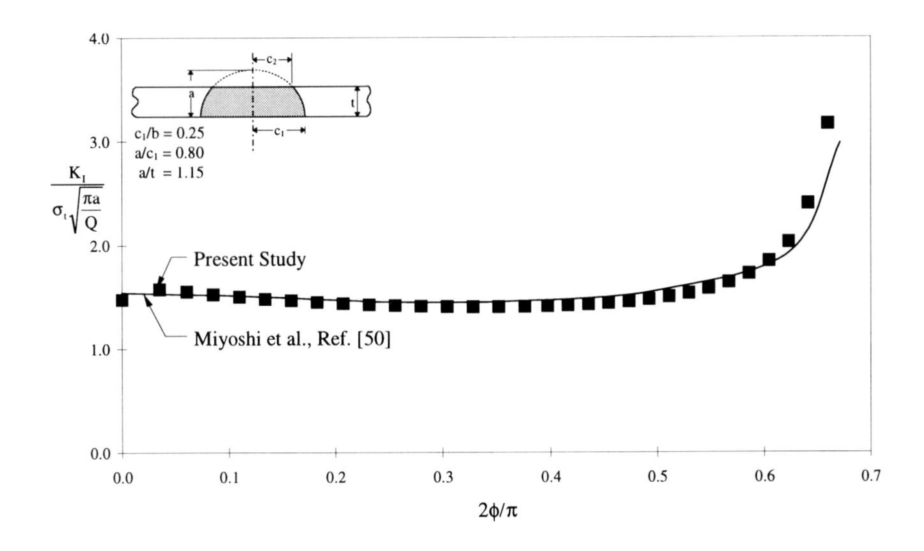

Figure 4.23 Comparison of Through Cracks with Oblique Elliptical Crack Front Subject to Remote Tension

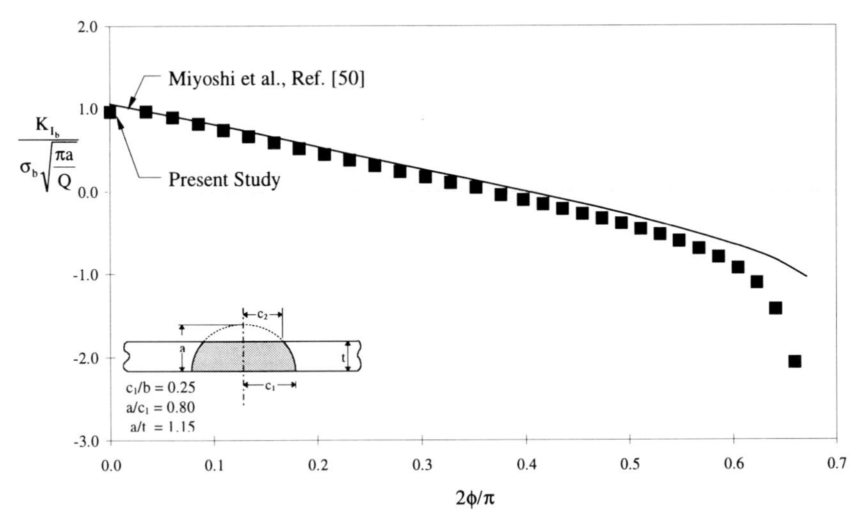

Figure 4.22 Comparison of Through Cracks with Oblique Elliptical Crack Front Subject to Remote Bending

Excellent correlation is seen at the free surface. The Miyoshi et al. results show the transition from positive to negative K's at $2\phi/\pi = 0.41$ ($z/t = 0.69$), whereas, the straight through crack subject to bending presented earlier transitioned at z/t = 0.5. Also note, the linear relationship is maintained until close to back surface penetration. The current results transition from positive to negative K's at $2\phi/\pi = 0.375$ (z/t = 0.64) and shows a stronger dependence on crack shape than Miyoshi et al. A steeper linear variation is visible through $2\phi/\pi= 0.5$, and a more pronounced non-linear gradient is apparent approaching back surface penetration.

The present results are lower than the Miyoshi et al. results for both the tension and bending cases; thus, when the tension and bending results are superimposed the difference increases between the present results and Miyoshi et al., see Figure 4.24. However, the front surface K's still exhibit excellent agreement. The difference in Miyoshi et al. and the current results is possibly due to the degree of mesh refinement toward the penetrated surface. Miyoshi et al. calculated K at only 6 locations compared to 31 in the current study. The pre-penetrated crack (surface crack) K's of Miyoshi et al. are compared to K values calculated from the Newman/Raju equations,[34] shown in Figure 4.25. Recall Miyoshi et al. used the boundary element method and Newman/Raju used the finite element method. The good agreement in the two surface crack solutions in Figure 4.25 indicate, as expected, the solution method has no effect on the stress intensity factor. Therefore, the differences between the current work and Miyoshi et al. is most likely due to modeling the crack boundary.

4.4.4 Influence of a Non Orthogonal Finite Element Mesh on Stress Intensity Factors Calculated Using the 3D VCCT.

Two final investigations are completed to further verify using the 3D VCCT and a non orthogonal, skewed, finite element mesh. Recall the force and COD methods require an orthogonal mesh with respect to the crack front in order to obtain accurate SIF solutions. The 3D VCCT makes no restrictions. The following two analyses aim to illustrate the effect of a skewed mesh on K. The first is a straight through crack that has been arbitrarily skewed (section 0), the second, an embedded elliptical crack (section 4.4.4.2) which represents the type and method of orthogonal mesh modification that is used to generate the meshes for calculating the K's for part-elliptical through cracks.

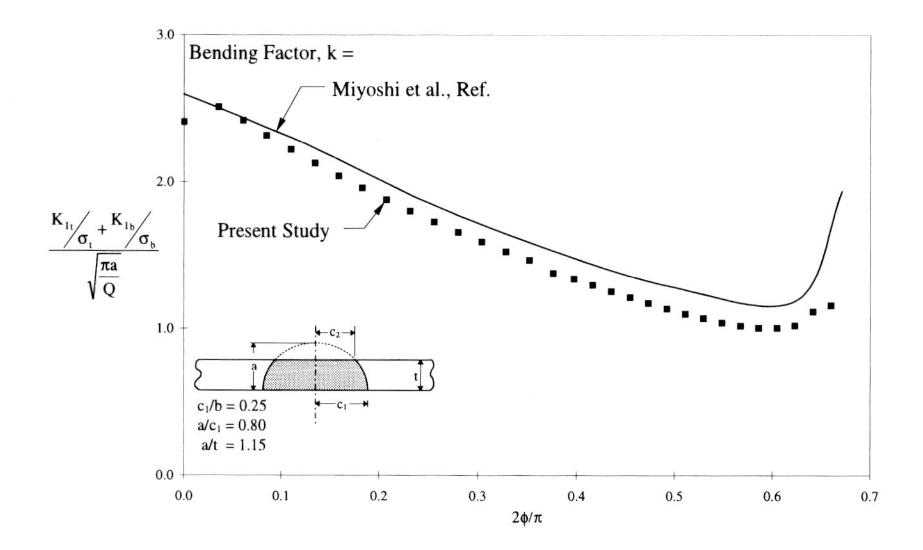

Figure 4.24 Comparison of Through Cracks with Oblique Elliptical Crack Front Subject to Remote Tension and Bending

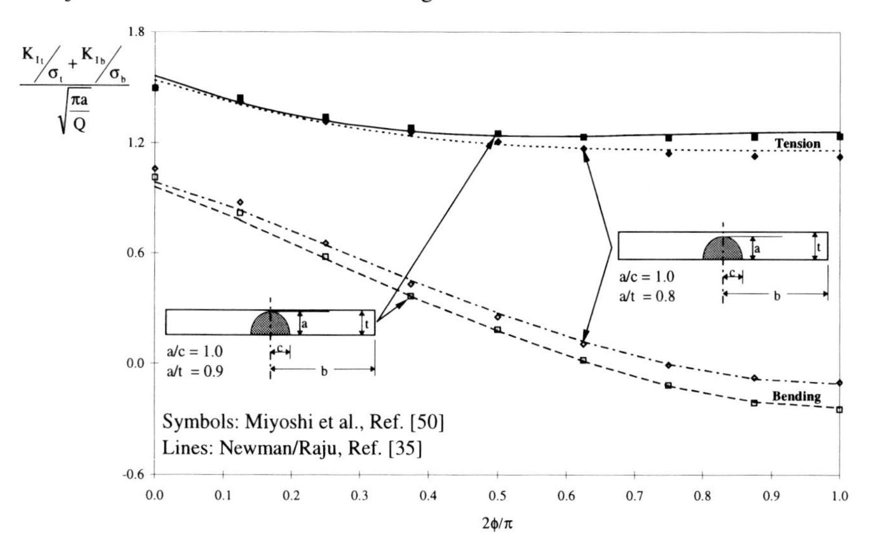

Figure 4.25 Comparison of SIF Solutions for a Semi-Elliptical Surface Crack Subject to Remote Tension and Bending

4.4.4.1 Center Cracked Tension with Skewed Mesh at the Crack Front Subject to Remote Tension

Using the straight through crack model described previously, several orthogonal and skewed meshes were analyzed to see the effect of the non-orthogonal mesh. The mesh around the crack front was skewed in two different ways by translating of nodes only in the thickness direction (z-direction). First, increasing the z coordinate linearly scaled the nodes in the crack wake, one element behind the crack front, by one half the original z dimension of the particular element. Then the nodes one element ahead of the crack front were linearly scaled by decreasing the z coordinate by one half the original z dimension of the particular element. The nodes one element behind, on, and one element ahead of the crack front now form a chevron as shown in Figure 4.27.

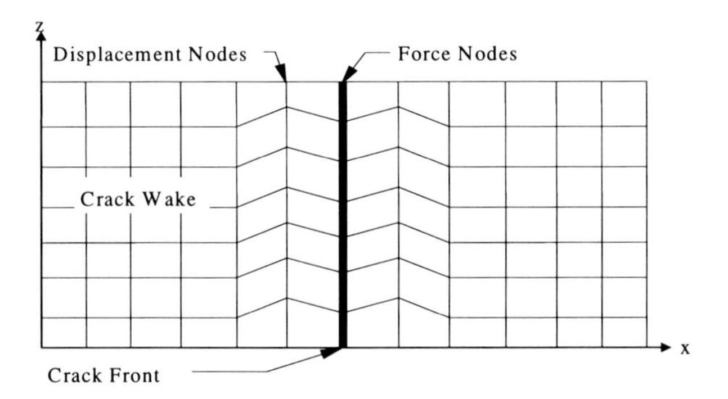

Figure 4.27 Diagram of Crack Plane of Chevron Skew Mesh with the Force and Displacement Nodes Used to Calculate K

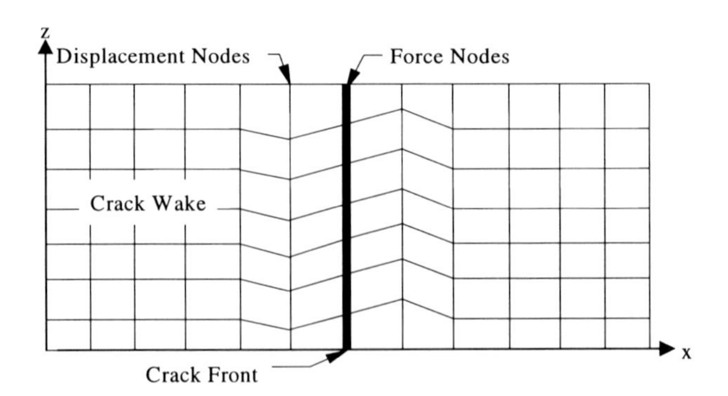

Figure 4.26 Diagram of Crack Plane of Linear Skew Mesh with the Force and Displacement Nodes Used to Calculate K

The second method of skewing the mesh uses the same process in the crack wake, but for the nodes one element ahead of the crack, the nodes are translated by increasing the z coordinate by one half the original z dimension of the particular element. The nodes one element behind, on, and one element ahead of the crack front now form collinear parallelograms as shown in Figure 4.26. The analysis results for these two models are shown in Figure 4.28.

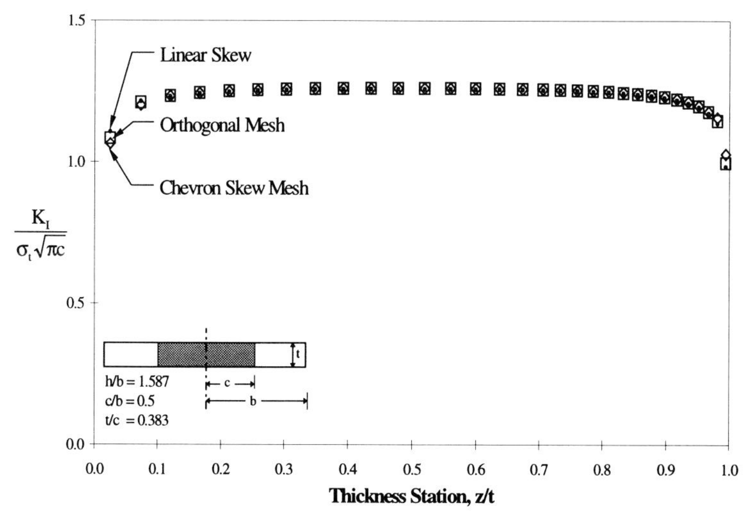

Figure 4.28 Effect of Mesh Pattern on the Mode I Stress Intensity Factor Solution for a Straight Through Crack Subject to Remote Tension

The small variation present is nearly uniform through the thickness; therefore, the mid-plane K's, z = t/2, are used for comparison. The difference is 0.14% for the Linear Skew Mesh and 0.008% for the Chevron Skew Mesh. From a geometric standpoint creating a non orthogonal mesh for a straight through crack with equal element area behind and in front of the crack front is a difficult task without changing the global dimensions of the plate being modeled. As a result, for both skewed meshes, the elements at the boundary, z = 0 and z = t, have different areas than the interior elements. At the boundary, the unequal area elements are symmetric for the Chevron Skew Mesh and asymmetric for the Linear Skew Mesh. The 0.14% error for the Linear Skew Mesh is attributable to the asymmetry of the elements with respect to the crack front, which is a requirement for the 3D VCCT. The 0.008% error for the Chevron Skew Mesh is due to the proximity of mesh transitions to the crack front which perturbates the local stress field thereby affecting the strain energy release rate calculations. The excellent agreement between the two skewed meshes and orthogonal mesh indicates no dependence on element orientation around the crack front for the calculation of K using the 3D VCCT.

4.4.4.2 Internal Elliptical Crack Embedded in an Infinite Solid Subject to Remote Tension

For an elliptical crack front, a non-orthogonal (skewed) mesh as shown in Figure 4.12 is quite easy to generate and modify. In addition, solution of this problem has been addressed by other researchers, which is useful for comparisons.[34,49] The purpose of this analysis was to determine to what extent the elements could be skewed and still have negligible effect on the K's. The normalized skew ratio (NSR) is defined in Eqn. (4.31) and α is shown in Figure 4.29.

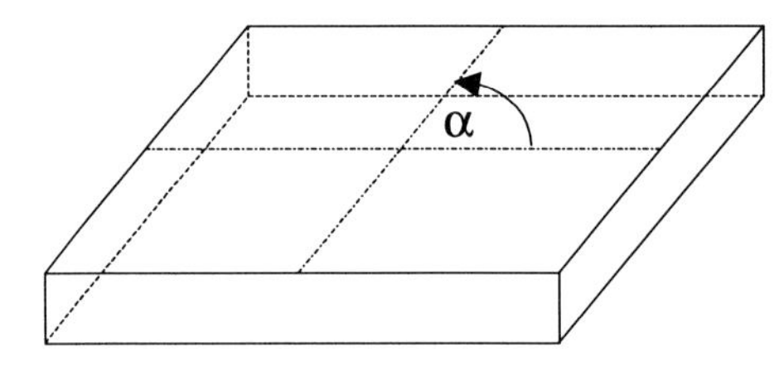

Figure 4.29 Definition of α

$$NSR = \frac{90 - \alpha}{90} \qquad (4.31)$$

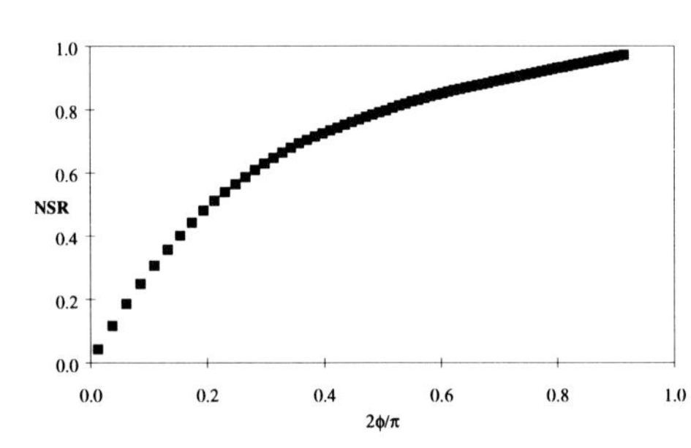

Figure 4.30 Normalized Skew Ratio for Finite Element Mesh of Internal Elliptical Crack Embedded in an Infinite Body

The normalized skew ratio for the elements approaching the minor axis of the ellipse was greater than 0.95 as seen in Figure 4.30. Skewing the elements any further results in a numerically unsolvable mesh. In other words, the integration points of the skewed elements are so close together, zero or negative volumes are calculated for them. Comparisons with known solutions are made in Figure 4.31. Even though the mesh is extremely skewed, the maximum error with respect to the Newman/Raju and Irwin solutions is 6%. Note, the Newman/Raju and Irwin solutions are coincident. The skew ratio of 0.95 is viewed as a limiting value. In view of these present results, orthogonality of the mesh with respect to the crack front is not required for the 3D VCCT.

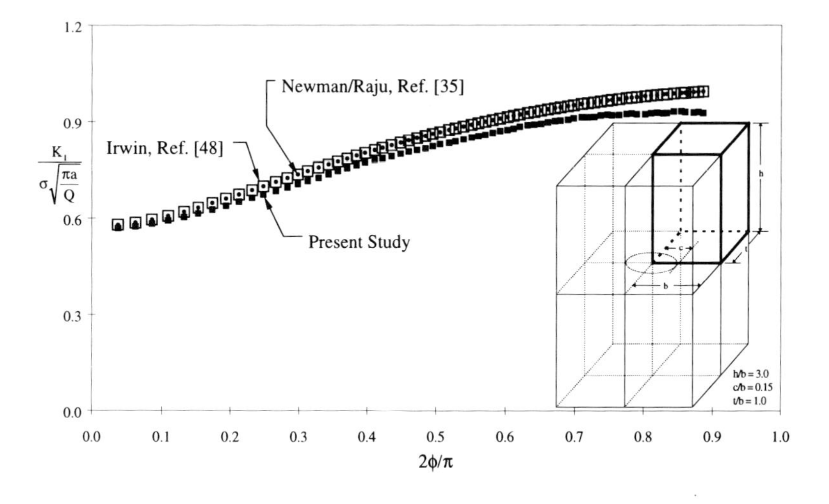

Figure 4.31 Comparison of Elliptical Internal Crack Embedded in an Infinite Solid Subject to Uniform Tension

4.5 Application Results

In the previous chapter the 3D VCCT was shown to provide accurate stress intensity factors for crack geometries and load conditions commonly found in riveted lap-splice joints in a transport aircraft fuselage. Now this technique is used to obtain a qualitative understanding of the influence of varying load conditions and crack geometries on the stress intensity factor. In section 4.5.1 K's are calculated for diametrically opposed through cracks growing from a centrally located hole in a plate subjected to three different rivet load distributions. In section 4.5.2, the focus is on the effect of the oblique, part elliptical crack front on K. A through crack in the center of a plate is used to calculate K for a straight crack and also an oblique, part elliptical crack. These qualitative results are assumed to apply for diametrically opposed through cracks with oblique elliptical crack fronts growing from a centrally located hole, reported in section 4.6, which is the main focus of analytical investigations.

4.5.1 Diametrically Opposed Through Cracks at a Pin Loaded Hole

Three-dimensional finite element models for diametrically opposed through cracks at a hole in a finite width plate are developed and analyzed. Again, only one quarter of the plate is modeled due to the available symmetry planes. The only published solutions for this crack geometry are two dimensional and obtained via conformal mapping[43,44,50] or boundary collocation.[37] In addition, referenced solutions are for infinite plates with only one crack; therefore, finite width[18] and Shah[40] corrections are superposed in the literature.

4.5.1.1 Pin Loaded Hole

The same finite element model used in the section 4.4.3.4 is also used to generate stress intensity factors due to the rivet loading. One of the fundamental issues to resolve in this part of the investigation is which load distribution to assume on the bore of the hole. Several researchers have addressed this question in developing stress concentration factors, K_t's.[51] Three different pin load distributions have been used in the literature; point, cosine and cosine squared loading. A parametric study has been accomplished to illustrate the effect of the pin load distribution on K. A priori, the point load results will be questionable since using point loads in a finite element model yields poor load introduction. This effect should be magnified for the smaller crack lengths were the point of load application is in close proximity to the crack front.

In order to make use of symmetry in a finite element model, the model must be symmetric with respect to geometry, load conditions, and material properties. Assuming the material is homogeneous and isotropic, a simple representation of a lap-splice joint, case I in Figure 4.32, is geometrically symmetric but is not loaded symmetrically. Case I can be decomposed into two parts that are loaded symmetrically. In equation form, the K solutions needed in Figure 4.32 are

$$K_{P\sigma} + K_{P\sigma} = K_w + K_{2P} \tag{4.32}$$

Case I: Stress Intensity Factor for Pin Loading (Bearing)

$$K_{P\sigma} = \sigma_{brg}\sqrt{\pi a}\beta_p \tag{4.33}$$

where

$$\sigma_{brg} = \frac{P}{2rt}$$

Case III: Stress Intensity Factor for Remote Tensile Loading

$$K_w = \sigma\sqrt{\pi a}\beta_w \tag{4.34}$$

Case IV: Stress Intensity Factor for Wedge Loading

$$K_{2P} = \sigma_{brg}\sqrt{\pi a}\beta_{2P} \tag{4.35}$$

By substituting Eqns. (4.33) - (4.35) into Eqn. (4.32) gives

$$2\sigma_{brg}\sqrt{\pi a}\beta_{P\sigma} = \sigma\sqrt{\pi a}\beta_{w} + \sigma_{2P}\sqrt{\pi a}\beta_{2P} \tag{4.36}$$

For case I, equilibrium requires $2bt\sigma = 2rt\sigma_{brg}$. The boundary correction factor for the pin loaded hole with load transfer can be written as

$$\beta_{P\sigma} = \frac{\frac{r}{b}\beta_{w} + \beta_{2P}}{2} \tag{4.37}$$

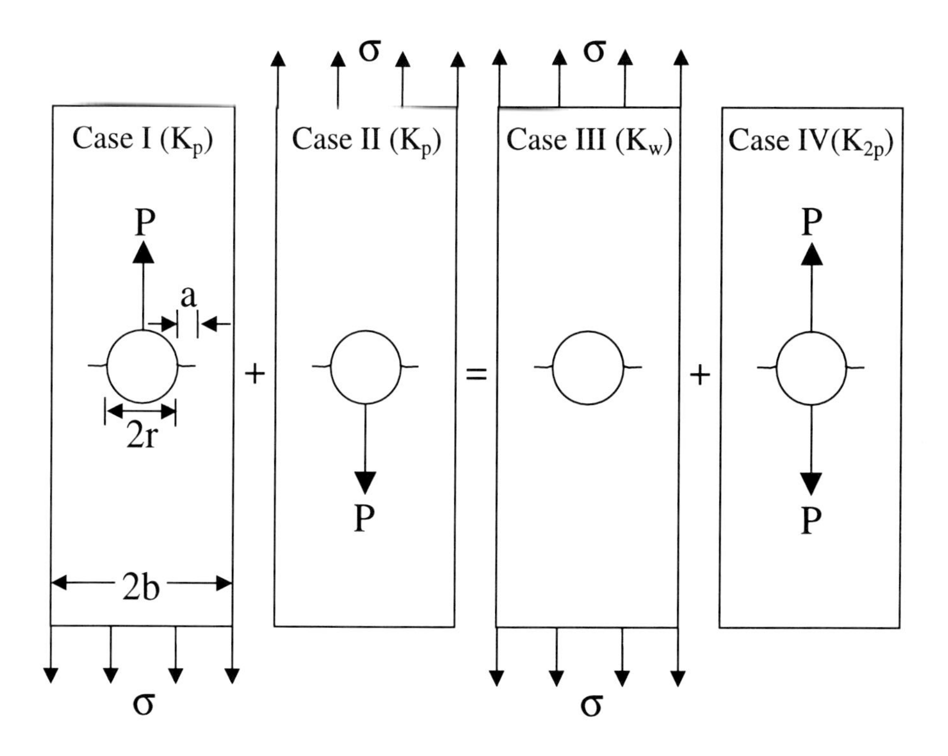

Figure 4.32 Decomposition of Loads in Typical Single Shear Lap-Splice Joint

Eqn. (4.37) is the most convenient form for calculation of K by the finite element method. In practice, for a 1/4 plate model of case I shown in Figure 4.32, two analyses must be run; one for remote tension and the other for the pin load.

4.5.1.2 Rivet Loading

Three different rivet load distributions were chosen to evaluate the effect of the rivet load distribution on K. Traditionally, the rivet load is modeled as a concentrated load located at the top of the hole, $\theta = 0°$, but due to recent investigations on the stress concentration factor of a loaded hole a cosine or cosine squared distribution may be more appropriate.[52-55] The effect of the assumed rivet load distribution is quite evident as seen in Figure 4.33 in addition to a comparison with reference [38]. As should be expected, the influence of the rivet load distribution is large for small cracks and diminishes with increasing crack length. The cosine distribution gives the highest K for the smallest crack length due to the proximity of the applied load to the crack tip. This behavior can be easily explained by the K solution for a through crack in a finite width plate subject to eccentric concentrated loads on the crack face as shown in Figure 4.34 and in functional form by the following equation.

$$K_{IA} = \frac{P}{\sqrt{\pi c}} \sqrt{\frac{c + x}{c - x}} \qquad (4.38)$$

As the point of application of load approaches the crack tip, the stress intensity increases sharply. Even though the magnitude of the applied load for both the cosine and cosine squared distributions approaches zero at the hole edge, the cosine distribution has a higher magnitude relative to the cosine squared; thus resulting in an higher K for the cosine distribution at the smallest crack length. The concentrated load being applied at the top of the hole yields the smallest K for the smaller crack length.

A similar trend of over estimating K for small cracks is apparent when comparing to reference [38]. However, for the large cracks, [38] underestimates K by nearly 100%. The large difference between the present study and [38] is of little consequence for large cracks since the contribution of the pin loading K to the total K for a typical lap splice joint is approximately 5%. Furthermore, since the majority of the fatigue life occurs during crack nucleation and growth to a visible crack, using the reference [38] solution is acceptable.

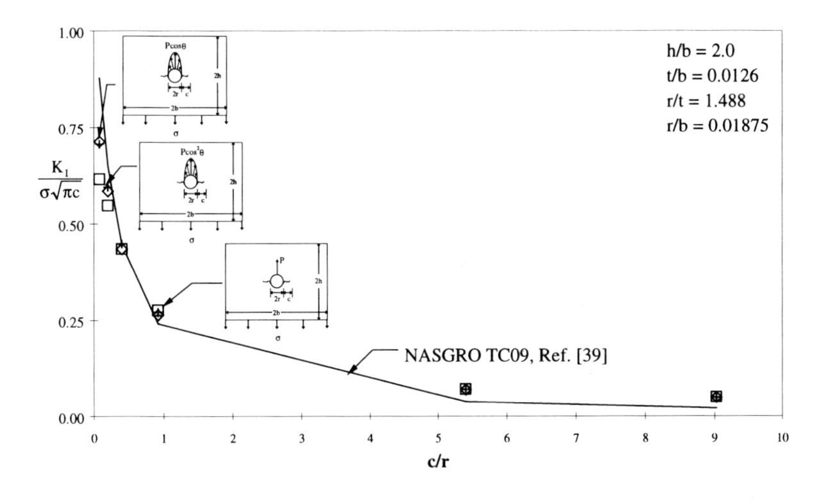

Figure 4.33 Effect of Rivet Load Distribution on Normalized K

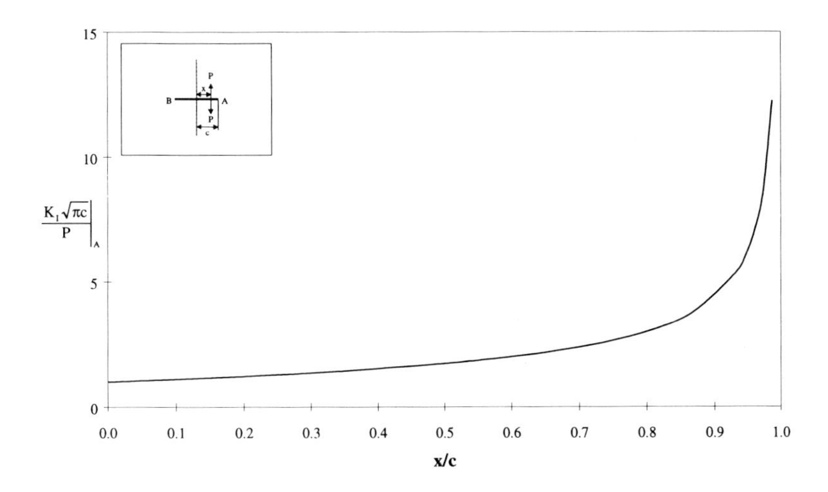

Figure 4.34 Effect of Load Location on Normalized K

4.5.2 Effect of Oblique Crack Shape on Stress Intensity Factor

From an engineering point of view, accounting for the oblique shape through cracks with an oblique elliptical crack front is traditionally accomplished by assuming the oblique crack has a crack front perpendicular to the sheet surface and a crack length c_1, recall definition of c_1 in Figure 4.21. This approach is undoubtedly conservative for those applications where the c_1 crack is visible. For riveted lap joints however, the visible crack is the penetrated crack, c_2, and the c_1 crack, faying surface crack, length is unknown without disassembling the joint; an impractical solution for transport aircraft fuselage skin. Using the same model as in the previous section, a sensitivity study is accomplished to determine the effect of the oblique shape. In Figure 4.35 and Figure 4.36, a comparison is made between an oblique crack with an elliptical crack front ($a/c_1 = 0.56$) and two through cracks with straight crack fronts ($c = c_1$ and $c = c_2$, respectively). As seen in Figure 4.35 for remote tension, an oblique crack subject to remote tension can almost be regarded as a straight through crack when comparing the results of the 3D FEMs. The c_1 crack length dictates failure; therefore, the higher K's at the penetrated surface are of no consequence. If 3D solutions are not available, using the secant approximation, Eqn. (4.39) with c_1 yields an overestimation of K by 5%.

$$K_I = \sigma\sqrt{\pi c}\sqrt{\sec\left(\frac{\pi c}{2b}\right)} \tag{4.39}$$

In the context of lap joints, the free surface is considered to be the surface where c_1 is measured. Using Eqn. (4.39) with c_2 results in an underestimation of K by 21%.

For remote tension, the higher K's at the penetrated surface supports the catch-up behavior reported by Grandt et al. However, for the same configuration as Figure 4.35 when remote tension and bending of are applied, catch-up of the penetrated crack is not likely as seen by Figure 4.36 $c_1 = c = 3.125$ underestimates K by only 0.2% when subject to both tension and bending. Recall the tension only analysis showed a 5% overestimation of K for this case. The 2D bending solution (NASGRO TC09) appears to be inadequate. The 3D FEM solution for the straight cracks both overestimate K at the free surface and could therefore be used to approximate the oblique crack without the extra resources required to model the latter.

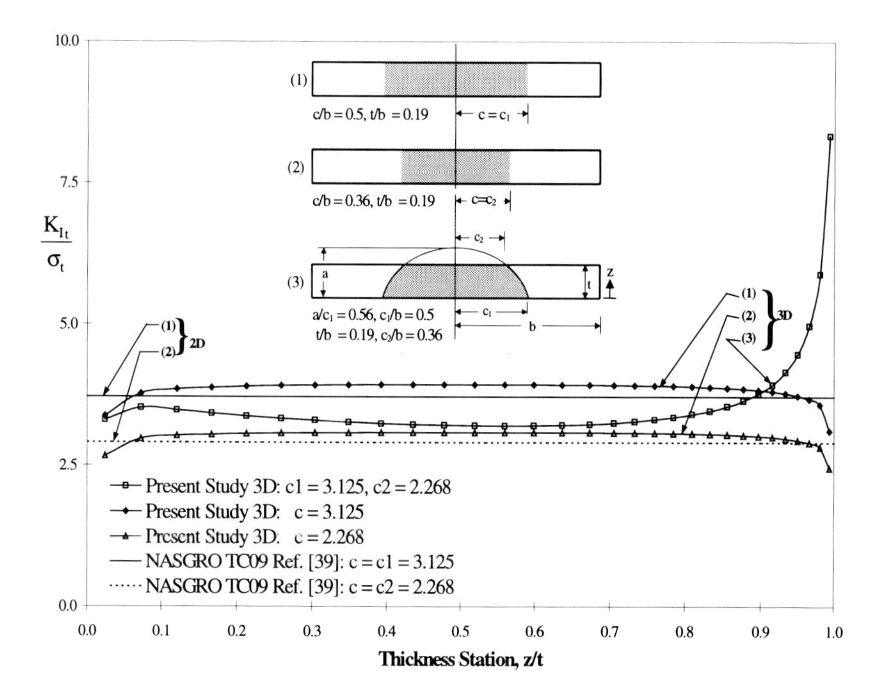

Figure 4.35 Sensitivity of K to Crack Shape Subject to Remote Tension

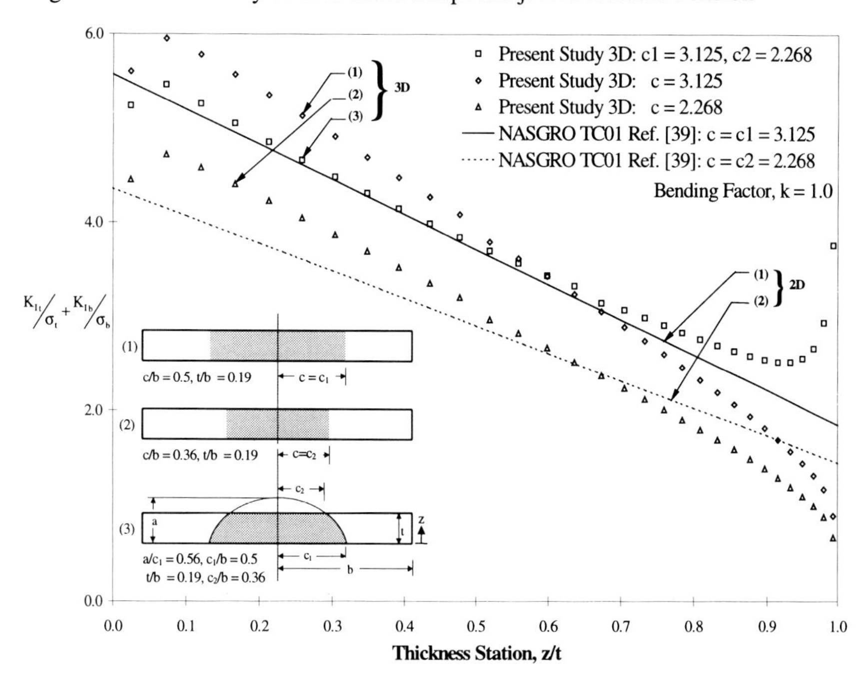

Figure 4.36 Sensitivity of K to Crack Shape Subject to Remote Tension and Bending

4.6 Stress Intensity Factors for Part Elliptical Through Cracks with an Oblique Front Under Combined Loading

The technique described above of using a non orthogonal finite element mesh to model a crack with a curved front is used to calculate stress intensity factors for through cracks with an oblique front under combined loading. The load conditions are chosen to represent the loads typically seen in a lap-splice joint; remote biaxial tension, remote bending, and pin loading. For each load condition, a unit stress ($\sigma = 1$) is applied. The remote tension is applied at y = h and four different values of the biaxiality ratio, B, are applied at x = b; B = 0.0, 0.25, 0.5, and 1.0 (the applied stress is $B\sigma$). Little difference in the K's is found between B = 0.75 and 1.0; therefore, the former is not considered. The remote bending is assumed to be linear through the thickness and is applied at the top of the model (y = h). Because the load distribution for the pin loading on the bore of the hole is not clear, three separate load distributions are used, point, cosine and cosine^2. The model used for all calculations is the same one used in the validation study in the previous sections, which has 9768 eight noded, isoparametric solid elements and 12033 nodes. The boundary conditions are simple, symmetry conditions at x = 0 and y = 0 where the *u* and *v* displacements, respectively, are suppressed. The dimensions shown in Figure 4.37 are h = 254 mm (10 in.), b = 127 mm (5 in.), and r = 2.38 mm (0.09375 in.). To calculate K's for more than one r/t ratio, the radius is kept constant and the plate thickness is scaled as needed. The plate thickness and hole radius dimensions chosen here are commonly used in lap joint designs. The plate height is determined to eliminate any finite height effect which has been well documented in references [9] and [56].

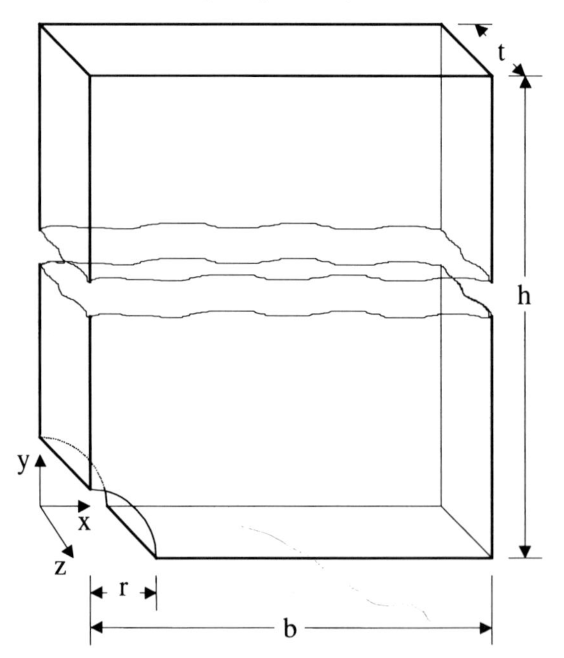

Figure 4.37 FEM Used for Part Elliptical Through Cracks with an Oblique Front Under Combined Loading

Similarly, the width is chosen to avoid any finite width effects.[18]

The crack shapes for which stress intensity factor solutions are calculated are listed in Table 4.2 and were chosen to bound all the crack shapes found during the experimental investigation discussed in Chapter 3. For a/c_1 values above 2.0, the crack front is rather straight, thus these crack fronts are not shown in Figure 4.38. Such cracks with a relatively large crack length are obtained with the higher a/t values. The hole edge is on the far left side of Figure 4.38.

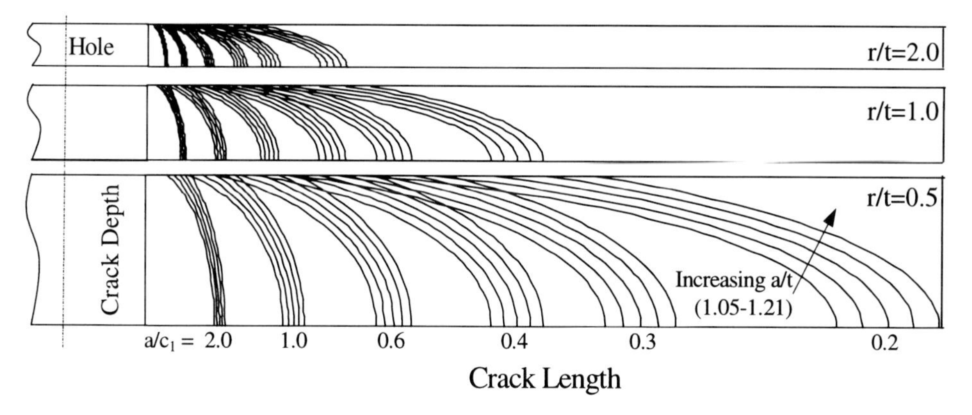

Figure 4.38 Crack Shapes for $a/c_1 \leq 2.0$, $a/t \leq 1.21$, and $r/t \leq 2.0$

Table 4.2 Crack Shape Geometries

a/t	1.05, 1.09, 1.13, 1.17, 1.21, 2, 5, 10
a/c_1	0.2, 0.3, 0.4, 0.6, 1.0, 2
r/t	0.5, 1, 2

Although the K's for part elliptical crack fronts are usually normalized by Eqn. (4.27), the approximation to the second elliptical integral of the second kind, Q, changes with the a/c ratio; thus making comparisons troublesome. Also recall that the K solutions for various crack shapes are calculated using one finite element model with a constant hole radius, plate width and plate height; therefore, to accommodate all shapes in the range shown in Table 4.2; t, c_1, and c_2 are varied from which the crack depth, a, is calculated. Also, as the c_1/b ratio increases, the finite width effect is no longer negligible. Therefore, a similar K normalization to that used in Eqn. (4.25) is employed with the addition of the finite width.

$$\beta = \frac{K_I}{\sigma\sqrt{\pi a} f_w} \tag{4.40}$$

When comparing the various solutions, the normalized K (β is used interchangeably with 'normalized K') at c_1 is used. Unfortunately, the effects on β of changing the crack shape ratios, $a/c_1$and a/t, and sheet thickness, r/t, are not as definitive as desired due to the interdependencies amongst a/c_1, a/t, and r/t. All calculated β solutions are listed in tabular format in Appendix E. Several of the more prominent trends are discussed below.

Varying the crack shape, a/c_1, for given a/t and r/t yields an easily identifiable trend as seen in Figure 4.39 where x is measured from the hole edge. For thick sheets (r/t = 0.5), an increasing a/c_1 for an a/t = 1.05 yields a decreasing β and a more uniform β distribution along the crack front for straighter crack fronts. This trend is also indicated for the other two thicknesses and the other a/t values. There is an obvious sharp rise of β at the penetrated side of the sheet (z/t = 1) for more oblique crack fronts (decreasing a/c_1).

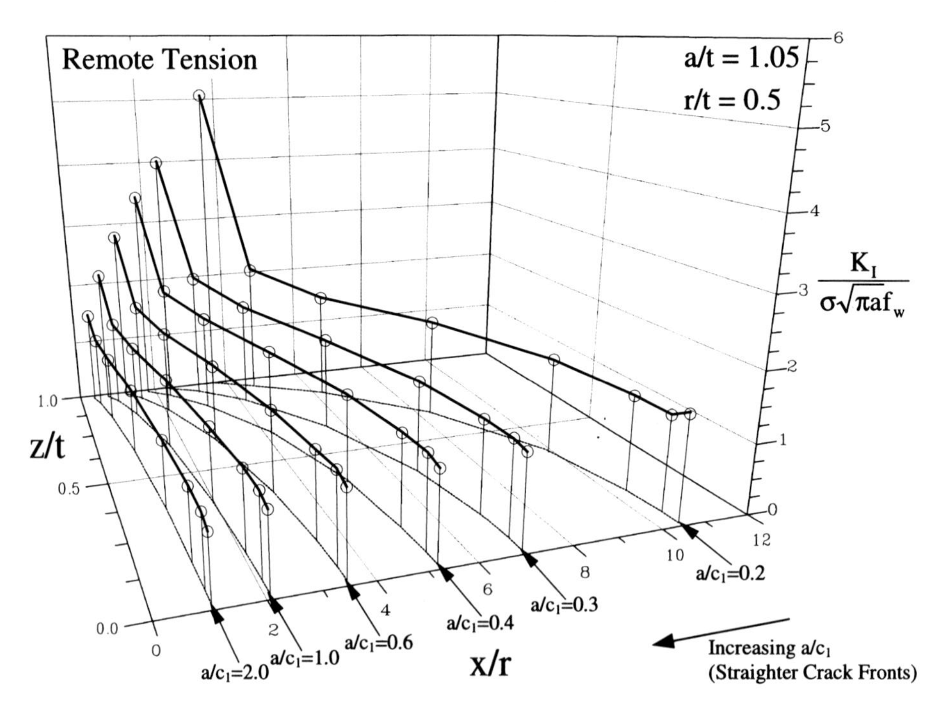

Figure 4.39 Effect of Changing a/c_1 on β for a/t = 1.05 and r/t = 0.5 (Thick Sheet)

Similarly, an increase in β along the crack front is seen when increasing the a/t ratio for a given a/c_1 and r/t which is shown in Figure 4.40. The increase in a/t with a constant a/c_1 and r/t is not a change in the shape of the crack, but an increase in the crack area. The β at the penetrated side of the sheet (z/t = 1) is slightly decreasing with increasing a/t, opposite of the front surface (z/t = 0) behavior. The increase in crack area decreases the variation in β along the crack front. For example, for a small crack (a/c_1 = 0.3, a/t = 1.05, r/t = 2.0) the ratio of β(z/t=1) to β(z/t=0) is approximately 3.0; whereas for a large crack (a/c_1 = 0.3, a/t = 10.0, r/t = 2.0), the same β ratio is only 1.4.

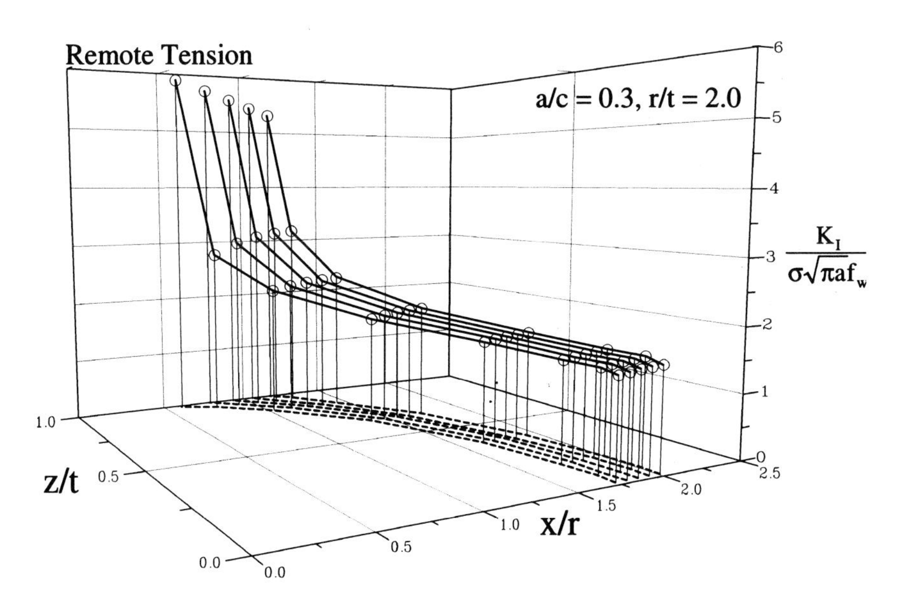

Figure 4.40 Effect of Changing a/t ratio (a/t = 1.05-1.21) on β for a/c_1 = 0.3 and r/t = 2.0

The interdependency of the crack shape parameters, a/c_1, a/t, and r/t, and their resulting effect on β make comparisons difficult amongst cracks through the entire range of crack shapes analyzed. The only trends seen for the entire range of crack shapes are:

- Pure Tension
  - shallow cracks, $a/c_1 \leq 0.6$, all r/t, increasing a/t → increasing β
  - deep cracks, $a/c_1 > 0.6$, all r/t, increasing a/t → decreasing β
- Remote Bending
  - all a/c_1, a/t, increasing r/t → decreasing β
- Pin Loading (Concentrated, Cosine, and Cosine2 Load Distributions)
  - all a/c_1, r/t, increasing a/t → decreasing β
  - all a/c_1, a/t, increasing r/t → increasing β

Although not shown graphically here, other trends in the β's are evident. For instance, in general the β's are decreasing for increasing biaxiality ratio for all a/c_1, a/t, and r/t with a few exceptions at the smaller a/c_1 and a/t ratios. Also, regardless of the crack geometry, pin loading with a concentrated load applied at the top of the hole is the most conservative yielding higher β's than both the cosine and $cosine^2$ distributions.

4.7 Conclusions

Non-Orthogonal Finite Element Mesh

It has been shown that the 3D virtual crack closure technique (3D VCCT) can be used with a non-orthogonal finite element mesh for the calculation of stress intensity factors. By generating one sufficiently fine finite element mesh, the 3D VCCT can be used to generate multiple stress intensity solutions of cracks with complex shapes by simply manipulating the crack plane geometry.

Three Dimensional Virtual Crack Closure Technique

K solution results were calculated with 3D VCCT for crack configurations and loading cases for which results were available in the literature. The configurations covered were circular internal crack embedded in an infinite solid subject to uniform tension, center crack tension, single edge crack tension, diametrically opposed through cracks at a hole subject to tension, bending, biaxial tension, and pin loading, semi-elliptical surface crack subject to tension and bending, and through cracks with an oblique elliptical crack front subject to tension and bending. In general the agreement was within 5% when comparing to 2D analytical solutions and 1% when comparing to published 3D finite element solutions.

Load Distribution on the Bore of the Rivet Hole

The assumed rivet load distribution on the bore of the rivet hole greatly influences K for small cracks, but has no measurable effect for large cracks. To remain conservative, a cosine-squared distribution should be assumed unless the rivet load distribution is known.

Oblique Crack Fronts

K solutions were calculated for through cracks with an oblique elliptical crack front subject to tension and bending. For the tension case, Figure 4.35, the oblique crack can be approximated as a straight crack having a crack front perpendicular to the sheet surface and crack length equal to the largest crack length of the oblique crack which has penetrated a free surface. Also, the high

K's on the penetrated surface would promote catch-up where the penetrated crack grows rapidly to the same length as the free surface crack. For the tension and bending case, however, the oblique crack cannot be approximated with a straight crack. No catch-up behavior seems possible since the K's for the penetrated surface crack are not only lower than those of the faying surface, but also the relative difference between the K's at the faying and penetrated surfaces becomes larger.

Crack Shape Effect

Increasing the a/c ratio, making the crack front more straight and less oblique, results in lower normalized K's caused by mitigating the perturbation of the stress field by the acute cusp at the intersection of the penetrated crack and back surface.

Loading Effects

Of the three assumed load distributions for the pin loaded hole analyses, the cosine squared load distribution is the most appropriate, conservative engineering solution giving a mean normalized K for small cracks and a moderately higher value for larger cracks when compared to the concentrated and cosine load distributions.

Stress Intensity Factors for Part Elliptical, Oblique Through Cracks Subject to Combined Loading

K-values have been calculated for crack shapes frequently seen in-service and in laboratory fatigue tests with a/c_1 ratios of 0.2, 0.3, 0.4, 0.6, 1, and 2; a/t ratios of 1.05, 1.09, 1.13, 1.17, 2, 5, and 10; and r/t ratios of 0.5, 1, and 2. Several load conditions are analyzed, biaxial tension (B = 0.0, 0.25, 0.5, or 1.0), remote bending, and pin loading (concentrated, cosine, or cosine^2 load distributions). Presented in tabular form, the K's can easily be incorporated into a crack growth prediction algorithm.

A strong interrelationship between the parameters describing the crack geometry, a/c_1, a/t, and r/t, makes comparisons between cracks of different shape difficult. However, some trends are clear.

- Pure Tension:
  - shallow cracks, $a/c_1 \leq 0.6$, all r/t, increasing a/t $\rightarrow$ increasing β
  - deep cracks, $a/c_1 > 0.6$, all r/t, increasing a/t $\rightarrow$ decreasing β
- Remote Bending
  - all a/c_1, a/t, increasing r/t $\rightarrow$ decreasing β

- Pin Loading (Concentrated, Cosine, and Cosine2 Load Distributions)
  - all a/c_1, r/t, increasing a/t → decreasing β
  - all a/c_1, a/t, increasing r/t → increasing β

Comparisons of the effect of biaxial loading and pin loading are in a broad sense the same as for straight cracks where increasing B decreases β and pin loading via a concentrated load at the top of the hole is more conservative than a cosine or cosine^2 distribution.

---

[1] Müller, Richard Paul Gerhard. An Experimental and Analytical Investigation on the Fatigue Behaviour of Fuselage Riveted Lap Joints, The Significance of the Rivet Squeeze Force, and a Comparison of 2024-T3 and Glare 3. Diss. Delft University of Technology, 1995. Delft, NL. ISBN 90-9008777-X, NUGI 834.

[2] Grandt, Jr., A. F., J. A. Harter, and B. J. Heath, "Transition of Part-Through Cracks at Holes into Through-the -Thickness Flaws," Fracture Mechanics: Fifteenth Symposium, ASTM STP 833, R. J. Sanford, Ed., American Society for Testing and Materials, Philadelphia, 1984, pp. 7-23.

[3] Piascik, Robert S., Scott A. Willard, and Matthew Miller. The Characterization of WideSpread Fatigue Damage in Fuselage Structure. NASA-TM-109142, 1994.

[4] Jeong, D. Y., D. P. Roach, J. V. Canha, J. C. Brewer, and T. H. Flournoy, "Strain Fields in Boeing 737 Fuselage Lap Splices: Field and Laboratory Measurements with Analytical Correlations," DOT/FAA/CT-95/25(DOT-VNTSC-FAA-95-10), FAA Technical Center, Federal Aviation Administration, US Department of Transportation, June 1995.

[5] Bartelds, G. and A. U. de Koning, "Application of Finite Element Methods to the Analysis of Cracks, (Phase I - Evaluation of Methods)," National Aerospace Laboratory of The Netherlands, NLR TR 78138 U, Dec. 1987.

[6] de Koning, A. U., Electronic Communication, The National Aerospace Laboratory of The Netherlands, 19 June 1996.

[7] de Koning, A. U., Personal Communication, The National Aerospace Laboratory of The Netherlands, 7 June 1996.

[8] Raju, I. S. and J. C. Newman, Jr., "SURF3D: A 3-D Finite-Element Program for the Analysis of Surface and Corner Cracks in Solids Subjected to Mode-I Loadings," Technical Report NASA TM-107710, February 1993.

[9] Raju, I. S. and J. C. Newman, Jr., "Three-Dimensional Finite-Element Analysis of Finite-Thickness Fracture Specimens," Technical Note NASA TN D-8414, May 1977.

[10] Dixon, J. R. and L. P. Pook. "Stress Intensity Factors Calculated Generally by the Finite-Element Technique." NATURE. 224 (1969): 166.

[11] Hellen, T. K., "On the Method of Virtual Crack Extensions," International Journal of Numerical Methods in Engineering. 9 (1975): 187-207.

[12] Parks, D. M., "A Stiffness Derivative Finite Element Technique for Determination of Crack Tip Stress Intensity Factors," International Journal of Fracture. 10 (1974).

[13] Parks, D. M., "The Virtual Crack extension Method for Nonlinear Material Behavior," Computer Methods in Applied Mechanics and Engineering, Vol. 12. pp. 353-364, 1977.

[14] de Koning, A. U. and C. Lof, "K-Distributions Extrapolated on the Basis of Stress Intensity Rates," Proceedings of the Third International Conference on Numerical Methods in Fracture Mechanics, eds. A. R. Luxmore and D. R. J. Owen, Pineridge Press Limited, Swansea, UK, 1984, p. 195.

[15] Rice, J. R., "An Examination of the Fracture Mechanics Energy Balance from the Point of View of Continuum Mechanics," Proceedings of the First International Conference on Fracture, Sendai, Japan, September 12-17, 1965, T. Yokobori, T. Kawasaki, and J. L. Swedlow, Eds. p. 309.

[16] Rice, J. R., "A Path-Independent Integral and the Approximate Analyses of Strain Concentration by Notches and Cracks," Journal of Applied Mechanics, Vol. 35, 1968, pp. 376-386.

[17] Bittencourt, T. N., A. Barry, and A. R. Ingraffea, "Comparison of Mixed Mode Stress-Intensity Factors Obtained Through Displacement Correlation, J-Integral Formulation, and Modified Crack-Closure Integral," Fracture Mechanics: Twenty-Second Symposium, Vol. II, ASTM STP 1131, S. N. Atluri, J. C. Newman, Jr., I. S. Raju, and J. S. Epstein, Eds., American Society for Testing and Materials, Philadelphia, 1992, pp. 69-82.

[18] Irwin, G. R., "Analysis of Stresses and Strains Near the End of a Crack Traversing a Plate," Journal of Applied Mechanics, Trans. ASME.. 24 (1957): 361-364.

[19] Bakker, Ad. The Three-Dimensional J-Integral: An Investigation into Its Use for Post-Yield Fracture Safety Assessment. Diss. Delft University of Technology UP, 1984. Delft, NL.

[20] Pickard, A. C., The Application of 3-Dimensional Finite Element Methods to Fracture Mechanics and Fatigue Life Prediction, Engineering Materials Advisory Services, LTD., West Midlands, UK, 1986, p. 70.

[21] Raju, I. S. And K. N. Shivakumar. Implementation of Equivalent Domain Integral Method in the Two-Dimensional Analysis of Mixed-Mode Problems. NASA-CR-182021, 1991.

[22] Shivakumar, K. N. and I. S. Raju.. An Equivalent Domain Integral Method for Three-Dimensional Mixed-Mode Fracture Problems. NASA-CR-182021, 1991.

[23] Rybicki, E. F. and M. F. Kanninen. "A Finite Element Calculation of Stress Intensity Factors by a Modified Crack Closure Integral," Engineering Fracture Mechanics. 9 (1977): 931-938.

[24] Shivakumar, K. N., P. W. Tan, and J. C. Newman, Jr. "A Virtual Crack-Closure Technique for Calculating Stress Intensity Factors for Cracked Three Dimensional Bodies," International Journal of Fracture. 36 (1988): R43-R50.

[25] Shivakumar, K. N. and J. C. Newman, Jr. ZIP3D - An Elastic and Elastic-Plastic Finite-Element Analysis Program for Cracked Bodies. NASA-TP-102753, 1990.

[26] Zienkiewicz, O. C. The Finite Element Method, Third Edition. New York: McGraw-Hill, 1979.

[27] Logan, Daryl L. A First Course in the Finite Element Method. Boston: PWS Publishers, 1986.

[28] Cook, Robert D. Concepts and Applications of Finite Element Analysis. New York: John Wiley & sons, 1981.

[29] Hughes, Thomas J. R. The Finite Element Method: Linear Static and Dynamic Finite Element Analysis. New Jersey, Prentice-Hall Inc., 1987.

[30] Zienkiewicz, O. C., R. L. Taylor, and J. M. Too. " Reduced Integration Techniques in General Analysis of Plates and Shells." International Journal of Numerical Methods in Engineering, 3 (1971): 275-290.

[31] Boresi, Arthur P. and Omar M. Sidebottom, Advanced Mechanics of Materials 4th Edition, John Wiley and Sons, New York, 1985, p. 570.

[32] Sneddon, I. N., The Distribution of Stress in the Neighbourhood of a Crack in an Elastic Solid," Proceedings of the Royal Society London A 187, 1946, pp. 229 - 260, qtd. in Broek, D. Elementary Engineering Fracture Mechanics. Martinus Nijhoff Publishers, Dordrecht, 1986, p. 88.

[33] Newman, Jr., J. C., Electronic Communication, NASA Langley Research Center, 2 Jan 1996.

[34] Newman, Jr., J. C. and I. S. Raju. Stress Intensity Factor Equations for Cracks in Three-Dimensional Finite Bodies Subjected to Tension and Bending Loads. NASA-TP-85793, 1985.

[35] Green, A. E. and I. N. Sneddon "The Stress Distribution in the Neighbourhood of a Flat Elliptical Crack in an Elastic Solid." Proceedings Cambridge Phil. Soc. 46 (1950): 159-164.

[36] Schijve, J. "Brief Note on the Stress Intensity Factor for Elliptical Crack." Engineering Fracture Mechanics, 18 (1983): 1067-1069.

[37] Newman Jr., J. C. An Improved Method of Collocation for the Stress Analysis of Cracked Plates with Various Shaped Boundaries. NASA-TN-6376, 1971.

[38] NASGRO Fatigue Crack Growth Computer Program, Version 2.01, NASA JSC-22267A, 1994.

[39] Tweed, J. and D. P. Rooke. "The Distribution of Stress Near the Tip of a Radial Crack at the Edge of a Circular Hole." International Journal of Engineering Science, 11 (1973): 1185-1195.

[40] Shah, R. C., "Stress Intensity Factors for Through and Part Through Cracks Originating at Fastener Holes," Mechanics of Crack Growth, ASTM STP 590, American Society for Testing and Materials, 1976, pp. 429-459.

[41] Roberts, R. and T. Rich, "Stress Intensity Factors for Plate Bending," Transactions of ASME, Journal of Applied Mechanics, 34 (1967): 777-779.

[42] Roberts, Richard, and John J. Kibler. "Some Aspects of Fatigue Crack Propagation," Engineering Fracture Mechanics. 2 (1971): 243-260.

[43] Shivakumar, V. and Y. C. Hsu. Stress Intensity Factors for Cracks Emanating from the Loaded Fastener Hole. Proc. of the International Conference on Fracture Mechanics and Technology, Hong Kong, March 1977.

[44] Shivakumar, V. and R. G. Forman. "Green's Function for a Crack Emanating from a Circular Hole in an Infinite Sheet." International Journal of Fracture, 16 (1980): 305-316.

[45] Tweed, J. and D. J. Cartwright. Compendium of Stress Intensity Factors. Her Majesty's Stationary Office, London. 1976.

[46] Murakami, Y., Stress Intensity Factors Handbook, Volume 3, The Society of Materials Science, Japan, Pergamon Press, Tokyo, pp. 607-610.

[47] Miyoshi, T., K. Ishii, and S. Yoshida, "Database of Stress Intensity Factors for Surface Cracks in Pre/Post Penetration," Transactions of Japanese Society of Mechanical Engineers, Vol. 56, No. 527 (1990), pp. 1563-1569. qtd. in Murakami, Y., Stress Intensity Factors Handbook, Volume 3, The Society of Materials Science, Japan, Pergamon Press, Tokyo, pp. 607-610.

[48] Miyoshi, T. and S. Yoshida, "Analysis of Stress Intensity Factors for Surface Cracks in Pre/Post Penetration," Transactions of Japanese Society of Mechanical Engineers, Vol. 54, No. 505 (1988), pp. 1771-1777. qtd. in Murakami, Y., Stress Intensity Factors Handbook, Volume 3, The Society of Materials Science, Japan, Pergamon Press, Tokyo, pp. 607-610.

[49] Irwin, G. R. "The Crack Extension Force for a Part-Through Crack in a Plate." Journal of Applied Mechanics, Trans. ASME. 29 (1962): 651-654.

[50] Bowie, O. L. "Analysis of an Infinite Plate Containing Radial Cracks Originating at the Boundary of an Internal Circular Hole." Journal of Mathematics and Physics. 35 (1960): 60-71.

[51] Chang, Fu-Kuo, Richard A. Scott, and George S. Springer. "Strength of Mechanically Fastened Composite Joints" Journal of Composite Materials, 16 (1982): 470-494.

[52] Chang, Fu-Kuo, Richard A. Scott, and George S. Springer. "Strength of Mechanically Fastened Composite Joints" Journal of Composite Materials, 16 (1982): 470-494.

---

[53] Crews, John H., C. S. Hong, and I. S. Raju. Stress Concentration Factors for Finite Orthotropic Laminates with a Pin-Loaded Hole, NASA-TP-1862, 1981.

[54] de Jong, Theo. “Stresses Around Pin-Loaded Holes in Elastically Orthotropic or Isotropic Plates.” Journal of Composite Materials, 11 (1977): 313-331.

[55] Eshwar, V. A., B. Dattaguru, and A. K. Rao. Partial Contact and Friction in Pin Joints, Report No. ARDB-STR-5010, Department of Aeronautical Engineering, Indian Institute of Science, 1977.

[56] Tada, Hiroshi, Paul C. Paris, and George R. Irwin, The Stress Analysis of Cracks Handbook, 2ND Edition, Paris Productions Incorporated and Del Research Corporation, St. Louis, 1985.

5.

Crack Growth Predictions

5.1 Introduction

The crack growth prediction model developed here does not use a new crack growth law or incorporate any new phenomenological behavior witnessed during the experimental investigation. Simply stated, the crack growth model predicts crack growth of part through and through cracks with crack shapes typically found in longitudinal lap-splice joints of pressurized fuselage structure. Results of fatigue crack growth experiments on different types of specimens are used for the validation of the prediction model. It covers,

- Center cracked tension specimens,
- Center cracked tension/bending specimens,
- Specimens with a single open hole with edge cracks, and
- Riveted lap joint specimens

In view of the aims of the present research program on fatigue of riveted lap joints of fuselage lap splices, the predictions are restricted to constant-amplitude (CA) loading, but it includes part through cracks with a quarter elliptical crack front as well as through cracks with an oblique crack front. Moreover, combined tension and bending is addressed.

The predictions are compared to results obtained in Chapter 3. Stress intensity factors are partly borrowed from the literature. For the part through crack growth, the well known Newman/Raju stress intensity factor equations are used. The through crack portion of the fatigue life is modeled using the newly developed stress intensity solutions for a part elliptical through crack which is presented in the previous chapter. A question, which immediately comes to mind, is how will the transition from a part through to through crack be addressed? At the point in the part through crack growth life where the crack penetrates the back surface; i.e., crack growth in the crack depth direction, the crack has a particular a/c_1 and a/t ratio. These two crack shape ratios are then assumed for the first increment of through crack growth using the newly developed K solutions.

The sheet material considered is 2024-T3 Alclad. The basic da/dN - ΔK crack growth data used for predictions are presented first (section 5.2), followed by predictions for part through cracks in section 5.3 and oblique through cracks in section 5.4. Both types of cracks are considered under different types of loading. The chapter is summarized in a number of conclusions in section 5.5.

5.2 Basic da /dN -ΔK Relation Adopted for Crack Growth Predictions

The Forman-Newman-de Koning, FNK, crack growth equation is used to describe the basic da/dN - ΔK relation applicable to 2024-T3 thin sheet material. The FNK equation, Eqn. (5.1) is an extension of the Forman equation with added parameters, p and q, to better fit the material data in the extremal regions of the da/dN vs. ΔK curve.[1-3]

$$\frac{da}{dN} = \frac{C(1-f)^n \Delta K^n \left(1-\frac{\Delta K_{th}}{\Delta K}\right)^p}{(1-R)^n \left(1-\frac{\Delta K}{(1-R)K_c}\right)^q} \tag{5.1}$$

where

a	= Crack length
N	= Number of applied fatigue cycles
R	= Stress ratio
ΔK	= Stress intensity factor range
C, n, p, q	= Empirically derived material constants
f	= Crack opening function
ΔK_{th}	= Threshold stress intensity factor
K_c	= Critical stress intensity factor

The crack opening function, f for plasticity induced crack closure is defined by Newman as[4]

$$f = \frac{K_{op}}{K_{max}} = \begin{cases} \text{maximum of R or } A_0 + A_1 R + A_2 R^2 + A_3 R^2 & R \geq 0 \\ A_0 + A_1 R & -2 \leq R < 0 \end{cases} \quad (5.2)$$

with the A_i coefficients given by the following equations.

$$A_0 = \left(0.825 - 0.34\alpha + 0.05\alpha^2\right)\left[\cos\left(\frac{\pi}{2}\frac{\sigma_{max}}{\sigma_0}\right)\right]^{\frac{1}{\alpha}}$$

$$A_1 = \left(0.415 - 0.071\alpha\right)\frac{\sigma_{max}}{\sigma_0}$$

$$A_2 = 1 - A_0 - A_1 - A_3$$

$$A_3 = 2A_0 + A_1 - 1$$

The plane stress/plane strain constraint factor is α and σ_{max} and σ_0 are the maximum applied stress and flow stress, respectively. For the predictions where the crack opening function is included, α is set to 1.5, as recommended in reference [2], and the ratio of σ_{max}/σ_o to 0.3.[2,4] Although the equations appear overly complicated for constant amplitude loading, it degenerates to the closure corrected Paris equation by setting $p = q = 0$. If the material does not exhibit significant crack closure, the crack opening function can be bypassed by setting f equal to R; i.e., $K_{op} = K_{min}$. In addition, if p and q are again set to zero, Eqn. (5.1) reverts to the Paris equation, $da/dN = C\Delta K^n$. A more in depth review of the FNK equation is presented in references [1-4]. The full equation has been coded in the computer program to give the user a choice of the three crack growth relations. For most predictions here, the Paris equation is used unless otherwise stated. From constant amplitude fatigue test data of center crack tension specimens, shown in Figure 5.1, the da/dN - ΔK relation is established, from which the Paris constants, C and n, are calculated to be 1.67×10^{-12} and 3.07, respectively. The computer program calculates the crack growth increment cycle by cycle from an assumed initial flaw size to failure. A program flow chart is shown in Appendix F.

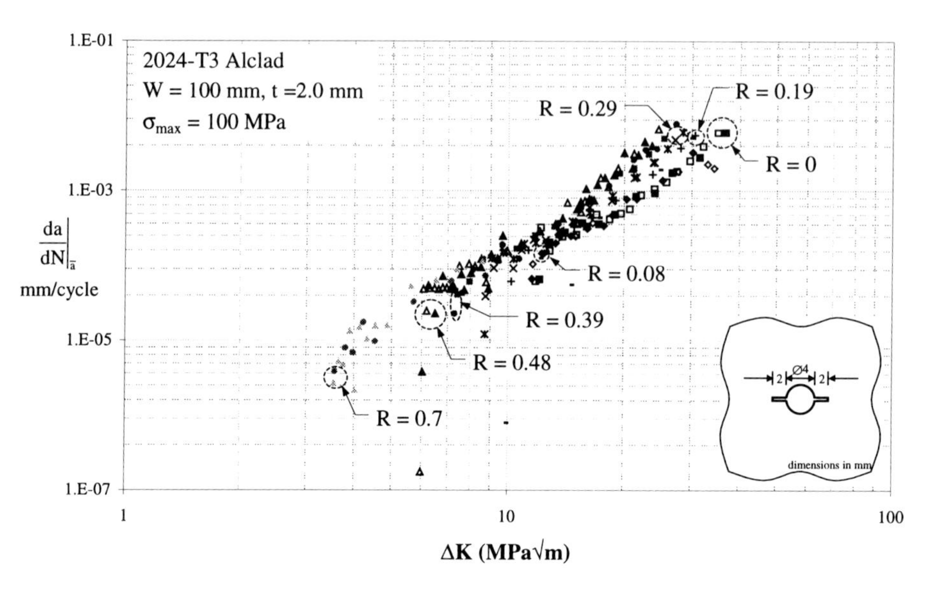

Figure 5.1 Crack Growth Rate Data Used for Predictions

5.3 The Growth of Part Through Cracks

Part through cracks initiated at the bore of a hole are generally supposed to have a quarter elliptical crack front. The size is then determined by the two semi-axis, the crack length "c" and the crack depth "a", see Figure 5.2. For the prediction of the fatigue crack growth life, an initial flaw assumption must be made, not only for the crack size, but also for the crack shape. For example, if the initial crack length "c_1" is assumed to be 1.27 mm (0.05 in., an initial crack size adopted by the USAF Damage Tolerance Requirements), an initial crack depth (a) must also be assumed. A parametric study is completed in section 5.3.1 to investigate the dependence of the part through crack growth on the assumed shape of the initial flaw. Predictions are made for the same value of "a" but three values of a/c_1 of the initial flaw.

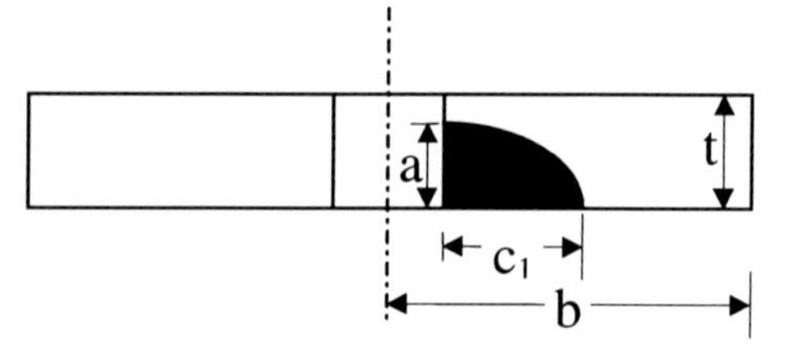

Figure 5.2 Part Through Crack Geometry

Predictions in section 5.3.2 are based on calculating da/dN and dc_1/dN in order to find the new locations of the semi-axis of the quarter elliptical crack. It thus is assumed that the crack front remains quarter elliptical and that the prediction can be restricted to two points of the crack front. This is the approach generally adopted in the literature. In section 5.3.2, predictions for crack extension are made for a large number of points distributed along the crack front. A quarter ellipse is then drawn through the predicted new points of the crack front. For that purpose a regression analysis is used. The purpose is to explore the sensitivity of the predicted crack shapes and observed crack shapes is made in section 5.3.3. It should be noted that the predictions are made for combined tension and bending. Such predictions for both the crack shape and fatigue life for combined tension and bending loading have not yet been published, as far as known to the author.

5.3.1 The effect of the Initial Flaw Shape a/c_l

The Newman/Raju surface and corner crack solutions for K-values remain as the primary point of reference for crack growth predictions of part elliptical crack geometries. For this reason, not to mention the ease of programming the numerous polynomial equations, the Newman/Raju solutions are the only solutions used for part through crack growth and can be reviewed in full in Appendix G. For the predictions that are completed here, an initial crack depth, a = 0.1 mm, has been used with three values of the a/c_l ratio: 0.5, 1.0, and 2.0, i.e. c = 0.2, 0.1, and 0.05 mm, respectively. The calculations are for a 100 mm wide specimen of 2024-T3, thickness 2.0 mm with a 4.8 mm hole in the center. The remote stress is 100 MPa of both tension and bending (bending factor k = 1.0). It is obvious that bending should affect the shape development of the fatigue crack. Results are shown in Figure 5.3 - Figure 5.5 for the three a/c_l ratios, respectively. The figures show crack fronts obtained after each 1000 cycles. The last crack front applies to static break through of the remaining ligament. Static break through occurs when K_{max} in the crack depth direction exceeds 1.4 times K_{Ic}, which is the same criterion used in the NASGRO Crack Growth Computer Program.[2] Although the initial a/c_l ratios are highly different in the three figures, the final a/c_l are similar ($a/c_l = 0.575$).

The figures show that the crack in all three figures is approximately quarter-circular ($a/c_l \approx 1.0$) at a crack depth of a = 0.5 mm. Subsequent crack growth then is also very much similar as should be expected, although the shape is changing from $a/c_l \approx 1.0$ to $a/c_l \approx 0.575$. The similarity of the crack growth in this period is confirmed by similar crack growth lives involved, see Table 5.1. As can be seen in the table, the initial K-value at the material surface, $K(c_l)$, is depending on the crack length "c_l". It explains why the first crack growth period until $c_l = 0.5$ mm is systematically depending on the initial c_l-value.

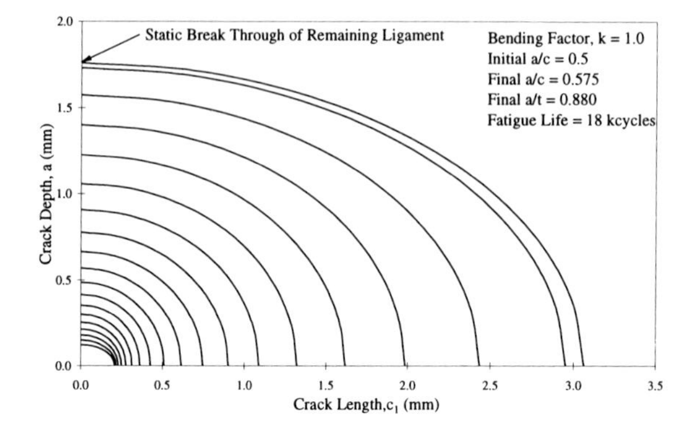

Figure 5.3 Crack Shape Development with Initial $a/c_l = 0.5$

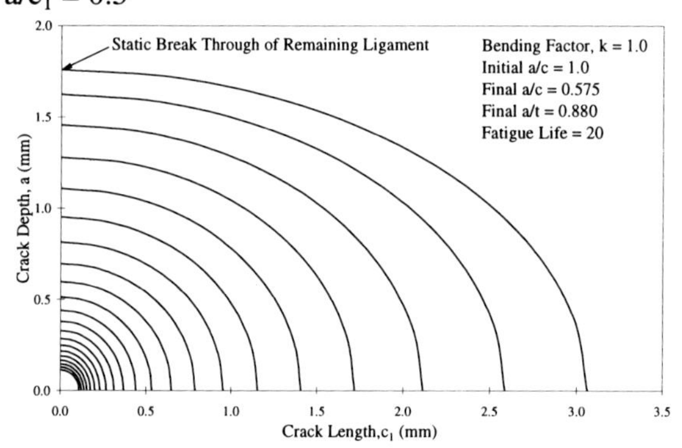

Figure 5.4 Crack Shape Development with Initial $a/c_l = 1.0$

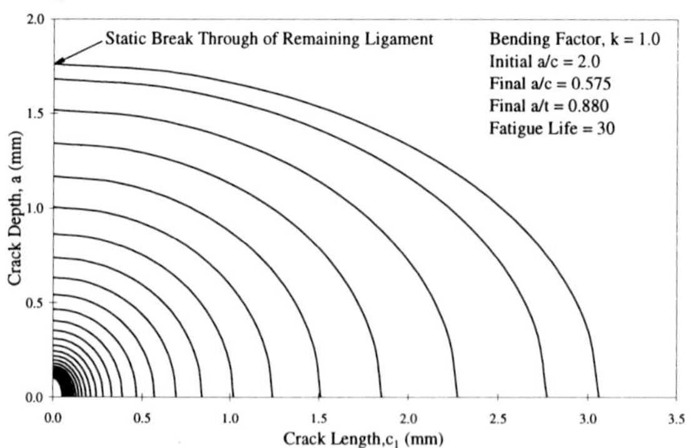

Figure 5.5 Crack Shape Development with Initial $a/c_l = 2.0$

Table 5.1 Crack Growth of Initial Small Corner Flaws of Different Shapes until Break Through

Initial Flaw Data					Crack Growth Life (kcycles)	
a/c_1	a (mm)	c_1 (mm)	K(a) $MPa\sqrt{m}$	$K(c_1)$ $MPa\sqrt{m}$	Life from initial flaw to c_1=0.5mm	Life from c_1=0.5mm to break through
0.5	0.1	0.20	8.31	6.75	8.88	9.32
1.0	0.1	0.10	6.90	6.71	11.64	9.26
2.0	0.1	0.05	3.87	4.18	21.32	9.22

It has also been noted in the literature that semi-elliptical surface cracks with highly different a/c_1 values of the initial flaw show a tendency to grow to crack shapes with approximately the same a/c_1 ratio. Ichsan recently discussed this.[5] However, the results of the semi-elliptical surface cracks were grown under cyclical tension, which resulted in stabilized a/c_1 ratios close to 1.0. The present observation of a similar $a/c_1 = 1.0$ for different initial flaw shapes, changing to a lower similar a/c_1 afterwards has not been noticed under cyclical tension. The present observation is obviously related to the occurrence of combined tension and bending.

5.3.2 Crack Extension Predictions along the Entire Crack Front

In the previous section, the crack extension was predicted for the ends of the two semi-axis, a and c_1, to arrive at the new crack front, assuming that it would remain quarter-elliptical. The Newman/Raju K-solution allows a calculation of K along the entire crack front. It implies that crack extension can be predicted for many points of the crack front. It leads to many new points of the moving crack front, which then can not be expected to be accurately a quarter elliptical curve with the same axes. However, it is possible to draw such a quarter ellipse through the new data points by adopting a regression analysis. The Newman/Raju equations can then be applied again for predicting the next crack extension. Ichsan adopted this procedure for a semi-elliptical surface crack loaded under remote tension only. He found that fitting an elliptical crack front for 32 points leads to a slightly longer fatigue life than predicted by the method of the previous section, i.e. predicting Δa and Δc_1 for the axes and assuming that the crack front remains elliptical. The difference was on the order of 10%. A similar comparison is made here for two corner cracks at a hole under remote tension and bending (bending factor, $k = 1$) with similar results. Using more than 32 calculation points along the crack front results in a negligible effect on the fatigue life. However using fewer calculation points results in a more significant difference between the curve fit and non-curve fit prediction. The crack growth for the fit data is less than the unfit since the linear regression is decreasing the a or c_1 dimension by a small amount after each cycle. As can be expected, the small systemic decrease in crack size has a cumulative effect. As a result, a prediction with a small initial flaw assumption will undergo more regression calculations thereby having a larger effect on the fatigue life. To eliminate the systemic error due to curve fitting, at least 32 calculation points should be used.

5.3.3 Comparison between Predicted and Observed Crack Shapes

Since in situ crack shape measurement is not currently possible, the only method of verifying the K solutions is by using the fatigue life or crack front shape. For the 7-open hole tests, several specimens were statically overloaded prior to failure by fatigue and the crack shapes were measured. Thus for a given number of cycles, the crack shape is known and comparisons can be made with the analytical predictions. Such a comparison is shown in Figure 5.6 for five separate cracks in the same specimen. The predictions are completed assuming each hole is located in a finite width strip with no interaction with

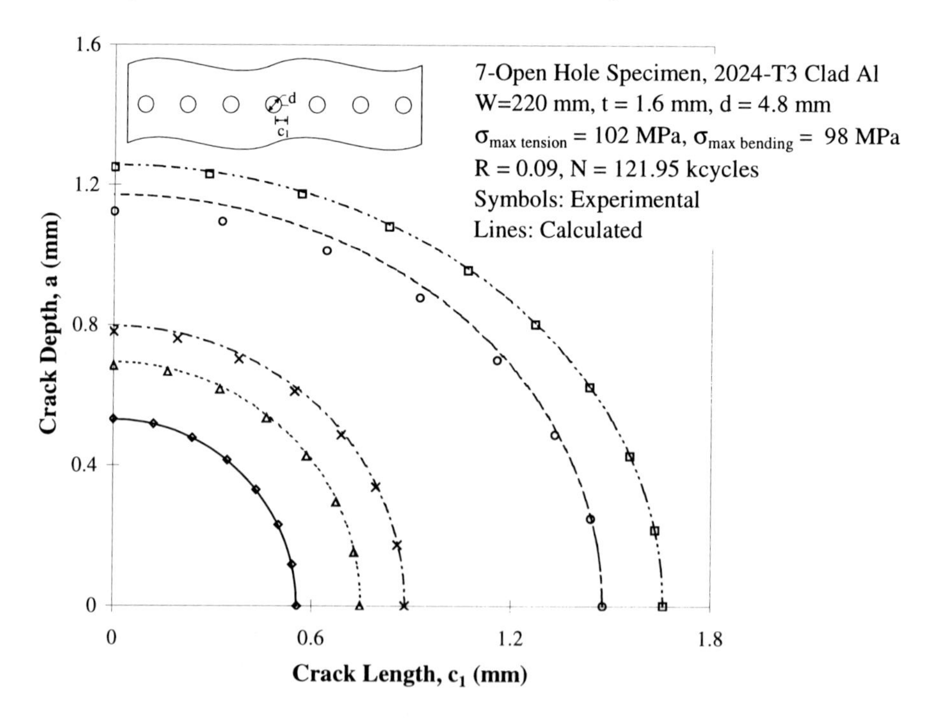

Figure 5.6 Measured vs. Predicted Crack Shape

adjacent cracks. Since the cracks are still small, this procedure should be allowed. The five cracks do not have the same dimension because the initiation has taken different numbers of cycles. The shapes are predicted for the crack length "c_1" as observed, starting from a quarter elliptical crack with $a/c_1 = 1.0$. The initial a/c_1 value is not very important for the present size of the crack as discussed in section 5.3.1. Figure 5.6 shows a very good agreement with the crack shape development as observed from the specimen fracture surface. Similar satisfactory results were obtained for specimens with a single open hole also loaded by combined tension and bending.

5.4 The Growth of Through Cracks

The only K-solution for through cracks under combined loading conditions now available is the library of the NASGRO Fatigue Crack Growth Computer Program.[2] It applies to the loading cases shown in Figure 5.7. It assumes that the crack has a crack front perpendicular to the sheet surface. Predictions with these K-solutions are compared in section 5.4.1 to the test results presented in Chapter 3. In section 5.4.2 and 5.4.3, the K-solutions obtained here in Chapter 4 are used.

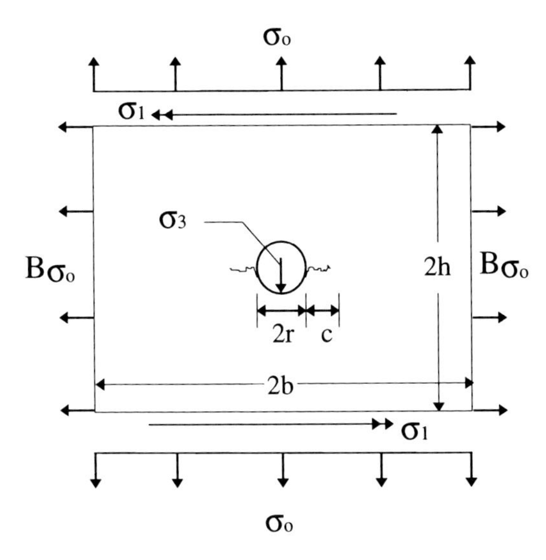

Figure 5.7 NASGRO TC09 Crack Case

5.4.1 Through Crack Growth: Published Stress Intensity Solutions

First predictions with the NASGRO K-solutions are made for through cracks starting from an open hole in a specimen loaded under tension only. The predictions are verified with test results obtained in the tests described in section 3.2.2. An assumption had to be made about the crack initiation life, which was made to predict not only the fatigue life, but also the entire crack history. To do so an initial flaw size is assumed for which a prediction is made. A comparison is then made between the actual and predicted crack histories. If the prediction is underestimating the crack growth, the initial flaw size is increase; conversely, if the prediction is overestimating the crack growth, the initial flaw size is decreased. An example of the comparison between prediction and tests results is shown in Figure 5.8, which shows a good agreement. Similar correlation was obtained for all open hole specimens loaded by remote tension only.

Unfortunately, the correlation is not as good for specimens subject to combined remote tension and secondary bending shown in Figure 5.9. Crack growth rates for cracks larger than 2 mm are highly overestimated. Although some aspects of the theoretical background of the NASGRO K-solutions maybe questioned, there is an obvious reason for disagreement. In the NASGRO concept, the crack is supposed to be a through crack with a straight crack front perpendicular to the plate surface. In reality, under combined tension and bending such cracks grow with an oblique part-elliptical front. In Figure 5.9, the crack length plotted is c_1 (see Figure 5.2), which is the largest length of the crack at the material surface where the bending stress has its maximum. Since the crack through the thickness is lagging behind this point the cracked area is smaller than assumed in the NASGRO solution. An overestimation of the crack growth should then be expected.

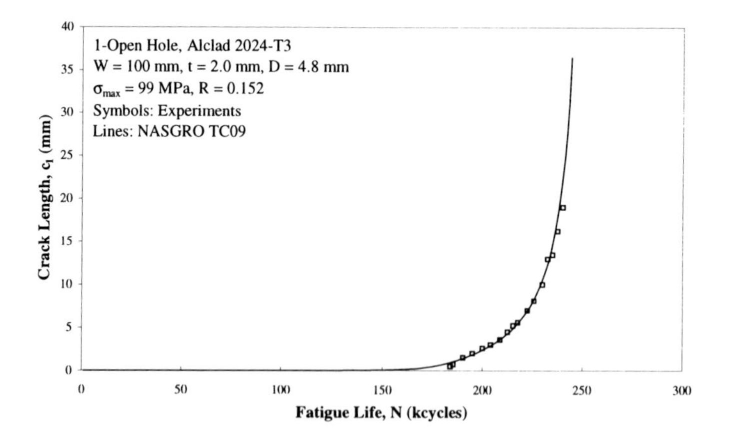

Figure 5.8 TC09 Verification Remote Tension

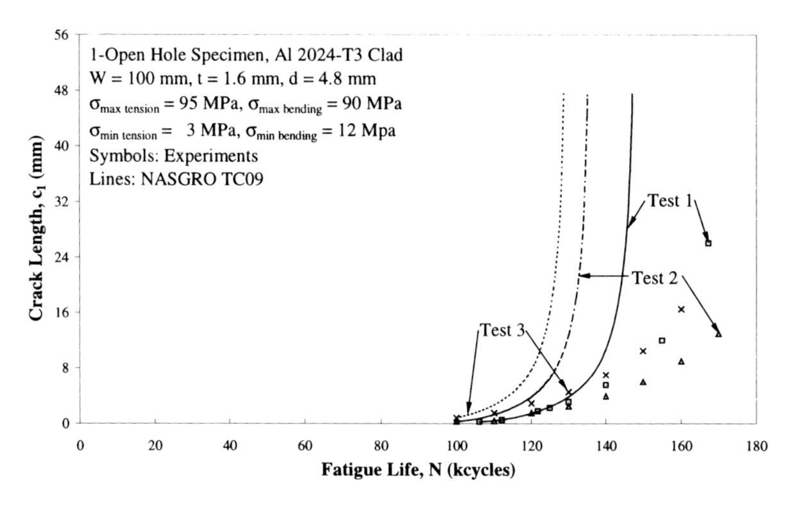

Figure 5.9 TC09 Verification Remote Tension and Secondary Bending

5.4.2 Through Crack Growth Predictions with the New K-Solutions: Open Hole Specimens

In Chapter 4, K-solutions for oblique cracks were obtained from the finite element analysis. The results were presented in tabular format. These values could be converted to a polynomial description of the results. However, with the computational power of current desktop personal computers, there is no need for the time consuming derivation of such polynomial equations. A computer program has been developed, which by interpolation obtains the K-values for the applicable crack shape and size, defined by values of a/c_1, a/t, and r/t. The interpolation is handled by a choice of three different interpolation routines; linear, higher order polynomial, or cubic spline. Although the linear interpolation is attractive in it's simplicity, it is not well suited for interpolating between the large ranges in the a/c_1 and a/t values. Both the higher order polynomial and cubic spline routines are used with no distinguishable difference. One benefit of the former is the error estimate in the interpolation where in the latter the error is not calculable. Conversely, the cubic spline appears to be more stable than the higher order polynomial in extrapolating K's from a table. Extrapolation maybe required in a/c_1 in the latter stage of the fatigue life where the crack is growing quite rapidly and the crack front is straightening out resulting in more of a slant than part elliptical crack.

The new K solutions for two part elliptical, oblique through cracks at a hole have been used to predict the fatigue life of an open hole specimen subject to tension and bending ($k \approx 1.0$). The cracks nucleate and grow naturally as corner cracks, thus the Newman/Raju K-solution is used until the crack grows through the thickness of the sheet. Once the crack is a through crack, the new solutions are used until final fracture.

The transition from a part through to through crack is difficult to examine experimentally; therefore an assumption of the transition behavior had to be made. Once the crack breaks through the back surface of the sheet, there is an instantaneous increase in the crack depth without a similar increase in the crack length. Although the crack depth, which is the minor axis of the ellipse, is now outside the sheet, it is convenient to continue to use the crack depth to describe the crack shape and size (a/c_1 and a/t ratios). The first crack depth after break through is calculated as 1.05 times the sheet thickness. Examination of the fracture surfaces of the fatigue specimens that were statically overloaded to failure when the crack was close to the transition area did not show a detectable change in the crack shape. As mentioned previously, since the fracture surfaces must be viewed by destructive inspection, the comparison is made of cracks from different specimens of the same geometry tested at the same remote stress level. Thus, the factor of 1.05 does not alter the crack shape significantly, but does account for the instantaneous increase in the crack depth at the moment of break through.

Additional assumptions that must be made in the prediction calculation are the initial flaw size and shape. Based on fractographic observations, the initial crack shape is assumed to be quarter circular, $a/c_1 = 1.0$. The initial flaw size is estimated, and the prediction algorithm iterates by varying the initial flaw size (constant quarter circular shape) until the crack length and number of cycles at failure are achieved.

The prediction of the open hole specimen subject to combined tension and bending shown in Figure 5.10 does not follow the entire crack growth history. The transition from a part through to a through crack is seen at approximately 104 kcycles where the crack length and depth (the latter not shown in the figure) is about 1.7 mm. The slight cusp in the prediction curve as the crack transitions from a part through to through crack appears to be artificial since the front surface K calculated for the last cycle as a part through crack is higher than the first K calculated for the through crack, implying a higher crack growth rate just before break through. The ligament width in the thickness direction is going to zero and such a behavior might occur in view of the questionable K-values at the moment of the transition. Grandt et al. reported a nearly constant crack growth rate for the c_1 crack length during the transition from a part through to through crack.[6]

The new K's do not account for a changing stress field which occurs as the crack becomes quite long. A long crack implies a reduction in the bending stiffness of the sheet, thus the crack is experiencing a more tensile stress field. In a tension dominant stress field it was shown for a center cracked sheet with a part elliptical oblique crack subject to pure tension (Figure 4.35) and separately combined tension and bending (k = 1.0, Figure 4.36) that the K's for the pure tension case are higher than that for the combined loading assuming the same applied remote stress.

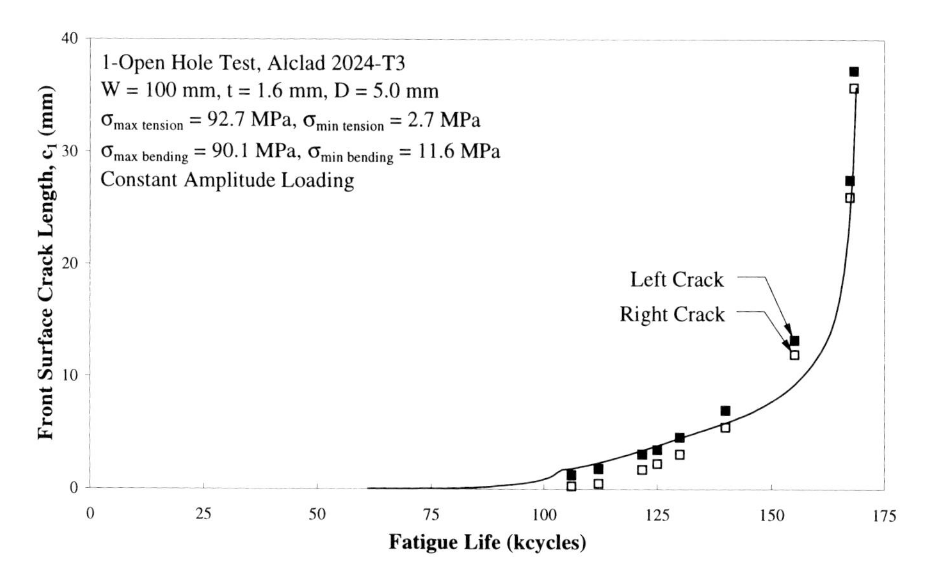

Figure 5.10 Crack Growth Prediction in 1-Open Hole Specimen Subject to Combined Tension and Bending

From the correlation between actual and predicted crack histories in Figure 5.10, the K solutions, both the Newman/Raju and new ones presented in Chapter 4, used for this complex crack shape subject to combined tension and bending loading are reliable. Within the scope of linear elastic fracture mechanics, K remains to be a good similitude parameter when the crack geometry and loading condition are well characterized.

5.4.3 Through Crack Growth Predictions with the New K-Solutions: Asymmetric Lap-Splice Joint Specimens

Predictions for the asymmetric lap splice joint are shown in Figure 5.12 and Figure 5.11. As was discussed in section 5.3.1, the initial flaw size and shape strongly affect the fatigue behavior of small cracks ($c_l \leq 0.5$ mm). Attempts have been made to try and predict the small crack history when the crack is still a corner crack using the Newman/Raju solution. However, since the flaw shape at these small crack lengths cannot be determined from the marker bands, predictions are made for several initial flaw shapes (a/c_l = 0.2, 1.0, and 2.0) using the known c_l crack length. From these predictions, the initial flaw shape did not affect the crack shape at break through which was previously shown in Figure 5.3 - Figure 5.5. Even though the life to break through slightly increases with a larger initial a/c_l, the Newman/Raju solutions do not adequately predict crack growth of an initial crack to break through. The reconstructed crack growth curves have regions, bounded by letters A – E and C –E in Figure 5.12 and Figure 5.11, respectively; which are used to explain the unsatisfactory predictions. The regions in each figure represent the same phenomena, thus, only Figure 5.12 is discussed. In region AB the crack is quite small, less than 200 μm. The fractographic examination of this fatigue crack nucleation point indicated the crack may have initiated as a surface crack along the bore of the hole a short distance from the faying surface. Since there is little material between the crack front closest to the faying surface and the faying surface itself, the stress intensity is locally quite high. As a result, the crack grows quite rapidly through this ligament as implied by the steep slope in region AB. The crack is growing slower than predicted in the region BC. The Newman/Raju solutions do not account for load transmission by friction between the sheets at the faying surface. Any portion of the remote load transferred by friction, results in a decrease of the load available to extend the crack. Clearly a shortcoming in the prediction model is the inability to account for load transmission by friction. As the crack enters the region CD, the effects of the rivet diminish. In this region, the dominant stresses are tension, bending, and pin loading for which the K solutions were calculated. Also, the crack is still relatively small, as are possible cracks at neighboring rivets, thus little crack interaction or load redistribution is occurring. In the final region, DE the crack transitions from a part through to a through crack. By the steep slope of region DE, the crack is growing quite fast with less than 10% of the fatigue life

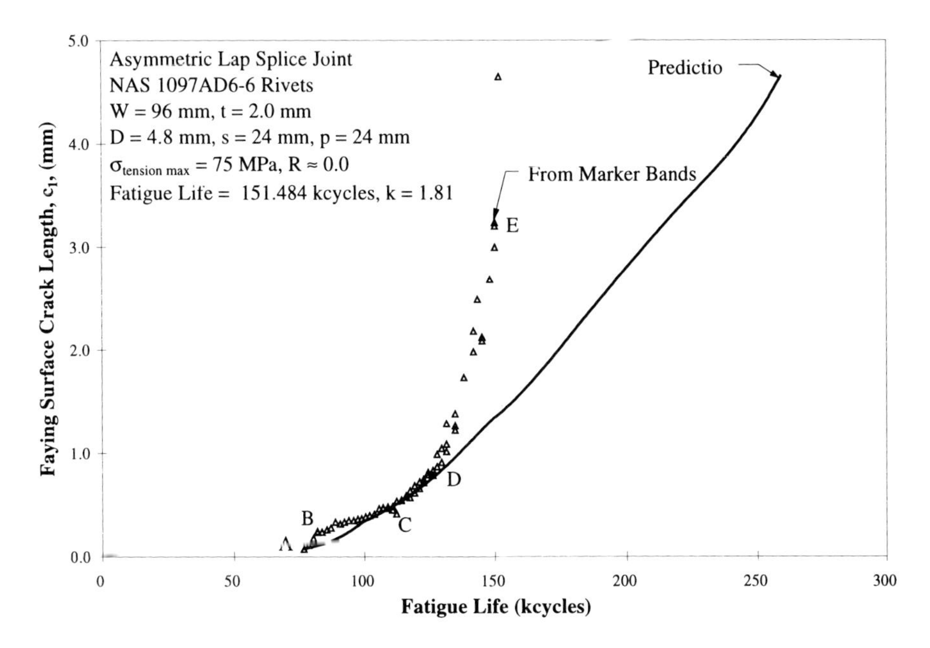

Figure 5.11 Crack Growth Prediction for Asymmetric Lap Splice Joint

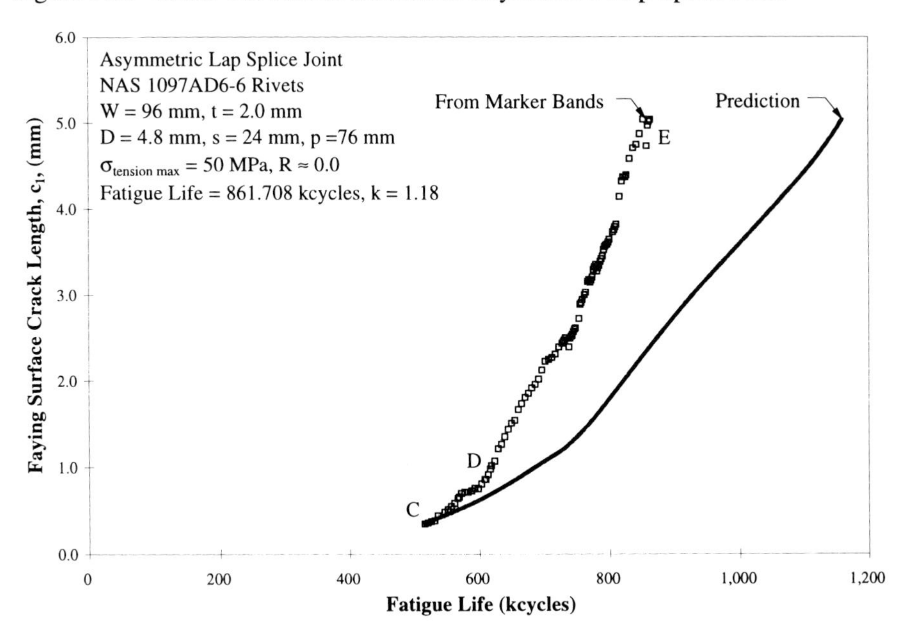

Figure 5.12 Crack Growth Prediction for Asymmetric Lap Splice Joint

remaining. Assuming the crack is growing from a rivet hole in a finite width sheet with no other cracks at other rivets is an oversimplification of the cracking scenario. Examination of the fracture surface of both critical rows showed extensive cracking. As adjacent cracks grow, the rivet is less effective in transferring the load; thus, the load is redistributed to the remaining rivets. Not accounting for load shedding is another weakness in the model. Regions CD and DE are also present in Figure 5.11 where the same arguments are applicable as used for Figure 5.12. The agreement in region CD is not as good as in Figure 5.12, but it is still reasonable. In region DE, the crack is again propagating faster than predicted, about twice as fast. Recalling that the Paris exponent is 3.07, it implies the crack driving force was enhanced by some 25% due to load shedding effects. For the experiment of Figure 5.11, the percentage should have been on the order of 70, which might well be possible because in this specimen, there was extensive cracking in both rivet rows.

The predictions for the asymmetric lap splice joint further illustrate the complexity of rivet joints. From these results, the need to include the effects of friction and crack interaction is essential.

5.5 Conclusions

Crack Growth Prediction Methodology: da/dN - ΔK Relation

In general, the same crack growth prediction methodology developed and used in the NASGRO Crack Growth Computer Program is adopted. The Forman-Newman-de Koning crack growth law is available in the crack growth prediction computer program, but due to the simplicity of the load spectrum (CA) and the well characterized sheet material (2024-T3), the FNK equation is degenerated to the closure corrected Paris Law for all predictions.

Crack Growth Predictions Using Published K Solutions

Double Corner Cracks at a Hole Subject to Tension and Bending,
Newman/Raju

- The Newman/Raju solutions accurately predicted both the crack shape and fatigue life for open hole specimens subject to remote tension and bending, (Figure 5.6).
- The initial flaw shape assumption has a negligible effect once the crack length at the material surface, "c", exceeds 0.5 mm in a 1.6 mm thick sheet. However, due to the dependence of the K-value at the material surface, K(c), on the "c" crack length, crack growth in the small crack regime ($c \leq 0.5$ mm) is affected.

- For studying the K variation along the crack front, a regression analysis must be preformed after each crack growth increment to fit the crack front back to an elliptical shape. To avoid regression analysis errors with the N/R solutions, 32 calculation points along the crack front must be used. Otherwise systematic errors occur in the crack shape (either in the crack depth or crack length) ultimately affecting the fatigue life.

Double Through Cracks at a Hole Subject to Tension and Bending (NASGRO TC09)

- The TC09 solution accurately predicted crack growth in open hole specimens loaded in pure tension (Figure 5.8), but under estimated the fatigue life consistently by at least 30% for combined tension and bending (Figure 5.9).

Crack Growth Predictions with Newly Developed K Solutions

The new K solutions for two part elliptical, oblique through cracks at a hole have been used to predict the fatigue life of an open hole specimen subject to combined tension and bending ($k \approx 1.0$). In addition, predictions have been completed for an asymmetric lap splice joint.

- Predictions of the *open hole* specimens subject to combined tension and bending ($k \approx 1$) show good agreement for a majority of the crack history. The predicted crack size is underestimated in the last 6% of the fatigue life.
- Predictions of the *asymmetric lap-splice joint* are inadequate with the fatigue life being overestimated for both reconstructed crack histories, Figure 5.12 and Figure 5.11. If crack growth in riveted joints is to be predicted, consideration of the effects of friction on the part through crack growth and crack interaction on the through crack growth is required.

---

[1] Forman, Royce G., Vankataraman Shivakumar, James C. Newman Jr., Susan M. Piotrowski, and Leonard C. Williams. Development of the NASA/FLAGRO Computer Program. Fracture Mechanics: Eighteenth Symposium ASTM STP 945. D. T. Read and R. P. Reed, Eds., American Society for Testing and Materials, Philadelphia, 1988. 781-803.

[2] NASGRO Fatigue Crack Growth Computer Program, Version 2.01, NASA JSC-22267A, 1994.

[3] Forman, R. G., and S. R. Mettu. Behavior of Surface and Corner Cracks Subjected to Tensile and Bending Loads in Ti-6Al-4V Alloy. Fracture Mechanics: Twenty-Second Symposium, Vol. 1, ASTM STP 1131, H. A. Ernst, A. Saxena, and D. L. McDowell, Eds., American Society for Testing and Materials, Philadelphia, 1992. 519-546.

---

[4] Newman Jr., J. C. "Crack Opening Stress Equation for Fatigue Crack Growth," International Journal of Fracture. 24 (1984): R131-R135.

[5] Putra, Ichsan, S. Fatigue Crack Growth Predicitons of Surface Cracks Under Constant-Amplitude and Variable-Amplitude Loading. Diss. Delft University of Technology, 1994. Delft:NL.

[6] Grandt, Jr., A. F., J. A. Harter, and B. J. Heath, "Transition of Part-Through Cracks at Holes into Through-the -Thickness Flaws," Fracture Mechanics: Fifteenth Symposium, ASTM STP 833, R. J. Sanford, Ed., American Society for Testing and Materials, Philadelphia, 1984, pp. 7-23.

6.

Finite Element Analysis of a Lap-Splice Joint

6.1 Introduction

The stress intensity factor solutions developed in Chapter 4 are used to make fatigue crack growth predictions of riveted joints. Unfortunately, these solutions do not account for some of the dominant behavior observed during the experimental program. Specifically, the effect of the rivet squeeze force could not be easily included in K solution analyses. The rivet squeeze force has been shown by Müller to effect the local stresses around the rivet holes, stress distribution in the joint overlap region, and behavior of the rivet. A three-dimensional finite element model is developed to explore these effects, the results of which can provide limits of applicability for the K solutions calculated in Chapter 4. The two models used are described in section 6.2 along with the boundary and loading conditions considered. The results of the preliminary three rivet model are discussed in section 6.3 and elucidated the more dominant behavior in the joint. The baseline model, an asymmetric lap joint with 3 rivet rows was then developed with the results presented in section 6.4. After examination of the displacements and stresses of the uncracked joint, a quarter circular part through crack was created at the hole edge of the critical upper rivet row. In section 6.5, the calculated stress intensity factors are presented and a comparison to published solutions is made. The conclusions are given in section 6.6.

6.2 Model Description

The three dimensional finite element model is an idealization of a longitudinal lap splice joint without longitudinal or circumferential stiffeners. Without using the asymmetry argument at the center of the middle pin, the model would appear as in Figure 6.1; however, by making use of the asymmetry condition the model size decreases, as shown in Figure 6.2, as does the computational time. Although several interesting results are obtained with the 3 rivet model shown in Figure 6.1; the asymmetric model shown in Figure 6.2 is the primary model used for this investigation. The model is composed of 11524 nodes, 2178 20 noded, isoparametric solid elements for both sheets and pins, and 270 gap elements define the interaction between sheets and pins. As evident from the relative element sizes in Figure 6.2, a one way bias is used in generating the mesh where the elements are smaller and more cubic in shape in the areas of interest but are larger with higher aspect ratios moving away from the areas of interest.

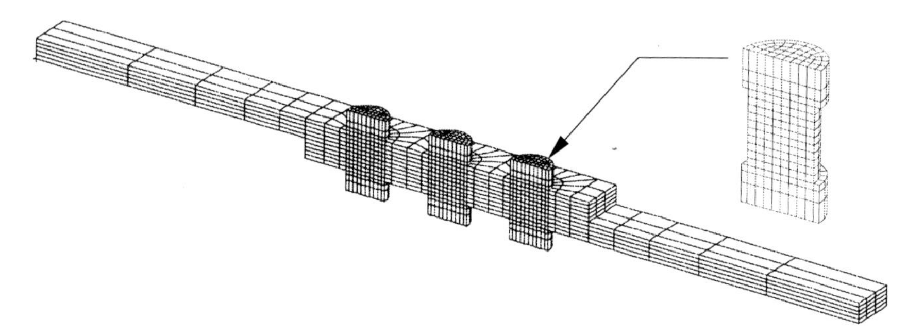

Figure 6.1 FEM without Asymmetry

The MSC/PATRAN pre/post processor is used for generating the mesh and all post processing of results. The MARC finite element analysis program is used for solving the model accounting for geometric but not material nonlinearity. Both the sheets and pins are assumed to be 2024-T3 aluminum with $E = 72.4$ GPa and $\upsilon = 0.33$. The model has two load conditions, one representing a low (LSF) and the other a high (HSF) squeeze force rivet installation. The LSF case has an incrementally applied load using a uniform surface tension in the negative x direction in the yz plane at $x = 0$ to a maximum tension of 70 MPa. The HSF case has two stress systems, the remote stress system created by applying the LSF case in addition to a residual stress system created by loading the bore of the rivet hole. The residual stress system is applied to represent the residual stresses created by expanding the rivet in the hole. Unless otherwise stated, HSF refers to the combination of the stresses caused by the remote load and residual stresses. The 1% radial expansion of the rivet in the hole is generated by applying a positive temperature, positive ΔT, to the rivet which creates an uniform expansion in all directions. Unfortunately, an undesired expansion along the longitudinal axis of the rivet also occurs; therefore a negative temperature, corresponding to a longitudinal rivet contraction of 0.5%, is applied in the longitudinal direction to negate the unwanted expansion. Calculating the required ΔT for thermal expansion and contraction is left until section 6.4.

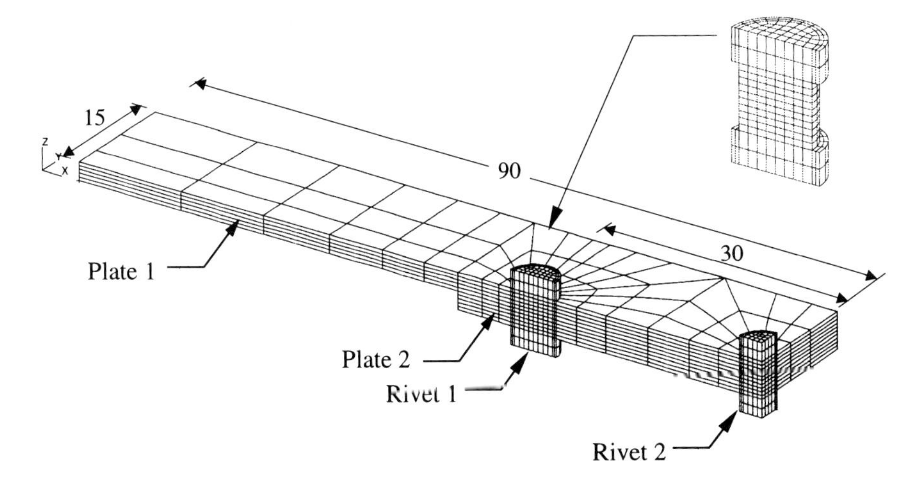

Figure 6.2 FEM Used for this Investigation using Symmetry

The boundary conditions for the model are listed in Table 6.1. Two clarifications on terminology, "tied displacements" represent a master-slave relation between one master node and all other slave nodes. Specifically, boundary condition 3 has the y displacements at (x,15,z) tied, thus if the one master node displaces say 0.01 mm, then all the slave nodes also displace this same amount. Physically, since the nodes are tied in the y direction at y = 15, boundary condition 3 represents an average stress (σ_y) in the xz plane at y = 15 approximating the behavior seen in an actual lap splice joint. "Coupled nodes," used at the interface of the rivet head and outer sheet surface, describes two nodes whose displacements are equal. The difference between coupled and tied nodes is that coupled nodes are a displacement relation between only two nodes; whereas tied nodes are between one node and many other nodes. Gap elements are used in two areas to define the contact behavior; between the two sheets at the faying surface and at the interface of the sheets and pins. At the latter location, the displacement component perpendicular to the contact surface must be degenerated from quadratic to linear to prohibit overlapping of the sheet and pin. For the nodes lying on the element face that creates the bore of the hole, the degeneration is accomplished by tying the mid-side node to the adjacent corner nodes. For numerical stability in the solution, a stiffness for the gap elements is required which was set at 10 GPa without this assumption convergence of the solution is only obtained if small load steps are used which is not computationally efficient.

Table 6.1 Boundary Conditions of Asymmetric Lap Splice Joint Model

Boundary Condition Number	Displacement (u,v,w)	Location (x,y,z)	Description
1	(0, 0, 1)	(0, y, 0)	Roller
2	(0, 1, 0)	(x, 0, z)	Symmetry Condition
3	(0, tied, 0)	(x, 15, z)	Average Stress Condition
4	u(z) = -u(-z)	(90, y, z)	Asymmetry Condition
5	v(z) = -v(-z)	(90, y, z)	Asymmetry Condition
6	w(z) = -w(-z)	(90, y, z)	Asymmetry Condition

0 → free boundary condition
1 → fixed boundary condition

6.3 Results and Discussion of Preliminary 3-Rivet Model

The preliminary three rivet model shown in Figure 6.1 originally was the primary model for this investigation; however, the excessive computational costs deemed further use of this model prohibitive. The high computation times for this model were due to the 364 gap elements that defined the interface not only between the rivet and sheets, but also between the sheets in the overlap region, the region between rivet 1 and rivet 3. The gap elements between the rivet and sheets were placed along the bore of the hole and also between the rivet head and outer surface of the two sheets. Two valuable results were obtained from the one analysis of the three rivet model which made for more efficient analyses of the asymmetric model shown in Figure 6.2. As seen in Figure 6.3 there is little relative displacement between the sheets in the overlap region; therefore, the gap elements in this area can be removed without significantly affecting the global response of the joint to the given load conditions. Furthermore, Figure 6.4 shows the contact stresses at the faying surface, σ_{zz}, are very small again supporting removal of the gap elements. The interaction behavior between the rivet head and outer surfaces of plates 1 and 2 was also insignificant (not shown) with all contact stresses less than 1.0 MPa.

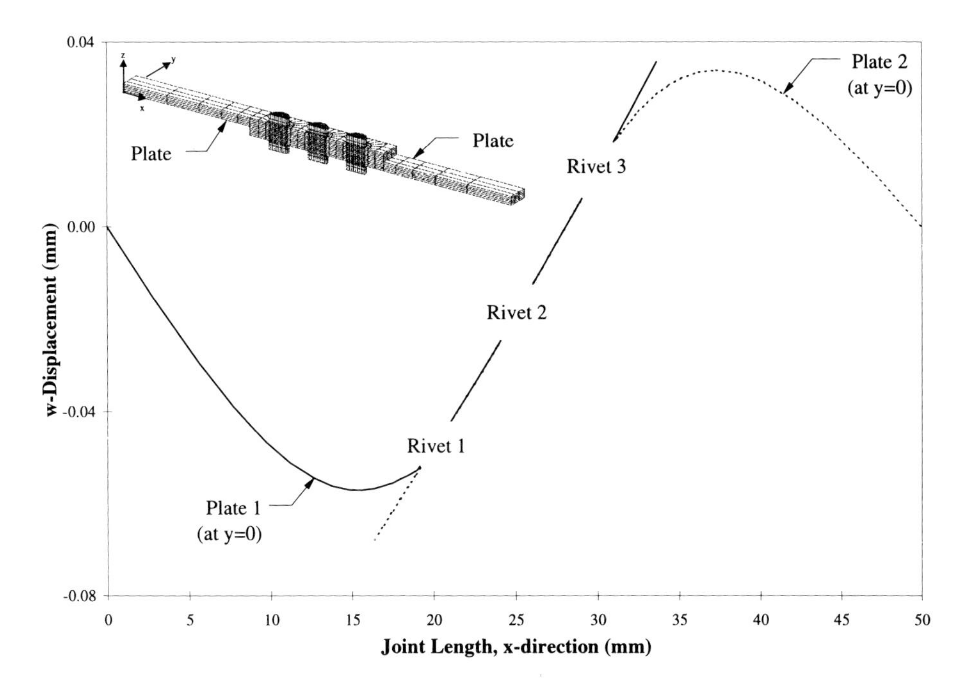

Figure 6.3 Faying Surface w-Displacement at the Symmetry Plane (y=0)

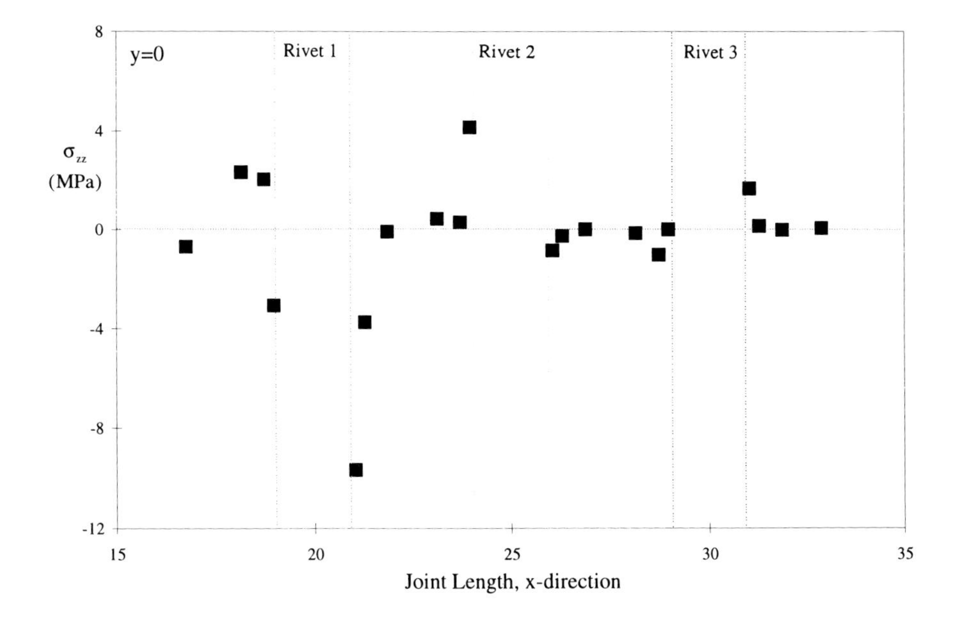

Figure 6.4 Faying Surface Contact Stresses in Overlap Region at the Symmetry Plane (y=0)

6.4 Results and Discussion of Asymmetric Lap Joint Analyses

The two analyses model two different values of the rivet squeeze forces, a low squeeze force (LSF) which has no residual stresses, and a high squeeze force (HSF) with residual stresses. Although valuable in their own right, the most interesting results are seen when comparing the two analyses to see the effect of the residual stress system. Determining the effect of the residual stress system (RSS) is the primary goal of this entire effort. Applying a temperature to the rivet thereby creating thermal stresses creates the RSS. The magnitude of the RSS due to the temperature is not known a priori; thus the applied temperature is estimated. The temperature estimate is based on the results of the LSF analysis. At the maximum load in the LSF analysis, the sheet loses contact with the pin at the top of rivet 1 at the faying surface. In the HSF analysis, the simulated expansion of the rivet and the resulting residual stresses are assumed to keep the rivet and sheet in contact which is most representative of fuselage lap-splice joints. Therefore, with the known gap between the sheet and rivet 1 in the LSF analysis, the temperature required to close this gap can be calculated using the following thermoelastic stress-strain relation.

$$\sigma_{ii} = \frac{E}{1-2\upsilon}\alpha\Delta T \qquad (6.1)$$

where

σ_{ii} = normal stress

υ = Poisson's ratio

α = Coefficient of thermal expansion

ΔT = Temperature difference

Due to the large amount of secondary bending in the LSF analysis, the gap is relatively large, approximately 0.046 mm. The ΔT required to close the gap is 121°C. Müller showed for a lap joint with the geometry used here, Figure 6.2, a 1% expansion of the rivet is realistic which equates to a ΔT of 132°C; therefore, the latter ΔT is used. In addition to the radial expansion of the rivet, an axial contraction of the rivet of 0.5% is applied to not only model the clamping forces associated with installing the rivet, but also to avoid numerical difficulties with the contact elements. As expected, the RSS is axisymmetric as seen in the Figure 6.5 where the Von Mises stresses around the perimeter of rivet 1 are normalized by the yield stress. The difference between the free and faying surface stresses in Figure 6.5 are due to the rivet head. Specifically, for

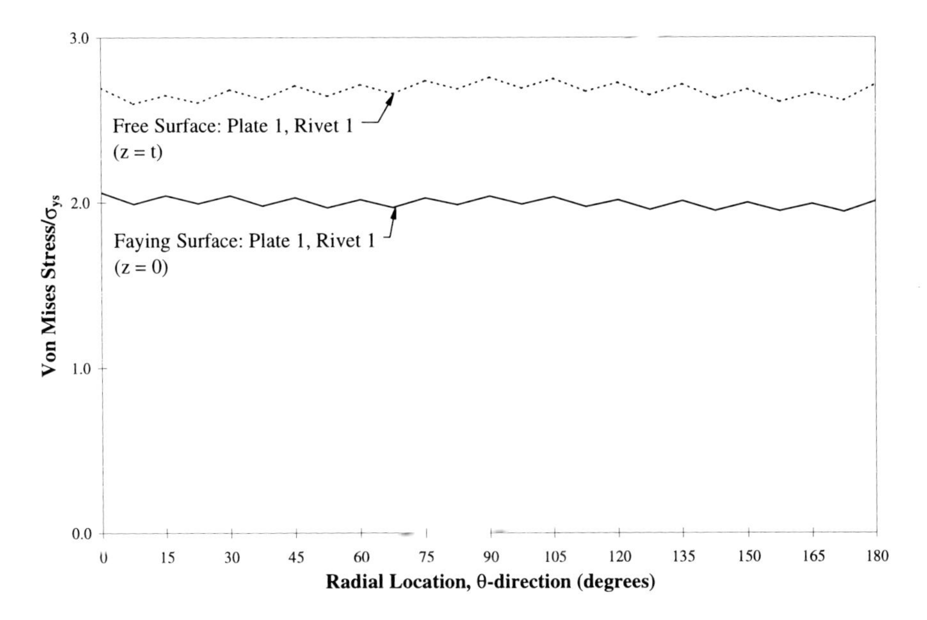

Figure 6.5 Residual Stress Around the Perimeter of Rivet 1 (No Remote Loading)

a constant radial expansion of the rivet, the locally higher stiffness at the interface of the rivet head and sheet free surface results in the higher stresses when compared to the faying surface. Material nonlinear behavior is not modeled, only geometric nonlinearity; therefore, with the residual stress well above the yield stress of the material, evaluating the effect of the RSS can only be qualitative.

The global effect of the RSS is quite evident in Figure 6.7. The compressive residual stresses are restricting the rivet tilting in addition to increased clamping of the sheets by the rivet head both of which result in a decrease in the secondary bending thereby decreasing the global deflection of the joint.

As previously shown with the preliminary 3 rivet model, the contact between the two sheets is small which also proves to be the case in the LSF and HSF analyses. Since the LSF and preliminary 3 rivet model results are similar, only the HSF results are shown in Figure 6.6. The displacements in the overlap region show the plates are intersecting one another slightly due to the secondary bending. As expected, the relative displacements between the plates are restrained in close proximity to each of the pins.

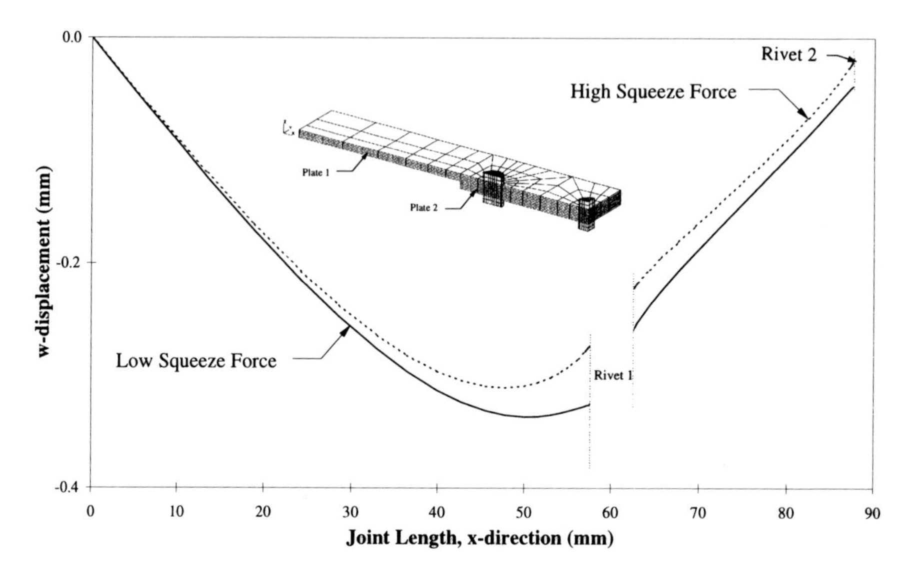

Figure 6.7 Faying Surface w-Displacement at the Symmetry Plane, y = 0

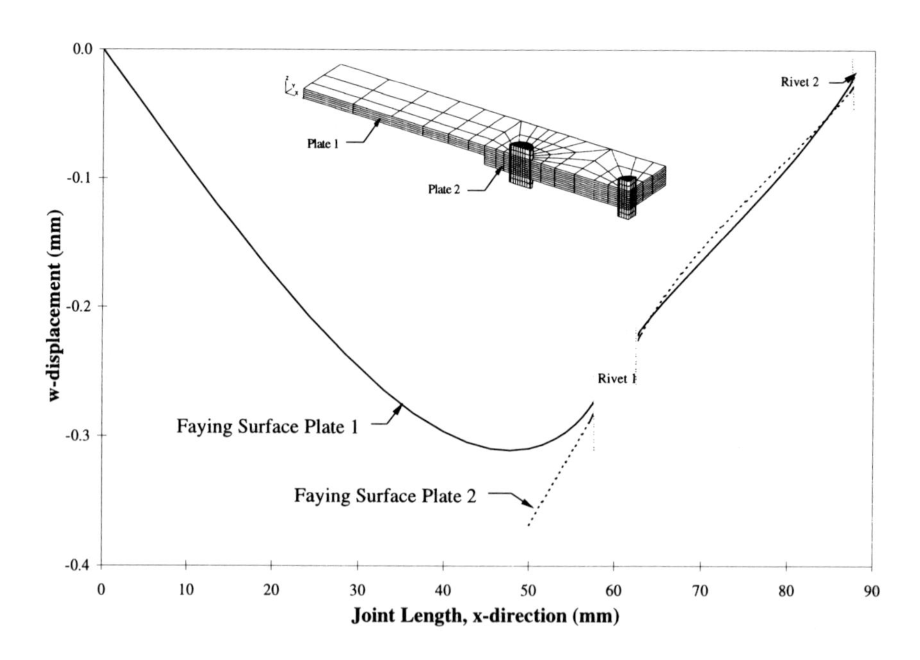

Figure 6.6 HSF w-Displacements at the Symmetry Plane, y = 0

As discussed previously, secondary bending plays a major role in defining the stress state in the joint. Müller experimentally and analytically showed the secondary bending is changing in the joint.[1] For the LSF analysis shown in Figure 6.8, the variation of the bending stress in the width direction is approximately 17 MPa. For this case the rivet is not offering much bending restraint; therefore, little change in σ_{xx} is expected through the width. Also notice in Figure 6.8 that the position of maximum bending shifted away from the joint net section which is contrary to the findings of Müller. The difference in the bending stress through the rivet pitch is small and there is a local increase in the stresses directly under the rivet head due to the coupling of nodes. Ignoring the artificially high stresses in close proximity to the rivet head, the location of maximum bending is moving toward the net section and is located at x = 53.17 mm for y = 0 and x = 54.53 mm for y = W.

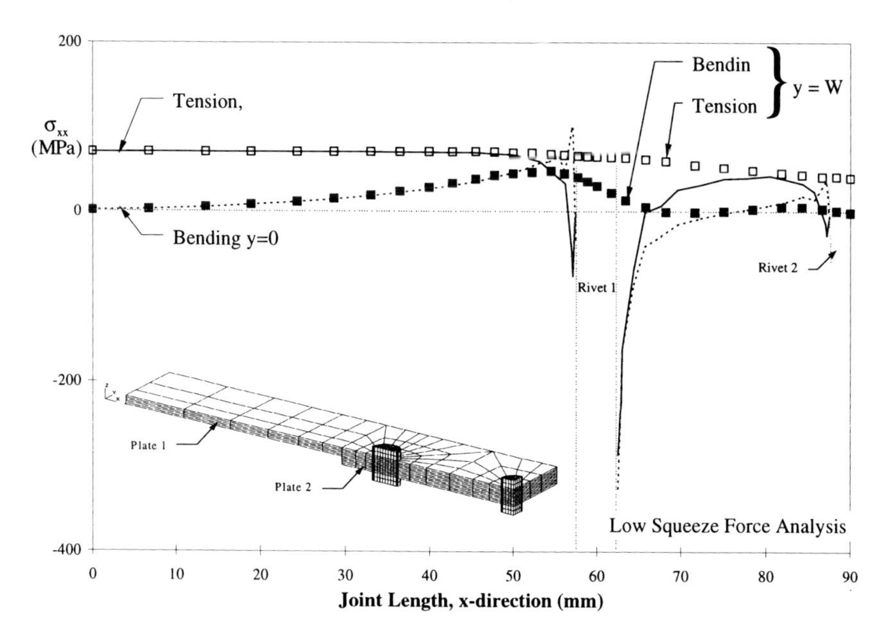

Figure 6.8 Variation of the Normal Stress (σ_{xx}) along the center of the Specimen (y=0) at the Sheet Surface (z=0). LSF Analysis

A more pronounced variation in the secondary bending stress is seen for the HSF analysis shown in Figure 6.9. The RSS dominates the stress state around the rivet hole resulting in the maximum secondary bending stress occurring at y = 0 at the top of rivet 1, which is 90° from the net section of the joint away from rivet 2. With increasing y, the bending stress decreases just as in the LSF case, but the location of maximum bending remains at x = 54.0 mm or may even shift slightly away from the net section which is opposite the behavior seen in the LSF analysis. Possibly the high RSS is masking the behavior in the

HSF analysis. If this is the case, the maximum bending at y = 0 is occurring at x = 51.75, the same location as the LSF analysis.

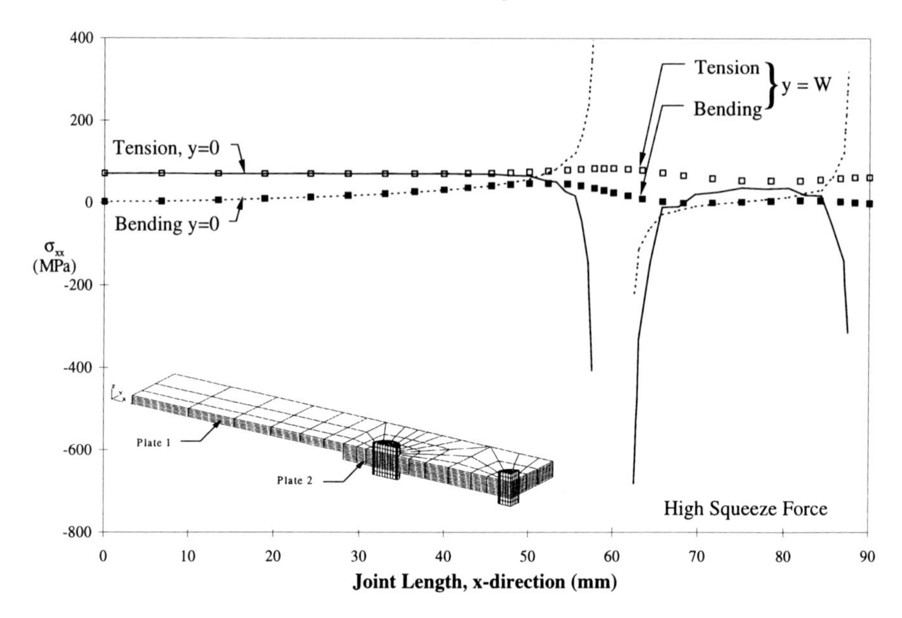

Figure 6.9 Variation of the Normal Stress (σ_{xx}) along the center of the Specimen (y=0) at the Sheet Surface (z=0). HSF Analysis

One of the most significant results of Müller is the increase in fatigue life with increasing squeeze force.[1] The increase in the fatigue life is due to the reduction of the cyclical stress caused by the RSS. Looking at Figure 6.10, the K_t for the LSF is 5.98 and for the HSF it is 6.05 including the RSS. However, from a fatigue loading point of view, only the cyclical stress is of importance that is the pressurization cycle for a fuselage structure. To determine the cyclical stress, the RSS must be subtracted since this is the equilibrium condition for the joint without any external loading due to cabin pressurization. As the cabin pressurizes, the stress at the hole edge does not increase from zero to maximum stress, but from the RSS stress to maximum stress which results in a smaller increase in the stress. This relation can be simply written as

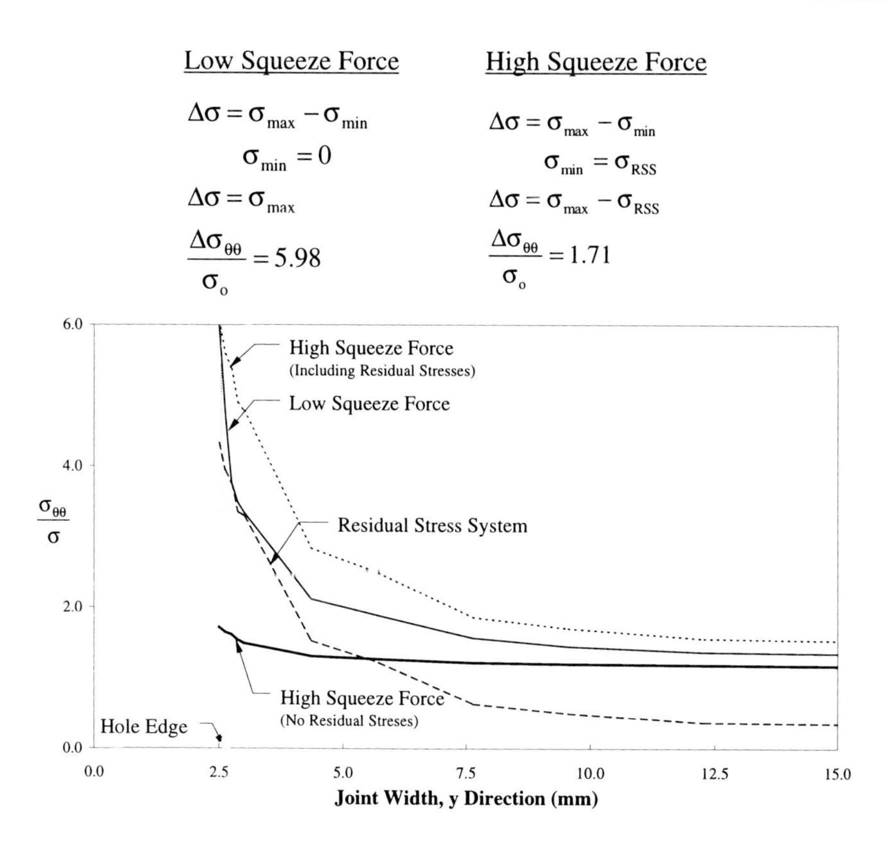

Figure 6.10 Effect of Residual Stress System on Fatigue Loading; Plate 1, Rivet 1

By increasing the squeeze force, the magnitude of the residual stress system is also increased and the cycle stresses are reduced more. Müller found an increase of the fatigue life by a factor of five or more depending on the squeeze force and joint geometry. The reduction in the cycle stress shown in Figure 6.10 is by a factor of 3.5 with the K_t being reduced to just 1.71 from 5.98 in the LSF case.

There is another interesting result with regard to the stress distribution at the hole edge and in the net section of the joint at rivet 1. The relationship between the tension and bending stresses at this location varies between the LSF and HSF analyses. In the LSF analysis, the maximum k, defined by Eqn. (3.1), is 0.56 at y = 4.36 mm and the minimum is 0.33 at the y = 2.75; whereas in the HSF analysis, the maximum k is 0.296 at y = 12.3 mm and the minimum is 0.016 at y = 2.75 mm. In both analyses, the maximum bending factor is away from the hole edge in the net section of the joint. Thus, a crack growing in the net section experiences non-negligible bending stress throughout its life.

Müller hypothesized a negligible bending stress at the hole edge due to the constraint of the rivet not allowing the sheets to bend which is supported by both analyses, more so by the HSF case having $k = 0.016$. The constraint offered by the rivet is less in the LSF case, thus the sheets are able to bend slightly as manifest by the $k = 0.33$. A further effect of the RSS is evident in the location of the maximum bending stress. For the LSF case, the maximum k and bending stress occur close to the hole edge; however, for the HSF case, the maximum k is at $y = 12.8$ mm and the maximum bending stress at $y = 12.6$ mm. More important is the shift of these maxima in the presence of the RSS further supporting a smaller stress intensity factor in the HSF than LSF analyses.

The line model derived by Schijve has been used for designing the lap splice joints tested in Chapter 3. The line model does not account for the stress concentration created by the rivet hole or the width of the joint. Since the rivets are assumed to be rigid, the model cannot describe any of the rivet behaviors discussed above.[2] Despite its simplicity, the line model is fairly accurate in predicting a k of 1.18, which is only 8.4% lower than the LSF analysis.

A second goal of this FE analysis is to determine the effect of the RSS on the rivet tilting behavior. As the rivet tilts, the contact boundary through the thickness may change which in turn changes the location of the maximum stress. The maximum stress for the LSF is found at the hole edge; however, for the HSF analyses, the maximum stress occurs between $40° < \theta < 90°$ as shown in Figure 6.11.

Several researchers have investigated the location of the maximum stress in a pin loaded hole using several different assumed load distributions which predict a location varying from 85° - 88° measured from the top of the hole.[3-6] In these previous studies, the bore of the hole is loaded uniformly through the thickness which is definitely not the case in a riveted connection since the rivet tilts and loses contact with the bore of the hole thereby changing the area over which the load is transferred. The LSF analysis represents such a case where the rivet tilts resulting in a changing stress distribution through the thickness as shown in Figure 6.12. Conversely, rivet tilting is restricted in the HSF cases as a result of the residual stress system, which yields a relatively even distribution of the stress through the thickness, again shown in Figure 6.12. Müller reports similar effects with regard to the rivet tilting behavior in low and high squeeze force joints. He measured the rivet flexibility that indicates the degree of rivet tilting; e.g., low rivet flexibility implies low rivet tilting; and found the rivet flexibility decreases with increasing squeeze force.[1]

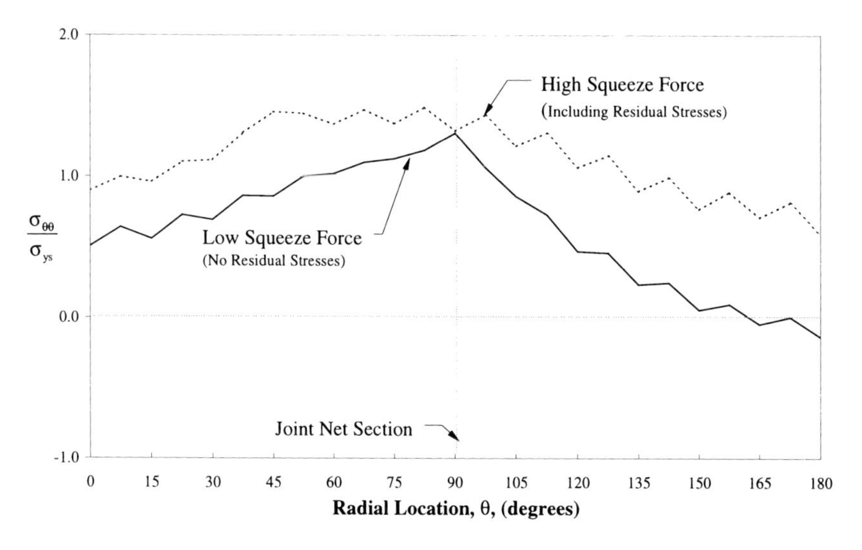

Figure 6.11 Stress Variation Along Perimeter of Rivet 1 Hole at the Faying Surface (z = 0)

From a crack growth prediction point of view, if the cracks are nucleating and growing at the hole edge then the K solutions calculated in Chapter 4 are applicable. If the cracks are not nucleating and growing from the hole edge, K solutions from Chapter 4 cannot be used and new solutions must be calculated. Therefore, in a joint whose rivets are installed using an average to high rivet

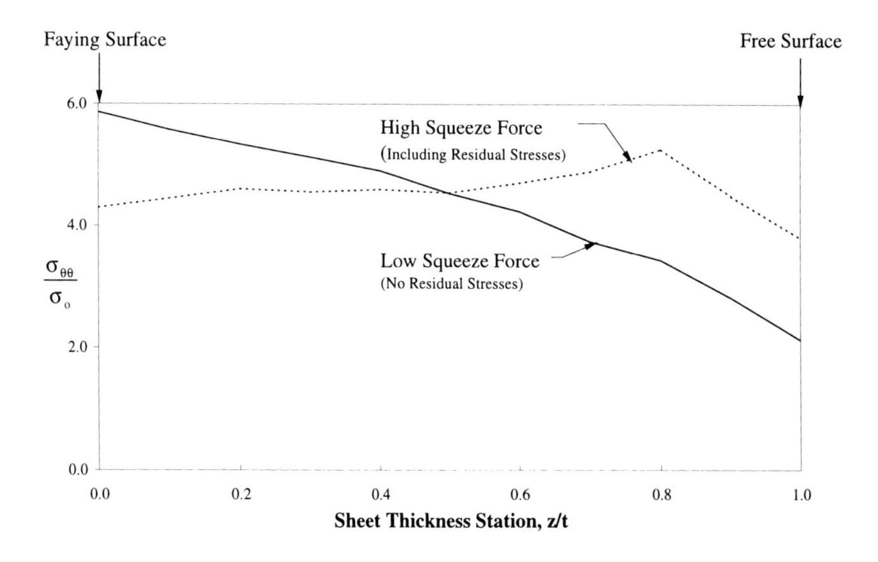

Figure 6.12 Stress Variation Through the Thickness at Rivet 1 Hole Edge (Joint Net Section, y = 2.5)

squeeze force, the new K solutions can be used, but this may be questionable for a low rivet squeeze force joint.

6.5 Stress Intensity Factor Calculation

A quarter circular corner crack was symmetrically created at the two hole edges of rivet 1 in the net section of plate 1 of the asymmetric lap-splice joint model. Both LSF and HSF cases of the cracked condition are analyzed. The purpose of these analyses is twofold; first, examine the effect of the RSS on the K-value, and second, compare the present results to the Newman/Raju[7] solutions (listed in Appendix G). The crack configuration analyzed is listed in Table 6.2. The K-values for the current analysis are calculated using the virtual crack extension method (VCE) developed by Bakker[8] which was discussed in section 4.1.2.3. The VCE method uses the finite element analysis results for the model with a crack. Without upsetting the equilibrium of the entire model, a small virtual crack extension is applied to the crack front nodes from which the J-integral is calculated. Recall J is related to G and hence K by Eqns. (4.16) and (4.18). The stresses, listed in Table 6.2, used for calculating the K-values using the Newman/Raju equations are most often determined from a simple, mechanics of materials approach. Due to the nonlinear relation between the applied load and secondary bending stress in addition to the contact behavior along the bore of the hole, the bending and bearing stresses are known a priori to be estimates of the real state of stress. The K-values for the crack depth and length locations are given in Table 6.3, and the full K distribution along the crack front is shown in Figure 6.13.

Table 6.2 Crack Configuration

Crack depth, a	0.8 mm
Crack length, c_1	0.8 mm
Plate width, W	30 mm
Plate thickness, t	2.0 mm
Rivet diameter, D	5.0 mm
$\sigma_{max\ bypass}$	40 MPa
$\sigma_{max\ bending}$	153 MPa †
$\sigma_{max\ bearing}$	120 MPa ††

† Schijve[2] Line Model
†† Eqn. (3.4), $\gamma = 40\%$

Table 6.3 Diametrically Opposed Corner Cracks at a Hole Subject to Combined Loading

Solution	$K_I(a)$ (MPa√m)	$K_I(c)$ (MPa√m)
Newman/Raju	10.24	9.57
LSF	11.86	8.71
HSF	3.95	3.75

The K variation along the crack front shows the effect of the secondary bending and pin loading. In the LSF case, the higher K's toward the bore of hole was seen previously in section 4.5.1.1 where K increased as the distance from the crack tip to the location of load application decreased. More importantly is the implication that contact is lost between the rivet and sheet where the K_t is increasing as a result of the rivet hole being partially open. The small variation in K for the HSF case implies contact is not lost. Just as in the stress analysis discussed in the previous section, when comparing the LSF and HSF cases, the effect of the RSS must be removed since the RSS represents the equilibrium condition for the HSF case. Thus, at the faying surface ($2\phi/\pi=0$) the LSF K's are 2.3 times higher than the HSF and 3.0 times higher at the bore of the hole ($2\phi/\pi=1$). The higher K-values for the LSF in comparison to the HSF (circles in Figure 6.13) explains the fatigue life increase Müller observed for joints manufactured with a average to high rivet squeeze force.

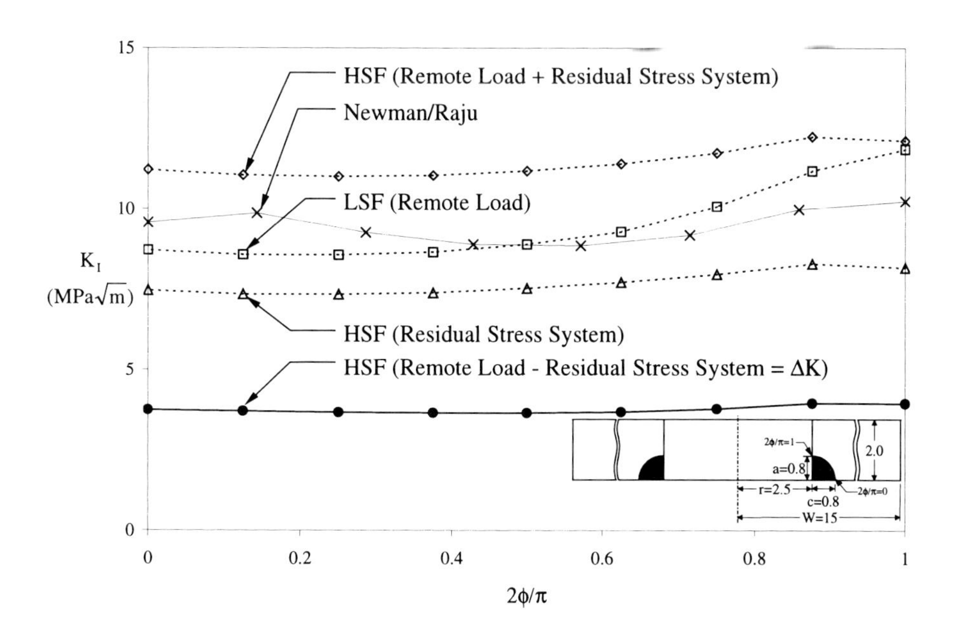

Figure 6.13 Comparison of K Distributions along Quarter Circular Part Through Crack Subject in a Lap-Splice Joint

Comparing the present results to the Newman/Raju solutions can only be qualitative in that the simplicity of the stress calculations in Table 6.2 can lead to errors in the K calculation. Although normalization of K in accordance with Eqn. (4.25) seems logical, the K's from the present lap joint FEA cannot be decomposed into the individual tension, bending, and pin loading components. Therefore, the distribution of K along the crack front is used for comparison between the Newman/Raju solutions and both the LSF and HSF solutions.

Most notable is the variation at the faying surface where the Newman/Raju solutions reach a maximum at $2\phi/\pi = 0.18$. The present results, both LSF and HSF, show a nearly constant K through $2\phi/\pi = 0.5$. The constant K over half of the crack front may account for cracks propagating with a self-similar shape. Similar behavior at the bore of the hole is seen between the Newman/Raju and LSF solutions where K is slightly increasing. In view of crack growth predictions in lap joints, the Newman/Raju solutions do not account for the residual stress system and will therefore overestimate ΔK. The results of the crack growth predictions in Chapter 5 (section 5.4.3, Figure 5.11 Region BC), also show the Newman/Raju solutions are producing K's that are too high at the faying surface. Therefore, the current FEA provides an essential second point of comparison for evaluating the performance of the Newman/Raju solutions for predicting crack growth in lap-splice joints.

6.6 Conclusions

Three Rivet Finite Element Model

- The three rivet finite element model has illustrated the more dominant behavior in the lap joint making subsequent analyses with the 1½ rivet model more computer and time efficient.
- Coupling the nodes between the rivet head and outer sheet surfaces of the joint has a negligible effect on the stresses around the rivet hole.
- The contact behavior between the rivet and sheets is accurately defined by contact elements connecting the rivet shank and hole bore.
- Contact between the sheets in the overlap region is minor and does not need to be modeled explicitly.

Asymmetric Lap Splice Joint

- A three rivet row lap splice joint was modeled with only 1½ rivets by employing symmetry and asymmetry boundary conditions in the joint width and joint length directions, respectively.
- The residual stress system as a result of rivet installation can be modeled by a thermal expansion in the radial direction and contraction along the longitudinal axis of the rivet.
- Effect of the Residual Stress System of a High Squeeze Force (HSF)
    - Rivet tilting is reduced decreasing the global joint deflection.

- The rivet hole remains filled decreasing the K_t by a factor of 4.
- The maximum tensile stress in the joint occurs at the hole edge in the net section of the joint and is nearly constant through the thickness. For the low squeeze force (LSF) analysis, the maximum tensile stress is at the faying surface (z = 0) and decreases to a minimum at the free surface (z = t).

- The secondary bending stress is changing through the joint width direction for both the LSF and HSF analyses being maximum at the symmetry plane (y = 0) and decreasing to a minimum at ½ the pitch (y = W).

Stress Intensity Factor Calculations

- A quarter circular part through crack was modeled at the hole edge of rivet 1 in the joint net section with an applied remote tensile stress of 100 MPa. K-values were calculated at 9 points along the crack front using the virtual crack extension method.
- The variation of K along the crack front is limited. The maximum occurs in the crack depth direction at the hole edge for both the LSF and HSF cases.
- The ΔK for a low squeeze force rivet installation is 3 times as high as that for a high squeeze force rivet installation thus leading to a longer fatigue life for the latter as experimentally shown by Müller.
- In view of the residual stresses around the hole, the Newman/Raju solution for double corner cracks at a hole under combined loading better represent the stress system for the LSF case where the residual stresses are negligible, but are not appropriate when the residual stresses are high and will overestimate K.
- Even though the residual stresses around the hole are negligible for the LSF case, the Newman/Raju solutions still overestimate K at the faying surface for the joint geometry, loading condition, and crack shape considered.

---

[1] Müller, Richard Paul Gerhard. An Experimental and Analytical Investigation on the Fatigue Behaviour of Fuselage Riveted Lap Joints, The Significance of the Rivet Squeeze Force, and a Comparison of 2024-T3 and Glare 3. Diss. Delft University of Technology, 1995. Delft, NL. ISBN 90-9008777-X, NUGI 834.

[2] Schijve, J. and F. A. Jacobs. Fatigue Crack Propagation in Unnotched and Notched Aluminum Alloy Specimens. NLR-TR M.2128. Amsterdam, NL: National Aerospace Laboratory, 1964.

---

[3] Chang, Fu-Kuo, Richard A. Scott, and George S. Springer. "Strength of Mechanically Fastened Composite Joints" Journal of Composite Materials, 16 (1982): 470-494.

[4] Crews, John H., C. S. Hong, and I. S. Raju. Stress Concentration Factors for Finite Orthotropic Laminates with a Pin-Loaded Hole, NASA-TP-1862, 1981.

[5] de Jong, Theo. "Stresses Around Pin-Loaded Holes in Elastically Orthotropic or Isotropic Plates." Journal of Composite Materials, 11 (1977): 313-331.

[6] Eshwar, V. A., B. Dattaguru, and A. K. Rao. Partial Contact and Friction in Pin Joints, Report No. ARDB-STR-5010, Department of Aeronautical Engineering, Indian Institute of Science, 1977.

[7] Newman, Jr., J. C. and I. S. Raju. Stress Intensity Factor Equations for Cracks in Three-Dimensional Finite Bodies Subjected to Tension and Bending Loads. NASA-TP-85793, 1985.

[8] Bakker, Ad. The Three-Dimensional J-Integral: An Investigation into Its Use for Post-Yield Fracture Safety Assessment. Diss. Delft University of Technology UP, 1984. Delft, NL.

7.

Summary, Recommendations, and Epilogue

7.1 Overview

The riveted lap joint in pressurized fuselages of transport aircraft remains difficult to understand despite enormous efforts since the historic Aloha Airlines catastrophic in-flight failure of 1988. This is not to imply the flying community is no better off than we were prior to the Aloha accident. Much has been learned in the last nine years regarding the fatigue performance of riveted connections. This chapter serves to wrap up the current research effort in section 7.2, summarize the major findings in section 7.3, give suggestions for future research in section 7.4 provide some prospective in the epilogue found in section 7.5.

7.2 Summary

Several questions were posed at the beginning of this research effort, Chapter 1, to provide a framework for a systematic approach to discovering the characteristics of fatigue crack growth from rivet loaded holes in thin 2024-T3 Alclad sheets. To give a sense of closure to these questions, they are answered in brief in this section with more complete summaries in section 7.3. The questions were:

1. Can an MSD cracking scenario be investigated in the laboratory with coupon sized test specimens?

    MSD was always observed in asymmetric lap-splice joint specimens with either two or three rows of rivets which represents the loading found in the more structurally complex fuselage lap-splice joints.

2. Why do some cracks nucleate at the top of the rivet hole and others at the hole edge in the net section of the joint?

A major discovery by Müller who showed the dependence of the crack nucleation site on the rivet squeeze force used to install the rivets. This finding was essential and exploited to ensure crack nucleation and growth from the rivet hole edge as commonly seen in-service.

3. What role does rivet tilting play?

   From the 3D finite element analysis of a three rivet row lap-splice joint, rivet tilting decreases with increasing squeeze force which affects the global deflection of the joint and the local stress at the hole edge in the net section of the joint in the critical outer rivet row.

4. Does the stress state vary between rivets in the same row?

   Again from the 3D finite element analysis of a three rivet row lap-splice joint, both the tensile and secondary bending stresses vary through the rivet pitch with the degree of variation dependent on the rivet squeeze force.

5. What are the crack growth rates of MSD cracks?

   The asymmetric lap-splice joint was loaded by remote cyclic tension to 75 MPa which created stresses in the critical outer rivet row similar to those is fuselage lap-splice joints. The small MSD cracks with a faying surface crack length of 75 μm grew at about 2×10^{-3} μm/cycle; whereas later in the life the larger MSD cracks, 4 mm in length, grew at nearly 1 μm/cycle.

6. Is small crack growth data obtainable?

   For open hole specimens, crack growth data was obtained via a fractographic examination using the scanning electron microscope for cracks approximately 10 μm in length. Due to fretting damage close to the hole edge, crack growth data for lap joints was obtainable down to 75 μm.

7. Can the crack shape be represented in a known geometric functional form?

   For open holed, 1½ dogbone, and asymmetric lap-splice joints, the characteristic crack shape is part elliptical for both part through and through cracks.

8. What effect does the local stress state have on the crack shape?

   Three effects were seen when increasing the secondary bending stress; the fatigue life decreases, the crack depth to sheet thickness ratio (a/t) is unaffected, and the crack depth to crack length ratio (a/c_1) decreases. Also, for low squeeze force rivet installations, fractographic investigations are difficult close to the rivet hole edge due to fretting damage.

9. Is the crack shape changing during its life?

   Crack shape development could not be monitored during the fatigue tests but information can be obtained by destructive testing. It shows that the cracks never propagate with a straight front but always with a part elliptical oblique front until just before unstable failure where the crack appears to have a slanted, oblique front.

10. Does the initial crack shape affect the fatigue life?

    This could not be experimentally verified, however from the new K solutions calculated there is a strong dependence on the initial crack shape (a/c_1), crack size (a/t), and sheet thickness (r/t).

11. Are the existing crack growth models sufficient?

    The Newman/Raju solutions for part through and NASGRO TC09 for through crack growth are conservative and underestimate the fatigue life.

12. Can the existing crack growth models be improved?

    The prediction model developed here is one improvement with consideration of part elliptical through crack growth. Additional work should focus on the quantifying the contribution of friction, fretting, and residual stresses around the rivet hole.

7.3 Survey of Results in More Detail

Existing Knowledge Base (Chapter 2)

- The design of the riveted lap joint is driven by the desire for a lightweight and easy to manufacture method of connecting the fuselage skin panels. In spite of its simplicity, the lap joint is geometrically complex with several contributions to a difficult analysis of the load transmission.

Joint Geometry Effects

- The eccentricity inherent in a lap joint creates secondary bending as an unavoidable result of the remotely applied tension stress. The circumferential stiffening elements (frames) reduce the hoop stress locally where the frames are attached to the joint; however, through the remaining width of the joint, the hoop stress distribution is homogeneous.

Rivet Squeeze Force Effects

- The work of Müller laid the foundation for the experimental work undertaken. His discovery of the importance of the rivet squeeze force lead to the design and testing of lap joint specimens which replicated the fuselage longitudinal lap-splice joint cracking behavior seen in-service.
- The residual stress system introduced as a result of rivet squeezing leaves radial and tangential stresses around the rivet and clamping stresses between the faying surfaces of the two sheets.

Fatigue Crack Growth

- Crack nucleation and growth at the faying surface imply that the cracks are small and invisible for a large portion of the fatigue life. The crack shape is part elliptical both before and after the crack has grown through the sheet thickness.
- Crack growth data for small and large cracks in riveted lap joints is not abundantly available in the literature. Not only is this data paramount to understanding the physics of the problem, but also the lack of data implies that the existing crack growth prediction models have not be adequately validated by empirical observations.

The Experimental Investigations (Chapter 3)

Fatigue Crack Growth

- The purpose of the present test series is to observe fatigue crack growth under combined cyclic tension and bending loads. The observation should include the development of size and shape of cracks starting as very small part through cracks up to oblique through cracks. Fractography has been essential for the observations. Tests were carried out on 3 types of specimens: (i) a newly developed combined tension and bending sheet specimen, (ii) a 1½ dogbone specimen, and (iii) an asymmetric riveted lap-splice joint. The major findings are summarized below.

Results of the combined tension and bending specimens

- The specimen yields a controllable stress field in the test area with a satisfactory agreement between secondary bending calculated with the Schijve line model and measured by strain gages.
- Tests were carried out on center cracked and 1, 5, and 7-open hole specimens. In the open hole specimens all cracks nucleated in the minimum net section, initially growing as corner cracks until they penetrate the back surface of the sheet to continue propagation as through cracks with oblique crack fronts.

Results of the 1½ dogbone specimens

- A modified 1½ dogbone specimens was adopted to obtain a bending factor k (= $\sigma_{bending}/\sigma_{tension}$) representative of fuselage lap joints. Strain gage measurements indicated an inhomogeneous secondary bending distribution in the width direction, not representative of fuselage lap joints.
- Specimens riveted with different squeeze forces confirmed that a higher squeeze force promotes crack initiation away from the minimum net section towards the top of the rivet hole.

Results of the riveted lap joints

- Asymmetric lap joints with two rivet rows were designed to different geometry's. All specimens revealed an MSD behavior similar to that found in the more structurally complex fuselage longitudinal lap-splice joints. Also secondary bending and crack growth appears to be similar.
- For an increasing sheet thickness the ratio a/t of the crack shape is not affected, but the a/c ratio decreases.
- The bending ratio k was varied in the range 1.2 to 2.0 by using different distances between the rivet rows. Increasing secondary bending reduced the fatigue life. It also led to lower a/c values for larger cracks.

Fractographic techniques

- Two types of marker loads were added to the constant-amplitude (CA) loading. The first type is using an under-load cycle (UL) to $\sigma_{min} = 0$ after every 50 cycles of the basic load cycles, which should have $\sigma_{min} > 0$. Distinct striations of the UL cycles could be observed in the electron microscope for small cracks, $c_1 \leq 1.0$ mm. For larger cracks such observations became difficult and marker bands could remain undetected.
- In the second approach a program loading spectrum was adopted by introducing blocks of 100 smaller cycles with maximum stress of 75 % of σ_{max} of the basic CA loading, but the same $\sigma_{min} = 0$ stress level. Clear marker bands were produced from a crack size of about 75 μm to specimen failure. The crack history can thus be reconstructed. This method is particular useful for small inaccessible cracks occurring in riveted lap joints.
- Observations on crack size and shape were also made by interrupting fatigue tests at certain percentages of the average fatigue life, followed by pulling the specimen to failure. The size and the shape of the crack nuclei could then be measured. It showed that crack nuclei in the open hole specimens developed relatively later than in riveted lap joints.

Crack front shapes

- The shapes of a large number of fatigue crack fronts obtained under combined tension and bending were examined. In general the crack front could well be approximated as a part elliptical shape with the two axes along the sheet surface and the edge of the hole. Values of a, c_1 and c_2 were then obtained by regression analysis

Calculating New Stress Intensity Factor Solutions for Part Elliptical, Oblique Through Cracks at a Hole Subject to Combined Loading (Chapter 4)

Non-Orthogonal Finite Element Mesh

- It has been shown that the 3D virtual crack closure technique (3D VCCT) can be used with a non-orthogonal finite element mesh for the calculation of stress intensity factors. By generating one sufficiently fine finite element mesh, the 3D VCCT can be used to generate multiple stress intensity solutions of cracks with complex shapes by simply manipulating the crack plane geometry.

Three Dimensional Virtual Crack Closure Technique

- K solution results were calculated with 3D VCCT for crack configurations and loading cases for which results were available in the literature. The configurations covered were circular internal crack embedded in an infinite solid subject to uniform tension, center crack tension, single edge crack tension, diametrically opposed through cracks at a hole subject to tension, bending, biaxial tension, and pin loading, semi-elliptical surface crack subject to tension and bending, and through cracks with an oblique elliptical crack front subject to tension and bending. In general the agreement was within 5% when comparing to 2D analytical solutions and 1% when comparing to published 3D finite element solutions.

Load Distribution on the Bore of the Rivet Hole

- The assumed rivet load distribution on the bore of the rivet hole greatly influences K for small cracks, but has no measurable effect for large cracks. To remain conservative, a cosine-squared distribution should be assumed.

Oblique Crack Fronts

- K solutions were calculated for through cracks with an oblique elliptical crack front subject to tension and bending. For the tension case, the oblique crack can be approximated as a straight crack having a crack front perpendicular to the sheet surface and crack length equal to the largest crack length of the oblique crack that has penetrated a free surface. Also, the high K's on the penetrated surface would promote catch-up where the penetrated crack grows rapidly to the same length as the free surface crack. For the tension and bending case, however, the oblique crack cannot be approximated with a straight crack. No catch-up behavior seems possible since the K's for the penetrated surface crack are not only lower than those of the faying surface, but also the relative difference between the K's at the faying and penetrated surfaces becomes larger.

Crack Shape Effect

- Increasing the a/c ratio, making the crack front more straight and less oblique, results in lower normalized K's caused by mitigating the perturbation of the stress field by the acute cusp at the intersection of the penetrated crack and back surface.

Loading Effects

- Of the three assumed load distributions for the pin loaded hole analyses, the cosine squared load distribution is the most appropriate, conservative engineering solution giving an mean normalized K for small cracks and a moderately higher value for larger cracks when compared to the concentrated and cosine load distributions.

Stress Intensity Factors for Part Elliptical, Oblique Through Cracks Subject to Combined Loading

- K-values have been calculated for crack shapes frequently seen in-service and in laboratory fatigue tests with a/c_1 ratios of 0.2, 0.3, 0.4, 0.6, 1, and 2; a/t ratios of 1.05, 1.09, 1.13, 1.17, 2, 5, and 10; and r/t ratios of 0.5, 1, and 2. Several load conditions are analyzed, biaxial tension (B = 0.0, 0.25, 0.5, or 1.0), remote bending, and pin loading (concentrated, cosine, or $cosine^2$ load distributions). Presented in tabular form, the K's can easily be incorporated into a crack growth prediction algorithm.
- A strong interrelationship between the parameters describing the crack geometry, a/c_1, a/t, and r/t, make comparisons between cracks of different shape difficult. However, some trends of shape and geometry effects on K could be indicated.
- Comparisons of the effect of biaxial loading and pin loading are in a broad sense the same as for straight cracks. Increasing the biaxiality decreases the geometry factor, β. Pin loading via a concentrated load at the top of the hole is more conservative than a cosine or cosine2 distribution.

<u>*Conclusions of the Crack Growth Predictions (Chapter 5)*</u>

Crack Growth Prediction Methodology: da/dN - ΔK Relation

- In general, the same crack growth prediction methodology developed and used in the NASGRO Crack Growth Computer Program is adopted.

The Forman-Newman-de Koning crack growth law is available in the crack growth prediction computer program, but due to the simplicity of the load spectrum (CA) and the well characterized sheet material (2024-T3), the FNK equation is degenerated to the closure corrected Paris Law for all predictions.

Crack Growth Predictions Using Published K Solutions

- Double Corner Cracks at a Hole Subject to Tension and Bending.
  - The Newman/Raju (N/R) solutions accurately predicted both the crack shape and fatigue life for open hole specimens subject to remote tension and bending.
  - The initial flaw assumption has a negligible effect once the crack length at the material surface, "c", exceeds 0.5 mm in a 1.6 mm thick sheet. However, crack growth in the small crack regime ($c \leq 0.5$ mm) is significantly affected by the initial crack size and shape.
  - A regression analysis must be preformed after each crack growth increment to fit the crack front back to an elliptical shape. To avoid analysis errors with the N/R solutions, a large number of points along the crack front must be used.
- The NASGRO TC09 solution accurately predicted crack growth in open hole specimens loaded in pure tension, but underestimated the fatigue life consistently by at least 30% for combined tension and bending.

Crack Growth Predictions with Newly Developed K Solutions

The new K solutions for two part elliptical, oblique through cracks at a hole have been used to predict the fatigue life of an open hole specimen subject to tension and bending ($k \approx 1.0$). In addition, predictions have been completed for an asymmetric lap splice joint.

- Predictions of the *open hole* specimens subject to combined tension and bending ($k \approx 1$) show good agreement until the last 6% of the fatigue life.
- Predictions of the *asymmetric lap-splice joint* are inadequate with the fatigue life being overestimated for both reconstructed crack histories. If crack growth in riveted joints is to be predicted, consideration of the

effects of friction on the part through crack growth and crack interaction on the through crack growth is required.

Conclusions from the 3D Finite Element Analysis of a Lap-Splice Joint (Chapter 6)

Three Rivet Finite Element Model

- The three rivet finite element model illustrated the more dominant behavior in the lap joint making subsequent analyses more computer and time efficient.
- Coupling the nodes between the rivet head and outer sheet surfaces of the joint has a negligible effect on the stresses around the rivet hole which are of prime importance in the fatigue performance of riveted connections.
- The contact behavior between the rivet and sheets is accurately defined by contact elements connecting the rivet shank and hole bore.
- Contact between the sheets in the overlap region away from the pins is minor and does not need to be modeled explicitly.

Asymmetric Lap Splice Joint

- A three rivet row lap splice joint was modeled with only 1½ rivets by employing symmetry and asymmetry boundary conditions in the joint width and joint length directions, respectively.
- The residual stress system as a result of rivet installation (high squeeze force, HSF) can be modeled by a thermal expansion in the radial direction and contraction along the longitudinal axis of the rivet.
- The residual stress system introduced by a high squeeze force preserves contact between the rivet shank and the rivet hole when the lap joint is loaded in tension and secondary bending. As a consequence, rivet tilting is reduced and the cyclic peak stress at the rivet hole edge is significantly smaller if compared to the case of a low squeeze force rivet installation. The peak stress distribution in the thickness direction is nearly constant, whereas the peak stress is clearly concentrated at the faying surface for a low squeeze force (LSF) rivet installation.

Stress Intensity Factor Calculations

- A quarter circular part through crack was modeled at the hole edge of rivet 1 in the joint net section with an applied remote tensile stress of 100 MPa. K-values were calculated at 9 points along the crack front using the virtual crack extension method.
- The variation of K along the crack front is very small for the HSF case, whereas it shows a maximum at the hole edge for the LSF cases, but for the latter case, it is a very weak maximum.
- The ΔK calculated for a low squeeze force rivet installation is 3 times as high as that for a high squeeze force rivet installation thus leading to a longer fatigue life for the latter as experimentally shown by Müller.
- In view of the residual stresses around the hole, the Newman/Raju solution for double corner cracks at a hole under combined loading better represent the stress system for the LSF case where the residual stresses are negligible, but are not appropriate when the residual stresses are high and will overestimate K.
- Even though the LSF case, the Newman/Raju solutions still overestimate K at the faying surface for the joint geometry, loading condition, and crack shape considered.

7.4 Recommendations

Experimental investigations remain the focal point of developing a viable crack growth prediction model for fuselage longitudinal lap-splice joints with or without MSD. The lap joints tested here need to evolve in complexity to physically represent the structure found in operational transport aircraft. Ideally, crack growth histories for curved, stiffened, three rivet row lap joints must be collected for verifying any prediction algorithm. With the known difficulties in collecting such data, the programmed loading spectrum used here to mark the fracture surface needs to be investigated further to optimize the quality of the marker bands in the different stages of crack growth. Also, to better characterize the geometric and loading parameters and their affect on the crack growth behavior, an in situ crack shape monitoring technique is needed.

The analytical work must also continue in step with the experimental investigations by developing stress intensity solutions for through cracks with oblique, part elliptical crack fronts for a collinear array of holes subject to combined loading giving insight in the crack interaction behavior under the various loading conditions and joint designs. Furthermore, multiple rows of rivets with random cracking at various locations may also be of interest. In addition, the effects of the longitudinal and circumferential stiffeners should be included in the stress intensity solutions.

Much work remains for the finite element analysis of the three rivet row lap-splice joint finite element analysis. The most needed improvement is the transition from straight shank to countersunk holes that must be done to accurately model real lap joints. Also, a more robust investigation into the residual stresses caused by installing the rivet would yield a more quantitative conclusion to the effect of the rivet squeeze force on the crack driving force. An investigation of the degree of load transmission by friction between the faying sheets and sheet/rivet interface is also needed. Lastly, a more complete investigation of three-dimensional cracks of various shapes at the hole edge in the net section is also of great interest.

7.5 Epilogue

Understanding of the fatigue crack growth mechanism in riveted lap-splice joints continues to grow every year and research in this area will proceed until the problem is solved or we stop building aircraft from fatigue sensitive materials. The present research was aimed at adding another piece to a puzzle for which the number of pieces is yet to be determined. Much attention was given to the through crack portion of the fatigue life where the crack front remains part elliptical. In the past, this portion of the life received little emphasis with the crack front assumed to be straight which is an oversimplification. Now that we have the tools to accurately represent the crack shape throughout its life, more effort can be directed toward comprehending the difficulties associated with the rivet; fretting, friction, and residual stresses. Although time consuming, a reliable method of obtaining crack growth histories has been presented which will serve as a means to validate any new prediction scheme. Empirical verification continues to be the final examination of our analysis tools.

Appendix A

Secondary Bending According to the Neutral Line Model

As an extension of the Schijve model[1], the neutral line approach is used for a three row riveted joint. The required equations to solve for the displacements and bending stresses are presented here. The equations are derived for a 3-row riveted joint. In Figure A.1, a 3-row riveted lap joint using the following notation is presented:

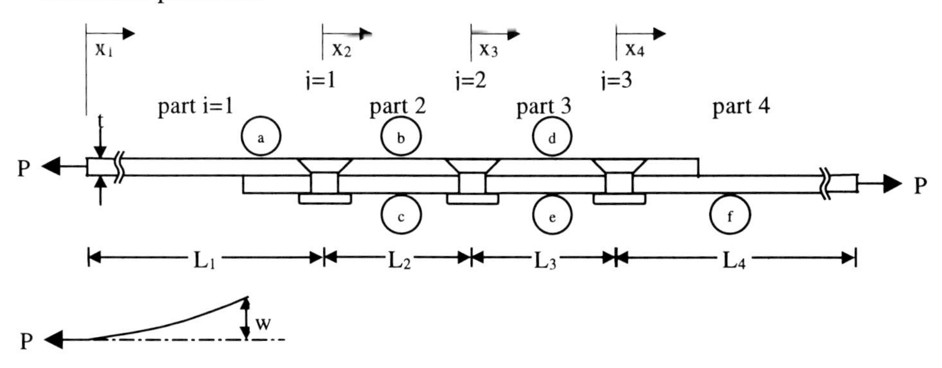

Figure A.1 Sign conventions and nomenclature for neutral line model

The four different parts in the model are referred by i=1 to i=4. The specimen ends are supposed to be loaded by P only, which presents a hinged connection. If L_1 and L_4 are sufficiently long as compared to L_2 and L_3, the secondary bending calculation gives practically the same results as for a clamped load introduction at the specimen ends.[1] For each part, equilibrium requires:

$$(M_x)_i = Pw_i = (EI)_i \left(\frac{d^2w}{dx^2} \right)_i$$

or:

$$\left(\frac{d^2w}{dx^2} \right)_i - \alpha_i^{\ 2} w_i = 0$$

$$\text{with } \alpha_i^{\ 2} = \frac{P}{(EI)_i}$$

solution:

$$w_i = A_i \sinh(\alpha_i x_i) + B_i \cosh(\alpha_i x_i)$$

where

$$\sinh(\alpha_i x_i) = S_i$$
$$\cosh(\alpha_i x_i) = C_i$$

$$\text{Part 1}: x_1 = 0 \rightarrow w(x_1 = 0) = 0 \rightarrow B_1 = 0 \tag{A.1}$$

At eccentricity j (j=1 to 3): $x_j=L_j, x_{j+1}=0$

Boundary conditions:

$$w_j + e_j = w_{j+1}$$
$$\left(\frac{dw}{dx}\right)_j = \left(\frac{dw}{dx}\right)_{j+1}$$

or :

$$A_j S_j + B_j C_j + e_j = B_{j+1} \quad (j = 1 \text{ to } 3) \tag{A.2}$$

$$A_j \alpha_j S_j + B_j \alpha_j C_j + e_j = A_{j+1} \alpha_{j+1} \quad (j = 1 \text{ to } 3) \tag{A.3}$$

$$\text{Part 4}: x_4 = L_4 \rightarrow w(x_4 = L_4) = 0 \rightarrow A_4 S_4 + B_4 C_4 = 0 \tag{A.4}$$

Eqns. (A.1) - (A.4) include 8 equations with A_j, B_j as the eight constants which must be determined. The shifts of the neutral lines at the rivet rows are the eccentricities e_j. Note that they are negative.

The lap joint to be considered here is symmetric in such a way that the two sheets have the same thickness (t), while $L_1 = L_4$ and $L_2 = L_3$. Moreover, in the neutral line model, the overlap between the two outer rivet rows is supposed to bend as a single sheet with double thickness (2t). The eccentricities then are $e_1 = e_3 = -t/2$ and $e_2 = 0$. In view of the symmetry only half the specimen has to be considered with the boundary condition,

$$X_2 = L_2 \rightarrow w_2 = 0 \tag{A.5}$$

With Eqns. (A.1) - (A.3),

$$\begin{aligned} B_1 &= 0 \\ A_1 S_1 + e &= B_2 \\ A_1 \alpha_1 C_1 &= A_2 \alpha_2 \\ A_2 S_2 + B_2 C_2 &= 0 \end{aligned} \tag{A.6}$$

where $S_1 = \sinh(\alpha_1 L_1)$, $C_1 = \cosh(\alpha_1 L_1)$, $S_2 = \sinh(\alpha_2 L_2)$, and $C_1 = \cosh(\alpha_2 L_2)$. The unknowns, A_1, A_2, and B_2, can be solved from these equations.

The bending moment at the first rivet row is:

$$M = Pw_{1,x_1=L_1} = PA_1S_1$$

The nominal bending stress at the first rivet row is:

$$\sigma_b \frac{M}{\frac{1}{6}bt_1^2} = \frac{6PA_1S_1}{bt_1^2}$$

where b is the specimen width. Since the remotely applied tensile stress, σ, is equal to P/bt the bending ratio, k, becomes:

$$k = \frac{\sigma_b}{\sigma} = \frac{6}{t_1}A_1S_1$$

Solving for A_1 from Eqn. (A.6) leads to:

$$k = \frac{-6\frac{e}{t_1}}{1+\frac{\alpha_1}{\alpha_2}\frac{T_2}{T_1}} = \frac{3}{1+\left(\frac{t_2}{t_1}\right)^{1.5}\frac{T_2}{T_1}}$$

with $T_1 = \tanh(\alpha_1 L_1)$ and $T_2 = \tanh(\alpha_2 L_2)$.

As an example, secondary bending and out of plane displacements, w, have been calculated for a specific three-row riveted lap joint. The dimensions are:

$t_1 = 2$ mm

$L_2 = 24$ mm

$L_1 = 100$ mm

$E = 72$ GPa

As shown by Figure A.2, the largest displacements occur at the outer rivet rows with a magnitude in the order of 0.5 mm for $\sigma = 100$ MPa. The secondary bending stress at the outer rivet rows in Figure A.3 illustrates the nonlinear behavior as a function of the applied stress, σ. Secondary bending is more nonlinear at the low σ-values. For the lap joint considered, the secondary bending stress is larger than the applied stress, $k > 1$.

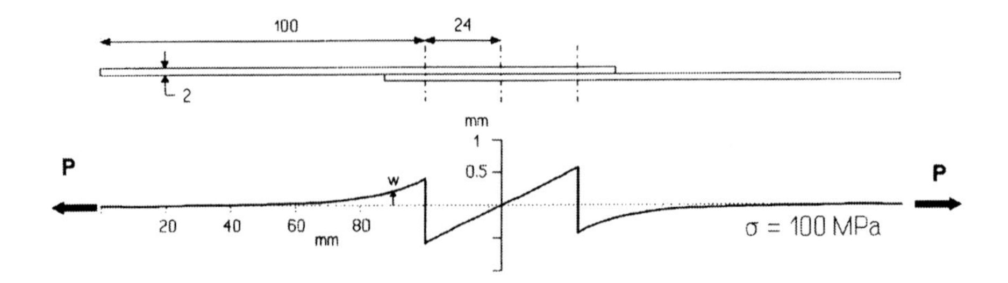

Figure A.2 Out-of-Plane Displacement (w) of a Riveted Lap Joint Calculated with the Neutral Line Model

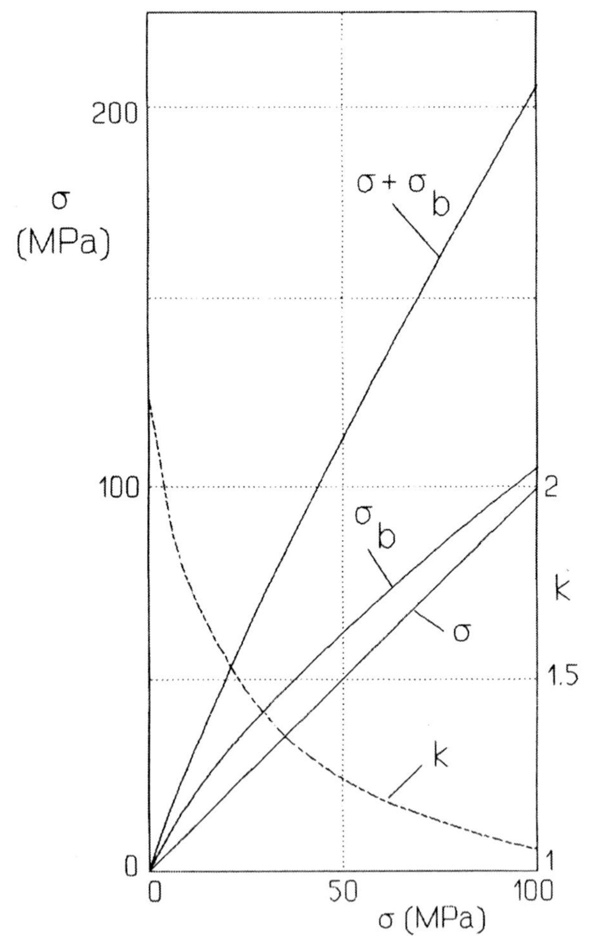

Figure A.3 The Secondary Bending Stress and Bending Ratio (k) as a Function of the Applied Stress on the Lap Joint of Figure A.2

[1] Schijve, J., Some Elementary Calculations on Secondary Bending in Simple Lap Joints, NLR-TR-72036. Amsterdam, NL: National Aerospace Laboratory, 1972.

Appendix B

Effect of a Marker Load Spectrum on Fatigue Crack Growth

The marker load spectrum is given in Figure B.1 below. The spectrum consists of 3 blocks of 1000 baseline cycles followed by programmed blocks of 100 smaller cycles and intermittent blocks of 10 baseline cycles. The number of these blocks was 6, 10, and 4 respectively, with the purpose to have some extra information of the load history on the fracture surface. The numbers of cycles in one spectrum program are:

Baseline cycles	3 x 1000 + (6 + 10 + 4) x 10	= 3200 cycles
Small marker load cycles	(6 + 10 + 4) x 100	= 2000 cycles
	Total number	= 5200 cycles

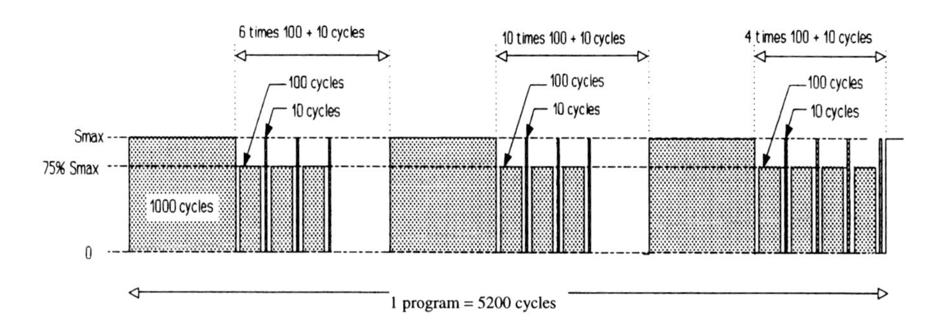

Figure B.1 Load History of Marker Spectrum

Adopting the plasticity induced crack closure concept and the simple Elber equation for calculating σ_{op} the results are:

$$S_{max} = 100\,MPa,\ R = 0 \rightarrow \Delta S = 100\,MPa$$

$$U(R) = \frac{\Delta S_{eff}}{S} = 0.5 + 0.4R$$

$$R = 0 \rightarrow U(R) = 0.5 \rightarrow \Delta S_{eff} = 0.5\Delta S = 50MPa$$

$$S_{op} = S_{max} - S_{eff} = 50\,MPa$$

For the smaller cycles, the S_{op} of the larger cycles applies (CORPUS concept). Thus:

$$(\Delta S_{eff})_{small\,cycles} = (S_{max})_{small\,cycles} - S_{op} = 75 - 50 = 25\,\text{MPa}$$

As a consequence:

$$\frac{(S_{eff})_{small\,cycles}}{(S_{eff})_{large\,cycles}} = 0.5$$

The Paris relation is assumed to be a good approximation:

$$\frac{da}{dN} = C\Delta K_{eff}^{m} = C\left(\beta \Delta S_{eff}\sqrt{\pi a}\right)^{m}$$

with β as the geometry factor.

During successive blocks of small and large cycles, the variation of the crack length can be ignored, and thus:

$$\frac{\left(\dfrac{da}{dN}\right)_{small\,cycles}}{\left(\dfrac{da}{dN}\right)_{large\,cycles}} = \left[\frac{(\Delta S_{eff})_{small\,cycles}}{(\Delta S_{eff})_{large\,cycles}}\right]^{m} = 0.5^{m}$$

Crack growth experiments were carried out on sheet specimens, material 2024-T3 Alclad, specimen width 100 mm, thickness 2 mm, pure tension with S_{max} = 100 MPa, R = 0, frequency 10 Hz. The specimens were provided with a central hole, diameter 4 mm, with two saw cuts of 2 mm each, leading to a $2a_o$ = 8 mm for the crack starter notch. The saw cuts are long enough to justify use of the Feddersen geometry correction factor ($\beta = \sqrt{\sec(\pi a)}$).

Two tests were carried out with CA loading only, and two tests included the marker load spectra. The crack growth rate data are compiled in Table B.1 and Table B.2 as the tip to tip crack length (2a) as a function of the number of applied cycles (N). The 2a-values are the sum of the crack length measured at the right hand side and the left-hand side. The symmetry was quite good. Note that the number of cycles (N) for the tests with the marker loads included the marker load cycles. Crack rates were calculated as $\Delta a/\Delta N$ for each interval in the tables. The corresponding crack length for calculating ΔK is the average length of the intervals.

Table B.1 Crack Growth Results of Two CA Tests

Test CA2				
2a (mm)	N (cycles)	Average a (mm)	da/dN (mm/cycle)	ΔK (MPa√m)
8.80	3823			
9.45	4813	4.56	0.00033	11.9
10.75	8872	5.05	0.00016	12.4
12.70	12801	5.86	0.00025	13.2
14.80	16808	6.88	0.00026	14.4
16.15	18806	7.74	0.00034	15.6
18.30	21010	8.61	0.00049	16.4
21.50	24813	9.95	0.00042	17.5
23.50	26805	11.25	0.00050	19.1
25.80	28807	12.33	0.00057	20.1
29.55	30948	13.84	0.00088	21.2
33.55	32826	15.78	0.00106	23.0
36.65	34002	17.55	0.00132	25.0
40.30	35007	19.24	0.00182	26.5
45.30	35998	21.40	0.00252	28.3
49.30	36506	23.65	0.00394	31.0
55.35	37011	26.16	0.00599	33.3

Test CA3				
2a (mm)	N (cycles)	Average a (mm)	da/dN (mm/cycle)	ΔK (MPa√m)
8.80	4000			
9.20	8002	4.56	0.00033	11.9
10.85	12006	5.05	0.00016	12.4
13.10	16010	5.86	0.00025	13.2
16.10	20141	6.88	0.00026	14.4
19.10	24008	7.74	0.00034	15.6
22.85	28011	8.61	0.00049	16.4
29.60	30009	9.95	0.00042	17.5
33.35	31999	11.25	0.00050	19.1
40.35	34003	12.33	0.00057	20.1
51.60	34501	13.84	0.00088	21.2
57.60	34796	15.78	0.00106	23.0

Table B.2 Crack Growth Results of Tests with Marker Loads

NASA5					
2a (mm)	N (cycles)	Average a (mm)	da/dN (mm/cycle)	ΔK (MPa√m)	da/dN* (mm/cycle)
8.72	10340				
9.48	15510	4.55	0.00007	12.0	0.00011
11.23	20680	5.18	0.00017	12.8	0.00026
12.73	25850	5.99	0.00015	13.8	0.00022
14.48	31020	6.80	0.00017	14.8	0.00026
16.98	36190	7.86	0.00024	16.0	0.00037
19.73	41360	9.18	0.00027	17.3	0.00041
22.98	46530	10.68	0.00031	18.8	0.00048
27.48	51700	12.61	0.00044	20.7	0.00067
34.23	56870	15.43	0.00065	23.4	0.00100
48.98	62162	20.80	0.00139	28.7	0.00213
53.73	62661	25.68	0.00476	34.1	0.00728
62.98	63043	29.18	0.01211	38.8	0.01852

*Corrected for the marker load

NASA6					
2a (mm)	N (cycles)	Average a (mm)	da/dN (mm/cycle)	ΔK (MPa√m)	da/dN* (mm/cycle)
8.67	5170				
9.42	10340	4.52	0.00007	12.0	0.00011
12.67	20680	5.52	0.00016	13.3	0.00024
15.67	31020	7.09	0.00015	15.1	0.00022
22.92	41360	9.65	0.00035	17.8	0.00054
34.42	51700	14.33	0.00056	22.4	0.00085
44.92	56870	19.83	0.00102	27.7	0.00155
55.67	58573	25.15	0.00316	33.5	0.00483
65.42	58891	30.27	0.01533	40.5	0.02345

*Corrected for the marker load

The results for the two CA tests without marker loads are plotted
Figure B.2. An average curve has been drawn though the data points. A first deviating data point is ignored for this purpose. The curve is not a linear one, as it should be if the Paris relation is applicable. However, the linear dotted curve does still approximate the empirical trend quite well. The slope of the line corresponds to a Paris exponent m = 3.3. The ratio of the crack rates during the marker load cycles and the baseline cycles becomes $0.5^{3.3} = 0.102 \approx 10\%$. In other words, the crack growth increment during a batch of 100 marker load cycles should be approximately equal to the crack extension during a batch of 10 baseline cycles. This was confirmed by the fractographic observations in the electron microscope (SEM). This is nice confirmation of the usefulness of considering crack closure during variable-amplitude loading.

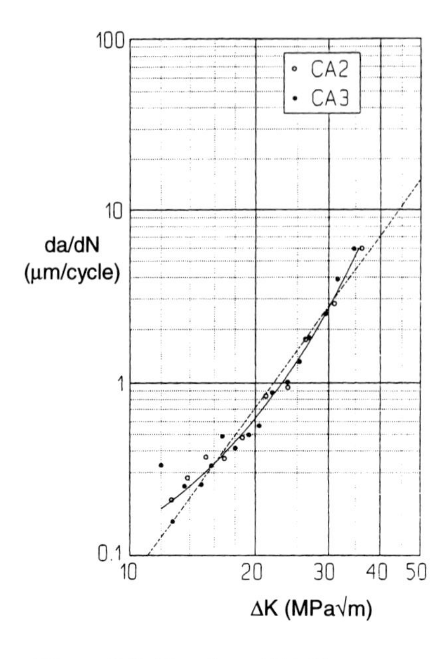

Figure B.2 Results of CA Tests

The crack growth rates during the tests with marker loads could now be corrected by replacing in the calculation every block of 100 marker load cycles by 10 baseline cycles. It implies that the 5200 cycles of the spectrum correspond to 3400 baseline cycles (reduction factor 0.654). The corrected crack rates are also given in Table B.2 and plotted in Figure B.3 together with the crack rates obtained in the CA test without marker loads. Apparently, the results of both types of tests are in the same scatter band. It is then concluded that the marker load cycles did not noticeably affect the growth rate during the baseline cycles.

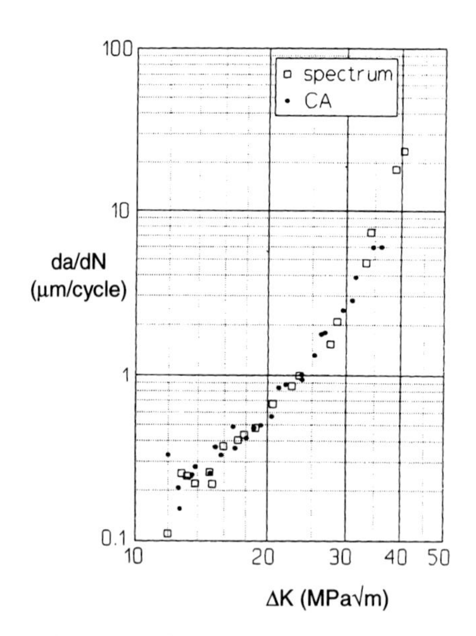

Figure B.3 Crack Growth Rates of the Spectrum Tests Corrected for the Marker Loads. Comparison to the Results of the CA Tests

Appendix C

Asymmetric Lap Splice Joint Crack Shapes and Thickness and Bending Stress Effect Investigations

The primary aim of the three asymmetric lap-splice joint tests was to determine the effect of changes in the joint geometry and load conditions on the crack shape. Specifically, does the crack shape change as the thickness of the adjoining sheets changes. Also, do changes in the secondary bending affect the crack front shape. Several locations along the crack front are measured and plotted in Figure C.2 - Figure C.5. In Table C.1 and Table C.2, the a/c_1 and a/t ratios are listed which are calculated from the a and c_1 values obtained from a least squares linear regression curve fit of the crack front measurements. In addition, the mean and standard deviations of a/c_1 and a/t for each lap joint are also given in the tables. The crack number system is shown in Figure C.1.

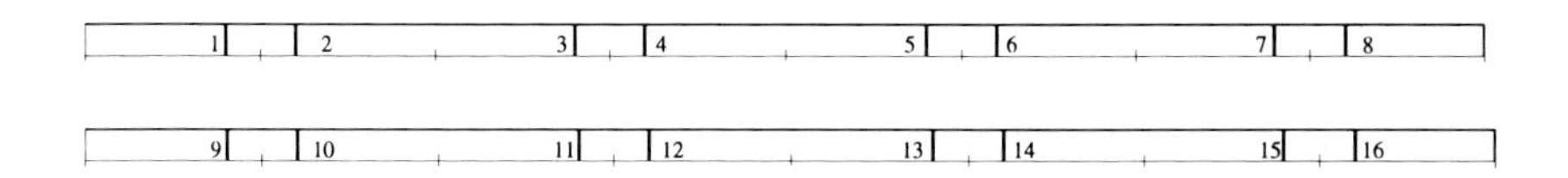

Figure C.1 Crack Number System

96W2T8N-4 (60%N)

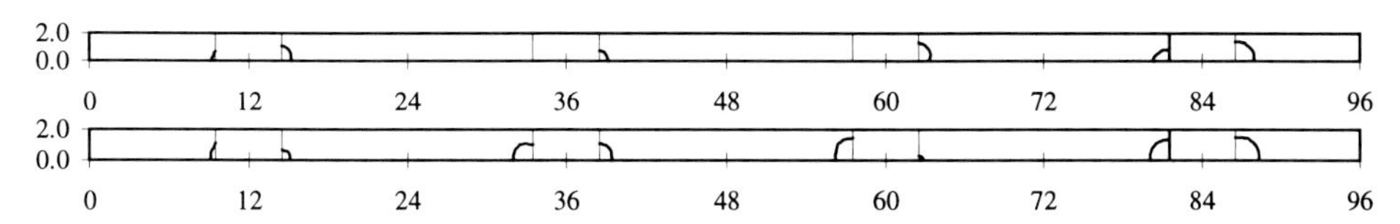

96W2T8N-5 (90%N)

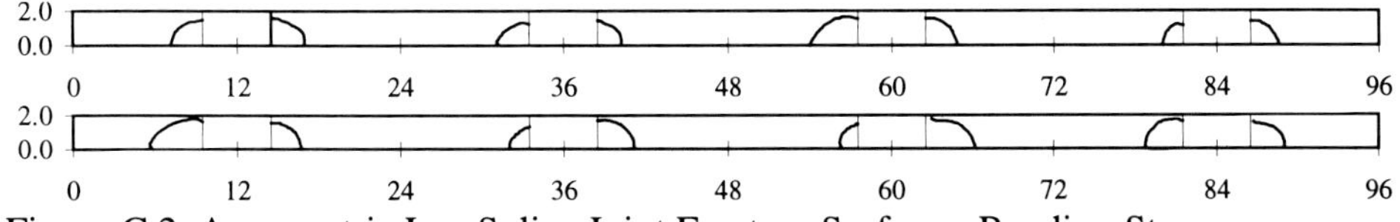

Figure C.2 Asymmetric Lap Splice Joint Fracture Surfaces, Bending Stress Effect - Test Series 3

96W2T12N-4 (60%N)

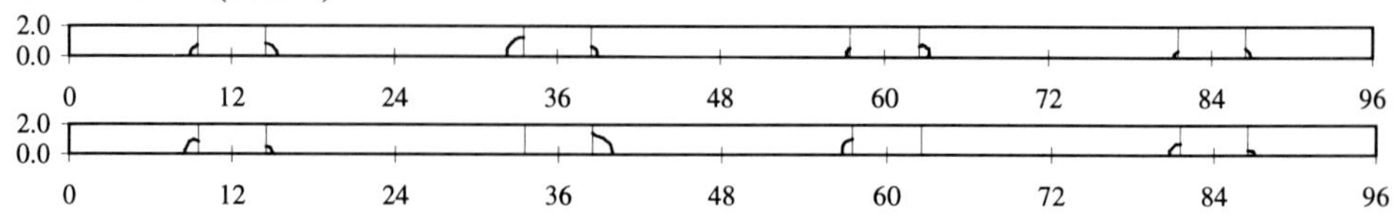

96W2T12N-5 (90%N)

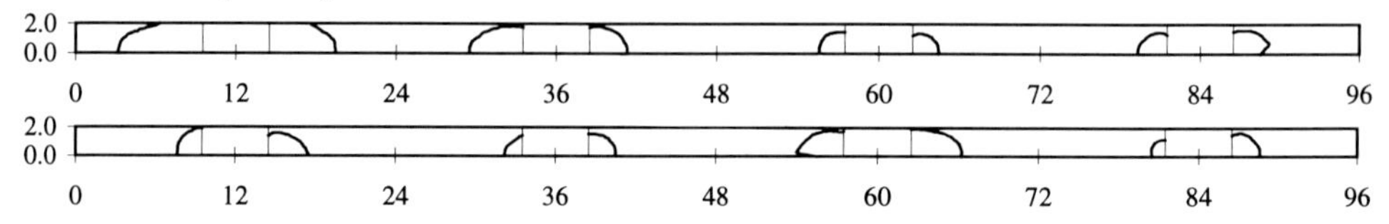

Figure C.3 Asymmetric Lap Splice Joint Fracture Surfaces, Bending Stress Effect - Test Series 3

96W2T16N-4 (60%N)

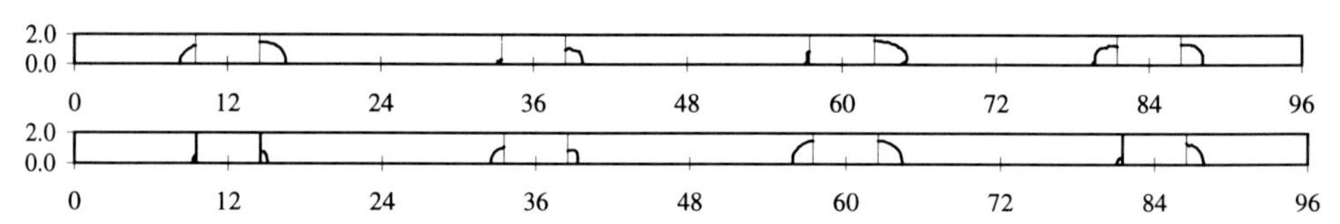

96W2T16N-5 (90%N)

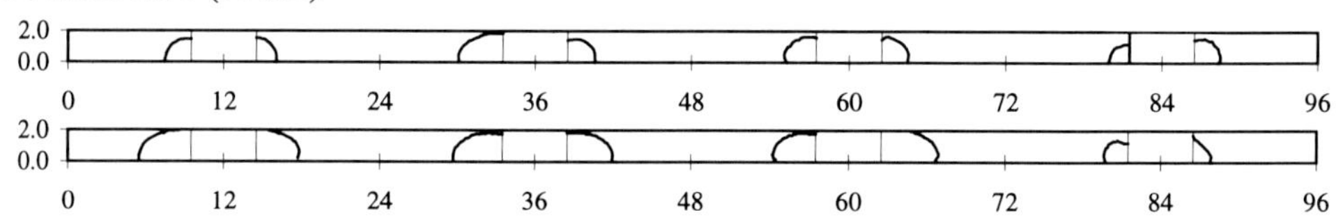

Figure C.4 Asymmetric Lap Splice Joint Fracture Surfaces, Bending Stress Effect - Test Series 3

96W2T24N-4 (60%N)

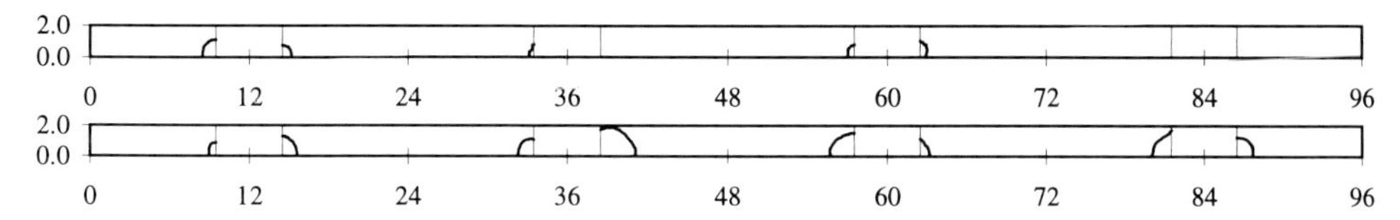

96W2T24N-5 (90%N)

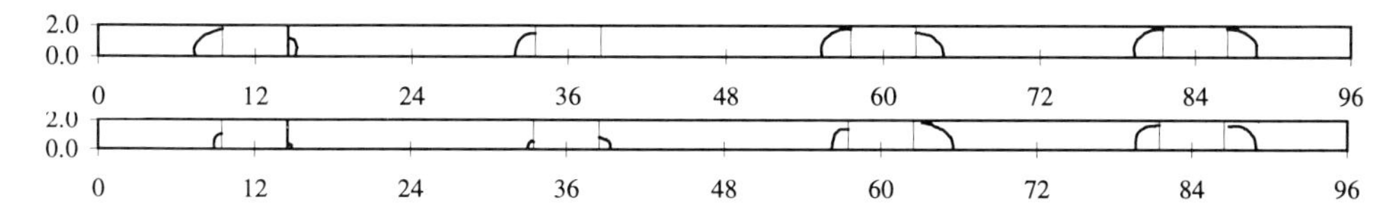

Figure C.5 Asymmetric Lap Splice Joint Fracture Surfaces, Bending Stress Effect - Test Series 3

Table C.1 Thickness Effect Investigation: Asymmetric Lap Splice Joint Test Series 1 and 2

Crack Location	Series 1						Series 2							
	64W1.2T		80W1.6T		96W2.5T		64W1.6T		80W2.0T		96W2.5T		128W3.2T	
	a/t	a/c	a/t	a/c	a/t	a/c	a/t	a/c	a/t	a/c	a/t	a/c	a/t	a/c
1	x	x	0.863	0.431	0.720	1.565	x	x	x	x	0.480	1.154	1.356	3.471
2	0.736	0.858	1.071	0.732	x	x	0.688	1.134	0.810	0.635	0.856	0.702	0.847	0.708
3	0.533	0.640	0.719	0.858	0.856	0.588	0.625	1.389	0.835	0.696	x	x	x	x
4	x	x	0.856	0.591	0.872	0.706	0.875	0.814	0.890	0.584	1.066	0.639	0.456	1.150
5	x	x	0.781	1.000	0.740	0.797	0.625	1.493	0.815	0.799	0.848	0.677	0.563	1.029
6	x	x	1.151	1.000	0.724	1.084	0.563	1.125	0.795	0.729	1.058	0.802	x	x
7	x	x	0.825	0.478	0.788	1.015	0.813	1.032	0.805	0.970	x	x	0.388	1.240
8	x	x	0.781	0.687	0.748	0.862	x	x	x	x	0.844	0.603	0.697	0.996
9	1.042	0.544	1.264	1.026	0.676	1.030	x	x	x	x	x	x	0.772	0.737
10	0.958	0.885	1.000	1.356	x	x	0.750	0.902	0.790	0.771	x	x	x	x
11	0.892	0.594	0.863	0.400	0.732	1.188	0.875	0.795	x	x	x	x	0.338	1.286
12	0.458	0.476	0.756	0.747	0.532	2.015	1.203	1.145	x	x	x	x	x	x
13	0.883	0.624	0.819	0.672	0.648	1.528	x	x	0.830	0.706	0.668	1.246	0.675	0.911
14	x	x	0.906	0.495	0.764	1.055	0.563	1.636	0.885	0.725	x	x	x	x
15	0.917	0.786	x	x	0.912	0.792	x	x	0.695	0.885	x	x	0.613	1.095
16	0.650	0.975	x	x	x	x	x	x	x	x	x	x	0.725	0.862
Mean	0.841	0.681	0.889	0.716	0.769	0.968	0.833	1.030	0.828	0.735	0.934	0.684	0.699	0.905
St. Dev	0.192	0.161	0.123	0.282	0.082	0.260	0.188	0.213	0.037	0.109	0.116	0.075	0.095	0.146

x – No Crack

Table C.2 Bending Stress Effect Investigation: Asymmetric Lap Splice Joint Test Series 3

Crack	96W2T8N-4		96W2T8N-5		96W2T12N-4		96W2T12N-5		96W2T16N-4		96W2T16N-5		96W2T24N4		96W2T24N-3	
Location	a/t1	a/c1	a/t1	a/c1	a/t1	a/c1	a/t1	a/c1	a/t1	a/c1	a/t1	a/c1	a/t1	a/c1	a/t1	a/c1
1	0.340	2.030	0.740	0.630	0.360	1.340	0.990	0.620	0.630	1.160	0.740	0.930	0.555	1.261	0.880	0.820
2	0.540	1.440	0.820	0.670	0.420	1.040	0.990	1.110	0.750	0.980	0.780	1.110	0.380	1.333	0.590	2.030
3	x	x	0.660	0.760	0.650	1.120	0.920	0.550	0.160	1.230	0.920	0.800	0.415	2.128	0.750	0.950
4	0.380	1.120	0.720	0.830	0.350	1.570	0.910	0.910	0.550	1.120	0.750	1.160	x	x	x	x
5	x	x	0.820	0.670	0.300	3.690	0.740	1.100	0.440	3.000	0.820	1.110	0.400	1.739	0.900	0.980
6	0.660	1.440	0.780	0.830	0.420	1.630	0.670	1.130	0.830	0.690	0.840	1.110	0.515	2.020	0.770	0.790
7	0.460	0.740	0.610	1.090	0.220	1.340	0.740	0.840	0.650	1.070	0.600	0.920	x	x	0.900	0.980
8	0.700	1.340	0.700	0.760	0.310	1.378	0.800	0.860	0.690	1.340	0.780	1.360	x	x	0.900	1.000
9	0.570	3.420	0.910	0.560	0.500	1.493	0.950	1.010	0.290	2.480	1.000	0.630	0.440	1.796	0.520	2.190
10	0.060	1.080	0.790	0.840	0.250	1.667	0.790	0.750	0.410	1.720	0.980	0.840	0.650	1.368	0.180	1.710
11	0.540	1.300	0.670	0.920	x	x	0.710	1.050	0.550	1.100	0.920	0.680	0.575	1.150	0.280	1.670
12	0.540	1.440	0.860	0.840	0.710	0.947	0.780	0.850	0.450	2.500	0.920	0.830	0.940	1.160	0.420	0.980
13	0.740	1.110	0.730	1.180	0.525	1.438	0.890	0.750	0.760	1.110	0.920	0.940	0.775	0.840	0.700	1.850
14	0.150	1.530	0.880	0.530	x	x	0.960	0.590	0.760	1.000	0.980	0.920	0.580	1.570	0.920	0.820
15	0.690	1.060	0.860	0.800	0.395	0.952	0.600	1.420	0.200	1.670	0.700	1.300	0.850	1.133	0.850	1.000
16	0.750	1.150	0.780	0.680	0.165	1.000	0.820	1.140	0.000	1.400	0.740	1.230	0.625	1.250	0.820	1.150
Mean	0.509	1.443	0.771	0.787	0.398	1.472	0.829	0.918	0.508	1.473	0.837	0.992	0.592	1.442	0.692	1.261
St. Dev.	0.216	0.645	0.089	0.175	0.162	0.713	0.116	0.231	0.251	0.666	0.117	0.221	0.183	0.394	0.244	0.485

x – No Crack

Appendix D

Crack Growth Rate Data from Open Hole Marker Load Investigation

The following eight crack growth rate curves, shown in Figure D.1, each have two sets of data. The open symbols are obtained by measuring the striation spacing from SEM fractographs. Five to seven striation measurements are used to calculate an average striation spacing and the number of cycles between striations is one for the CA loading. For the marker bands created using the single under-load spectrum, each marker band represents 50 cycles. The solid symbols are obtained from the crack length vs. cycles data measured during the fatigue test, and da/dN is calculated using the secant method in the following manner.

$$\bar{a} = \frac{1}{2}\left(a_{i+1} + a_i\right) \tag{D.1}$$

$$\left.\frac{da}{dN}\right|_{\bar{a}} = \frac{\left(a_{i+1} - a_i\right)}{\left(N_{i+1} - N_i\right)} \tag{D.2}$$

The computed da/dN in Eqn. (D.2) is an average rate and is plotted versus the average crack length given by Eqn. (D.1). Two representative fractographs for a small and large crack for each of the eight tests are shown in Figure D.2.

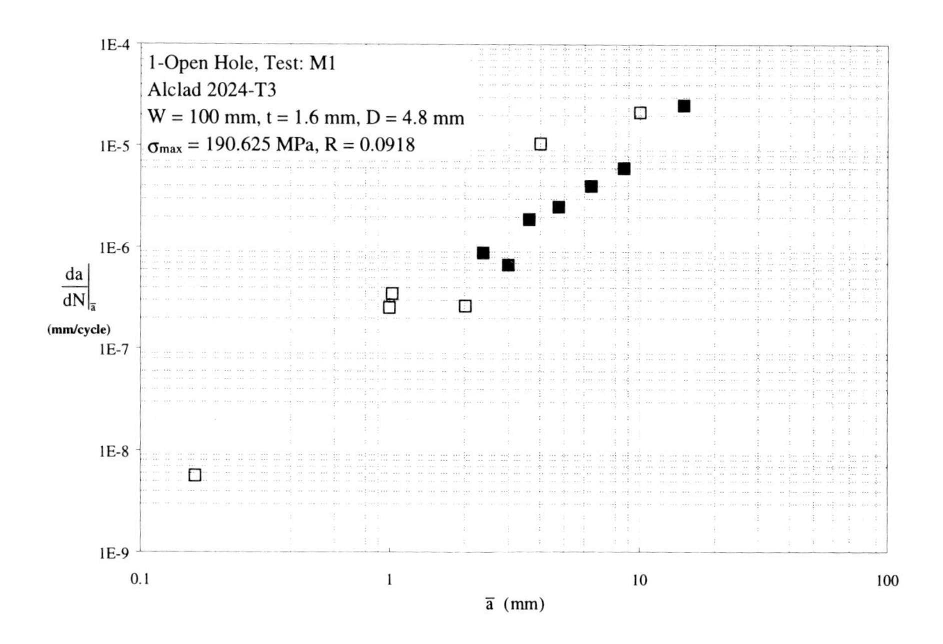

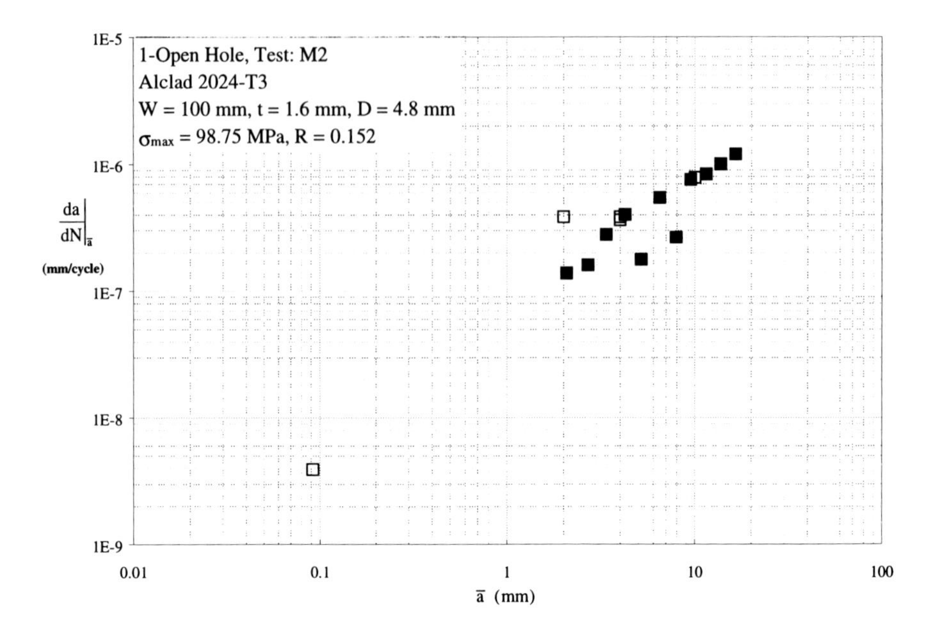
1-Open Hole, Test: M2
Alclad 2024-T3
W = 100 mm, t = 1.6 mm, D = 4.8 mm
σmax = 98.75 MPa, R = 0.152
da/dN|ā
(mm/cycle)
1E-5
1E-6
1E-7
1E-8
1E-9
0.01
0.1
1
10
100
ā (mm)

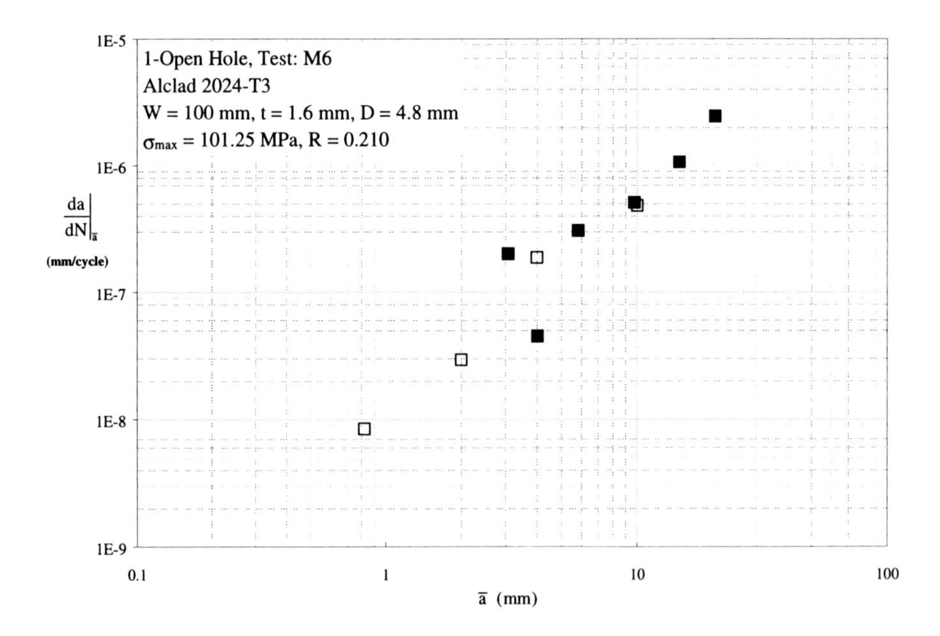
1-Open Hole, Test: M6
Alclad 2024-T3
W = 100 mm, t = 1.6 mm, D = 4.8 mm
σmax = 101.25 MPa, R = 0.210
da/dN|ā
(mm/cycle)
1E-5
1E-6
1E-7
1E-8
1E-9
0.1
1
10
100
ā (mm)

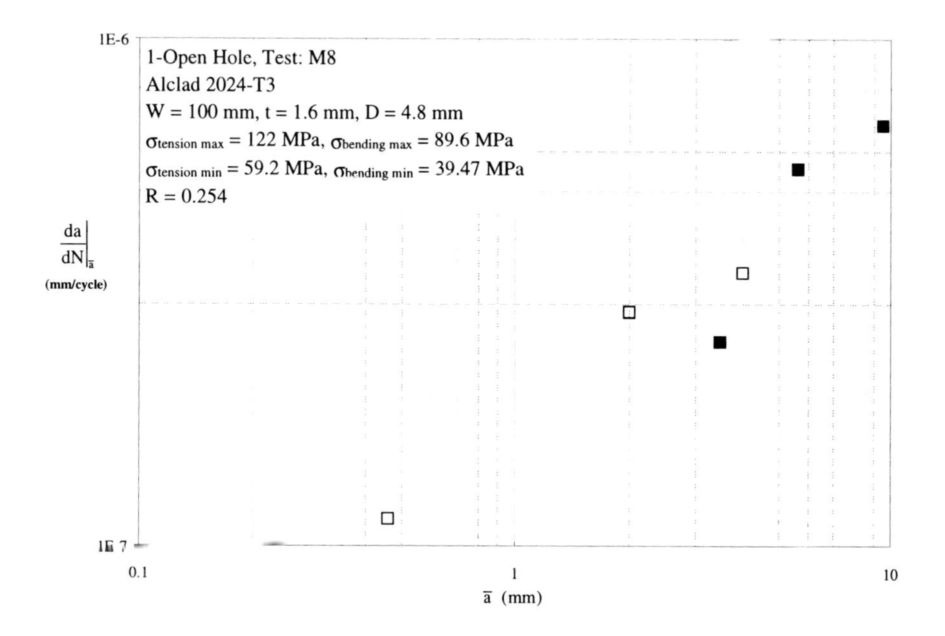
1-Open Hole, Test: M8
Alclad 2024-T3
W = 100 mm, t = 1.6 mm, D = 4.8 mm
σtension max = 122 MPa, σbending max = 89.6 MPa
σtension min = 59.2 MPa, σbending min = 39.47 MPa
R = 0.254
1E-6
1E-7
da/dN|ā
(mm/cycle)
0.1
1
10
ā (mm)

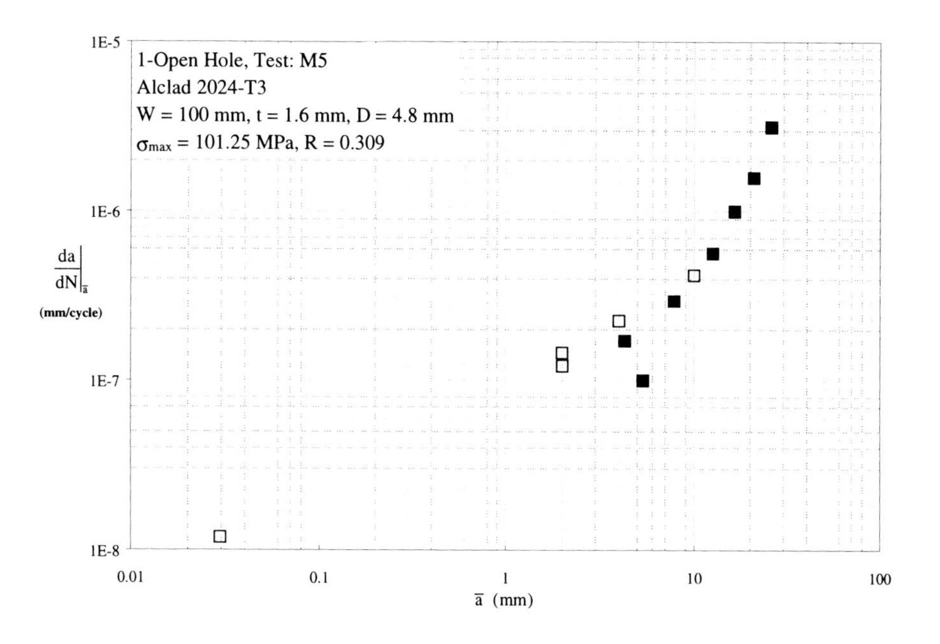
1-Open Hole, Test: M5
Alclad 2024-T3
W = 100 mm, t = 1.6 mm, D = 4.8 mm
σmax = 101.25 MPa, R = 0.309
1E-5
1E-6
1E-7
1E-8
da/dN|ā
(mm/cycle)
0.01
0.1
1
10
100
ā (mm)

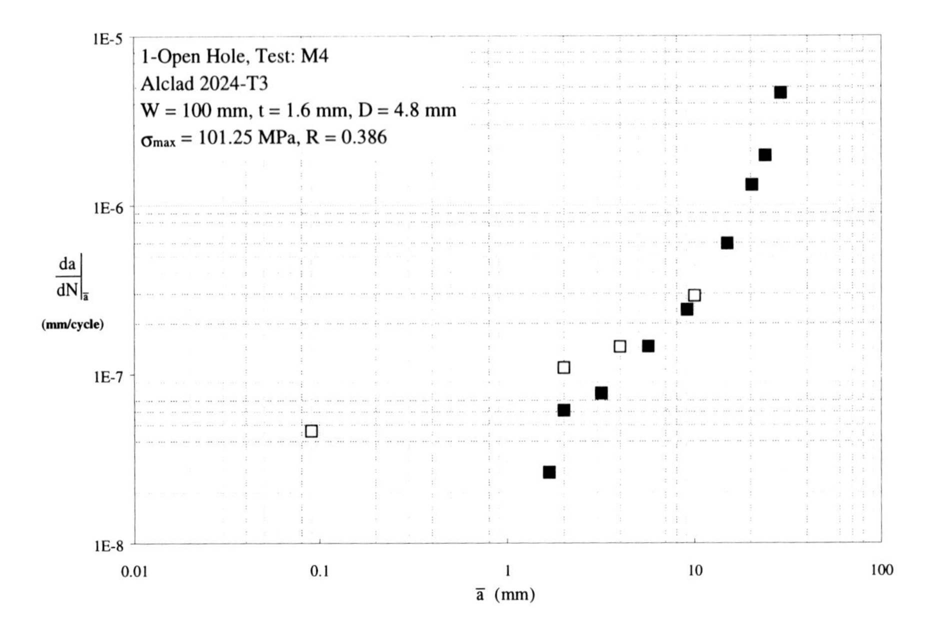
1-Open Hole, Test: M4
Alclad 2024-T3
W = 100 mm, t = 1.6 mm, D = 4.8 mm
σmax = 101.25 MPa, R = 0.386
1E-5
1E-6
1E-7
1E-8
da/dN|ā
(mm/cycle)
0.01
0.1
1
10
100
ā (mm)

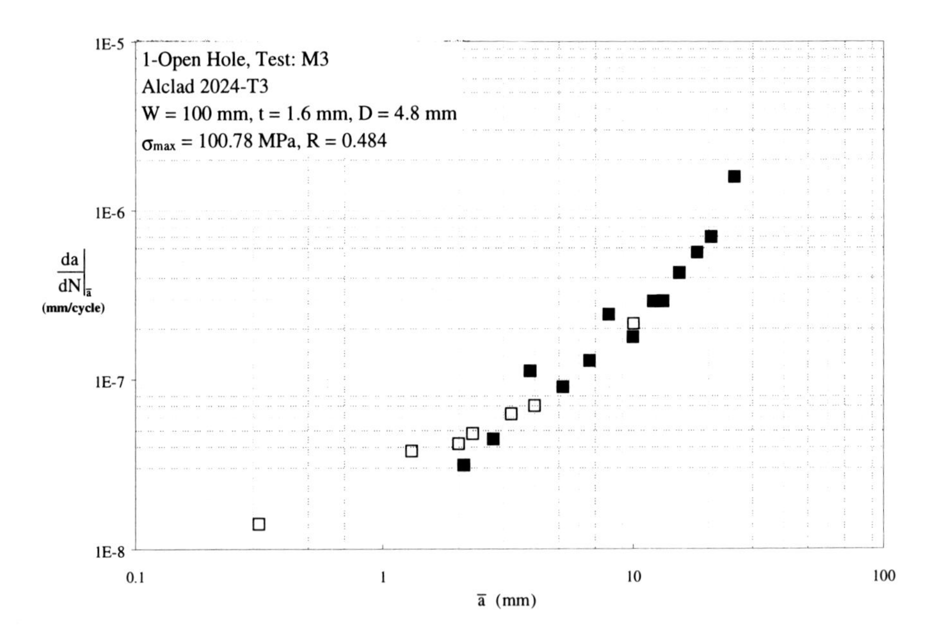
1-Open Hole, Test: M3
Alclad 2024-T3
W = 100 mm, t = 1.6 mm, D = 4.8 mm
σmax = 100.78 MPa, R = 0.484
1E-5
1E-6
1E-7
1E-8
da/dN|ā
(mm/cycle)
0.1
1
10
100
ā (mm)

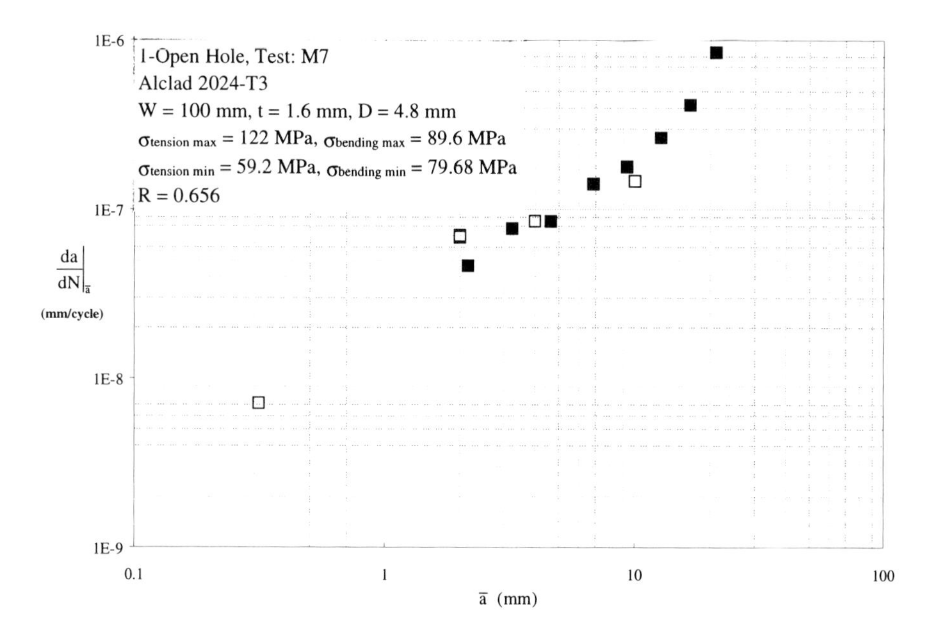

Figure D.1 Crack Growth Rates for 1-Open Hole Tension and Combined Tension and Bending Marker Load Test Specimens

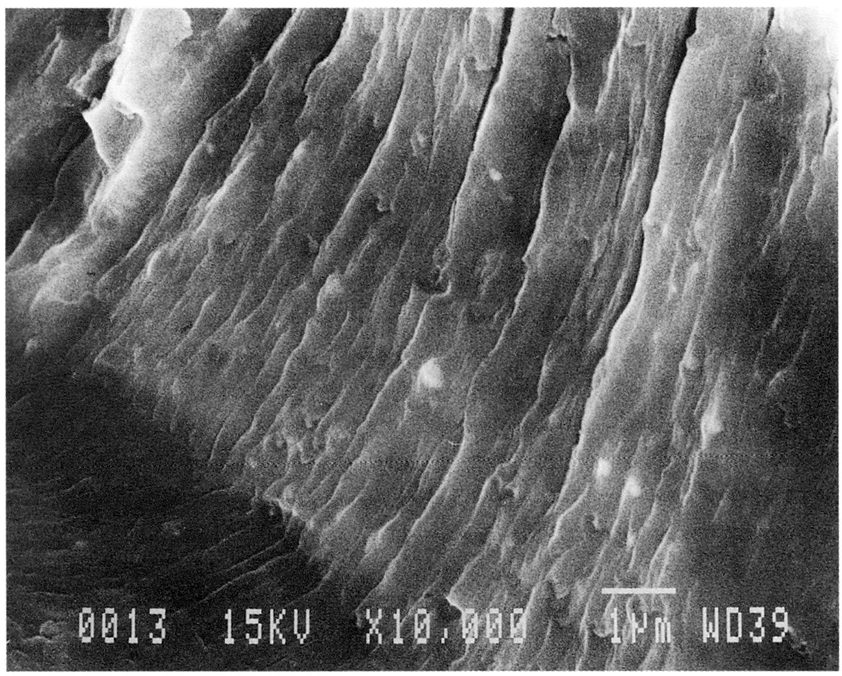

1-Open Hole, Alclad 2024-T3, Test: M1, W = 100 mm, t = 1.6 mm, D = 4.8 mm, $\sigma_{tension\ max}$ = 190.625 MPa, R = 0.0918, c_1 = 0.166 mm

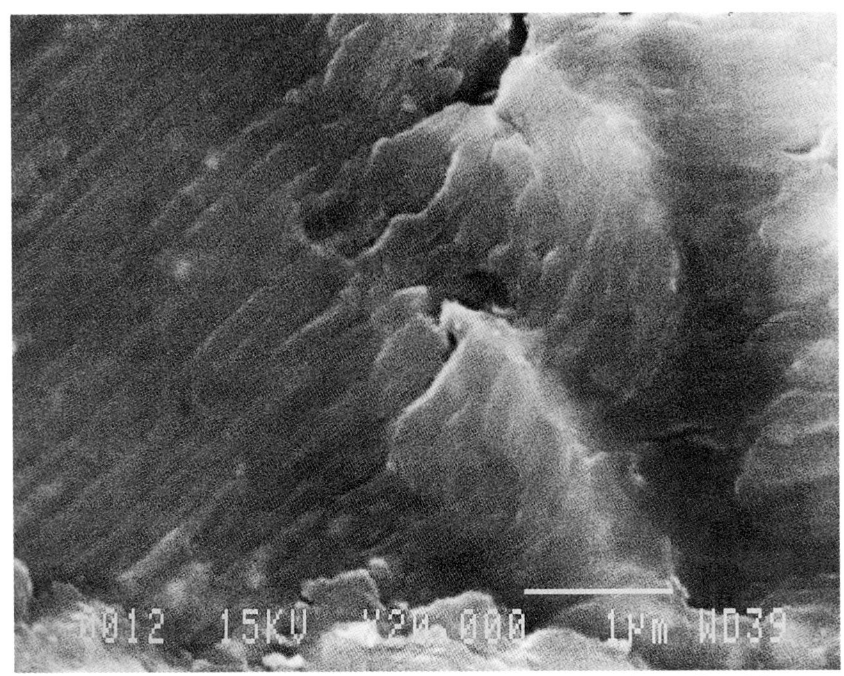

1-Open Hole, Alclad 2024-T3, Test: M1, W = 100 mm, t = 1.6 mm, D = 4.8 mm
$\sigma_{tension\ max}$ = 190.625 MPa, R = 0.0918, c_1 = 1.0 mm

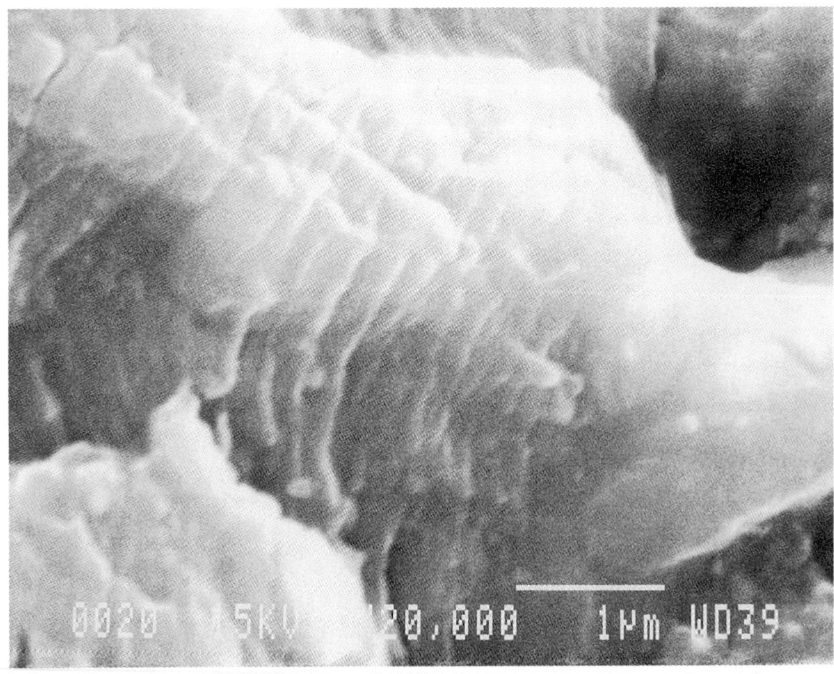

1-Open Hole, Alclad 2024-T3, Test: M2, W = 100 mm, t = 1.6 mm, D = 4.8 mm
$\sigma_{tension\ max}$ = 98.75 MPa, R = 0.152, c_1 = 2.0 mm

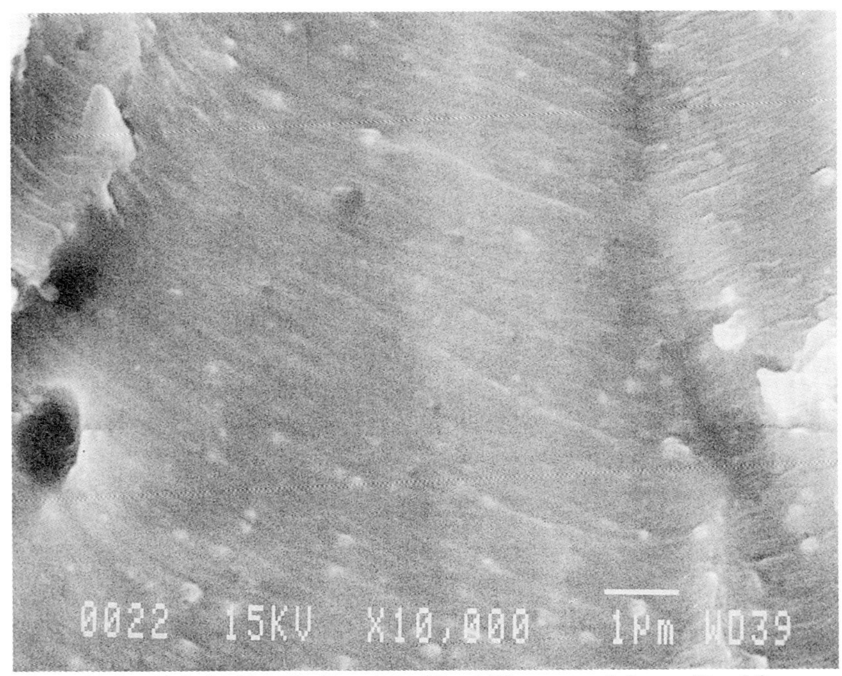

1-Open Hole, Alclad 2024-T3, Test: M2, W = 100 mm, t = 1.6 mm, D = 4.8 mm
$\sigma_{tension\ max}$ = 98.75 MPa, R = 0.152, c_1 = 2.0 mm

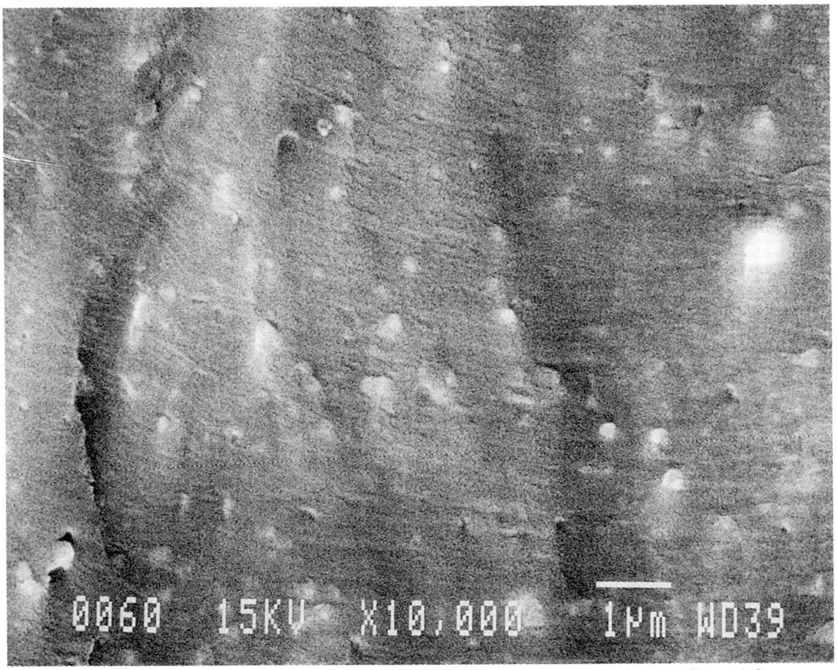

1-Open Hole, Alclad 2024-T3, Test: M6, W = 100 mm, t = 1.6 mm, D = 4.8 mm
$\sigma_{tension\ max}$ = 101.25 MPa, R = 0.210, c_1 = 0.821 mm

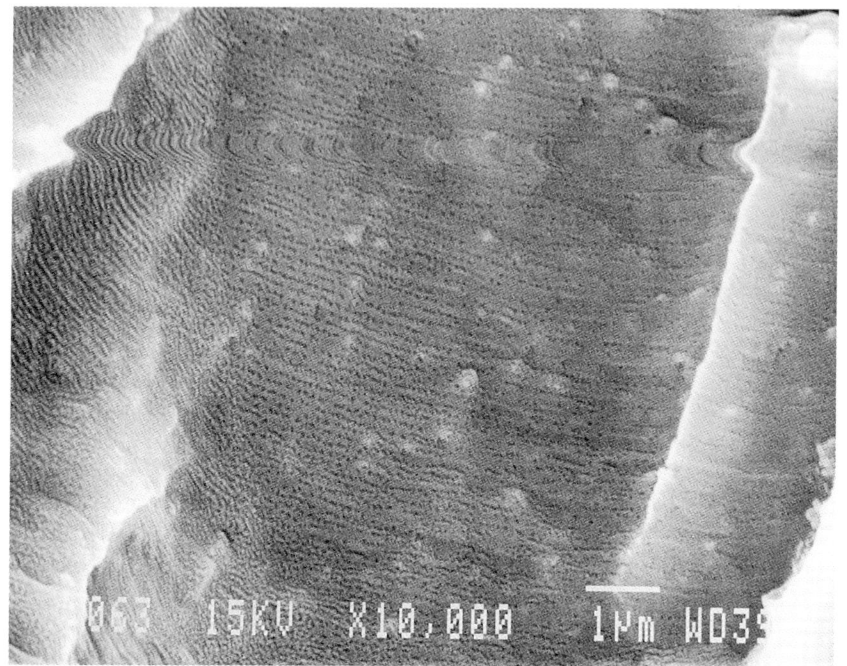

1-Open Hole, Alclad 2024-T3, Test: M6, W = 100 mm, t = 1.6 mm, D = 4.8 mm
$\sigma_{tension\ max}$ = 101.25 MPa, R = 0.210, c_1 = 2.0 mm

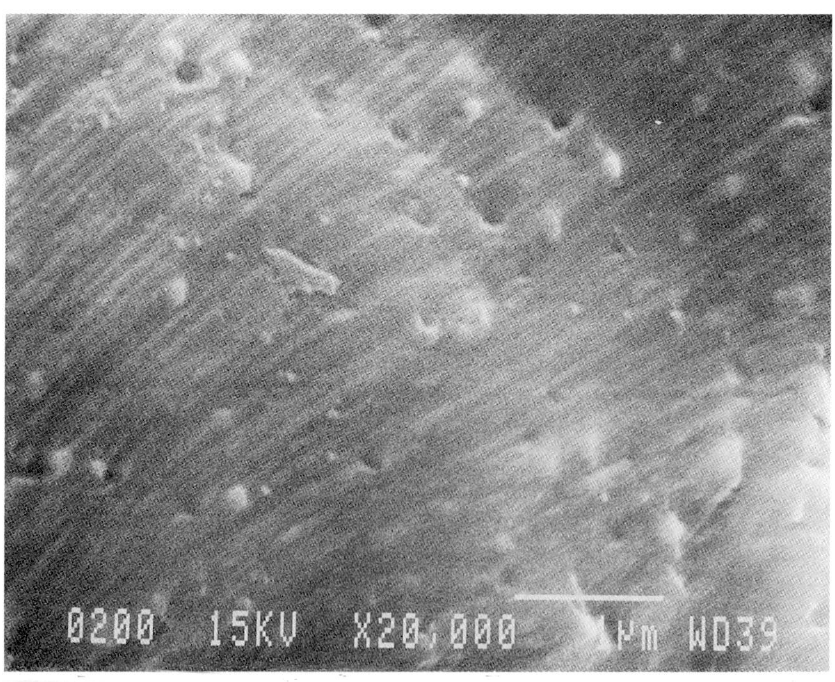

1-Open Hole, Alclad 2024-T3, Test: M8, W = 100 mm, t = 1.6 mm, D = 4.8 mm
$\sigma_{tension\ max}$ = 122 MPa, $\sigma_{bending\ max}$ = 89.6 MPa, $\sigma_{tension\ min}$ = 59.2 MPa,
$\sigma_{bending\ min}$ = 39.47 MPa, R = 0.254, c_1 = 0.459 mm

1-Open Hole, Alclad 2024-T3, Test: M8, W = 100 mm, t = 1.6 mm, D = 4.8 mm
$\sigma_{tension\ max}$ = 122 MPa, $\sigma_{bending\ max}$ = 89.6 MPa, $\sigma_{tension\ min}$ = 59.2 MPa,
$\sigma_{bending\ min}$ = 39.47 MPa, R = 0.254, c_1 = 2.0 mm

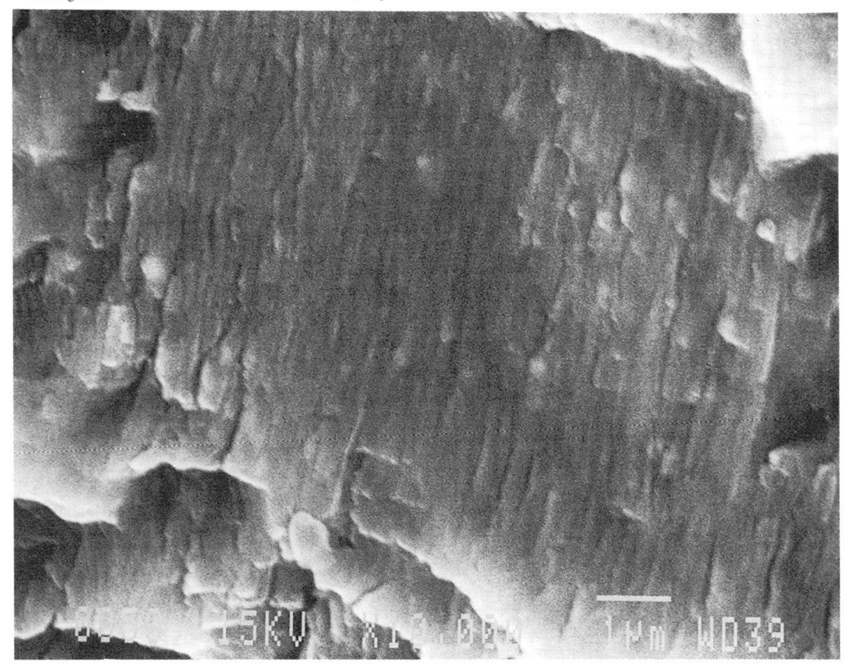

1-Open Hole, Alclad 2024-T3, Test: M5, W = 100 mm, t = 1.6 mm, D = 4.8 mm
$\sigma_{tension\ max}$ = 101.25 MPa, R = 0.309, c_1 = 0.089 mm

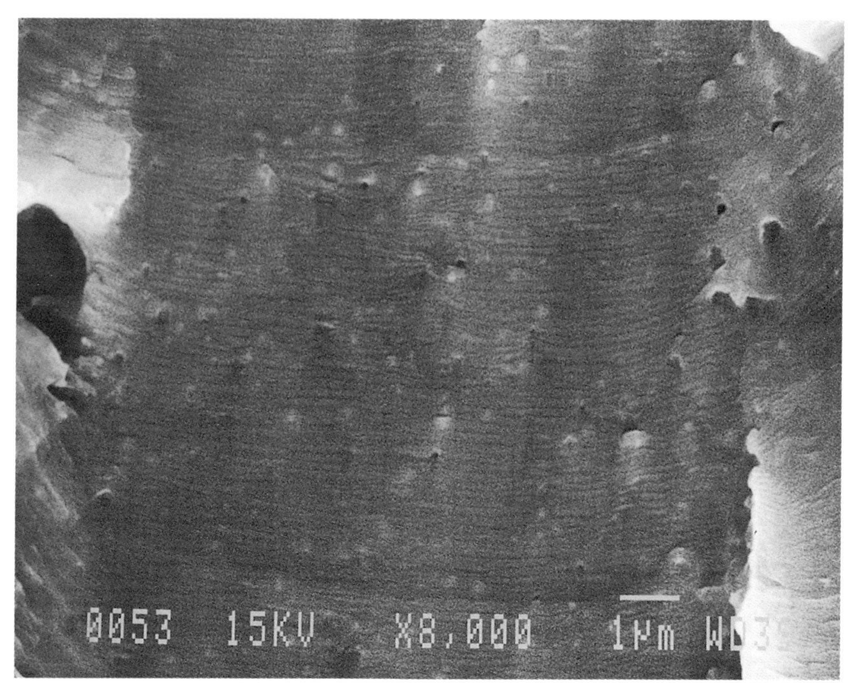

1-Open Hole, Alclad 2024-T3, Test: M5, W = 100 mm, t = 1.6 mm, D = 4.8 mm
$\sigma_{\text{tension max}}$ = 101.25 MPa, R = 0.309

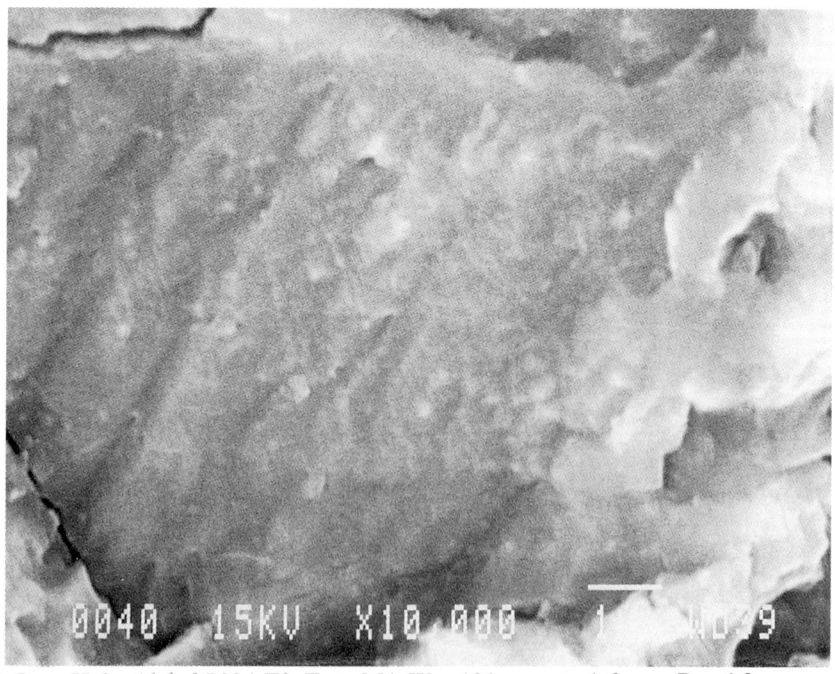

1-Open Hole, Alclad 2024-T3, Test: M4, W = 100 mm, t = 1.6 mm, D = 4.8 mm
$\sigma_{\text{tension max}}$ = 101.25 MPa, R = 0.386, c_1 = 0.089 mm

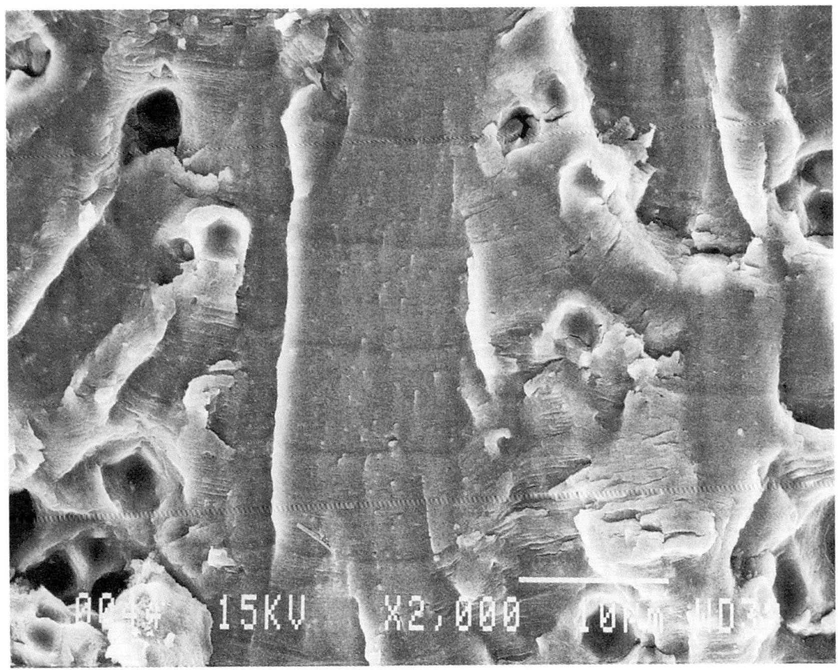

1-Open Hole, Alclad 2024-T3, Test: M4, W = 100 mm, t = 1.6 mm, D = 4.8 mm
$\sigma_{tension\ max}$ = 101.25 MPa, R = 0.386, c_1 = 4.0 mm

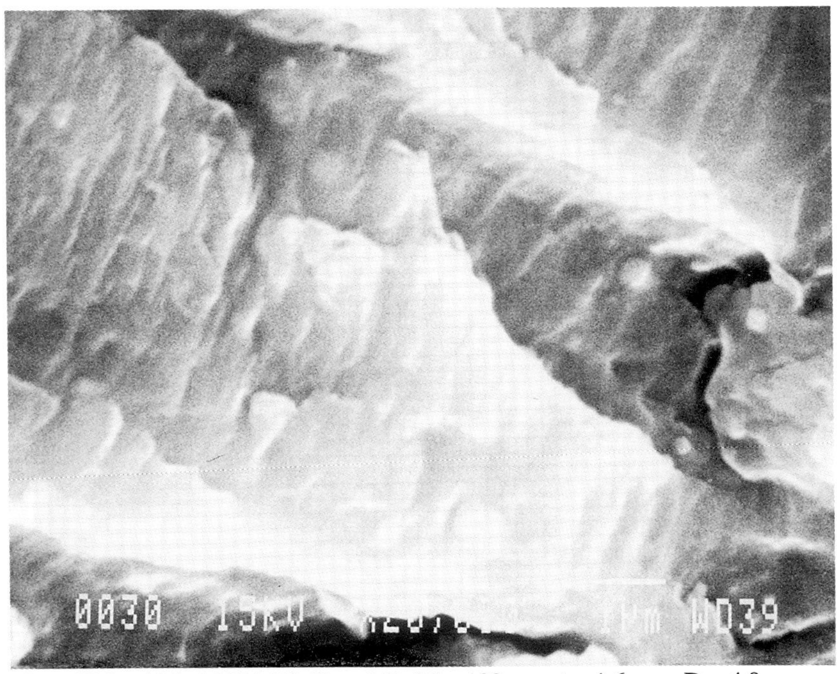

1-Open Hole, Alclad 2024-T3, Test: M3, W = 100 mm, t = 1.6 mm, D = 4.8 mm
$\sigma_{tension\ max}$ = 100.78 MPa, R = 0.484, c_1 = 0.318 mm

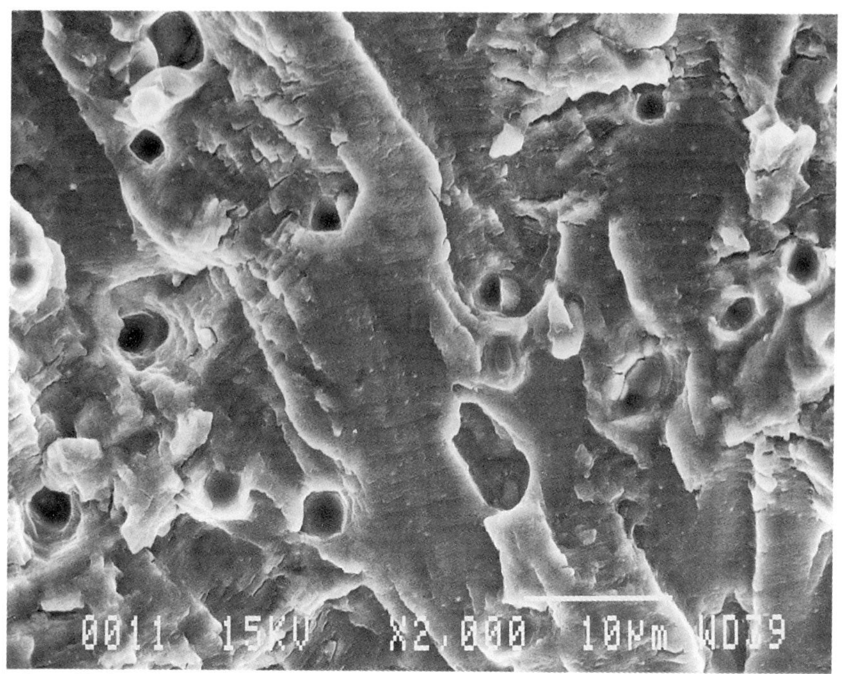

1-Open Hole, Alclad 2024-T3, Test: M3, W = 100 mm, t = 1.6 mm, D = 4.8 mm
$\sigma_{tension\ max}$ = 100.78 MPa, R = 0.484, c_1 = 1.3 mm

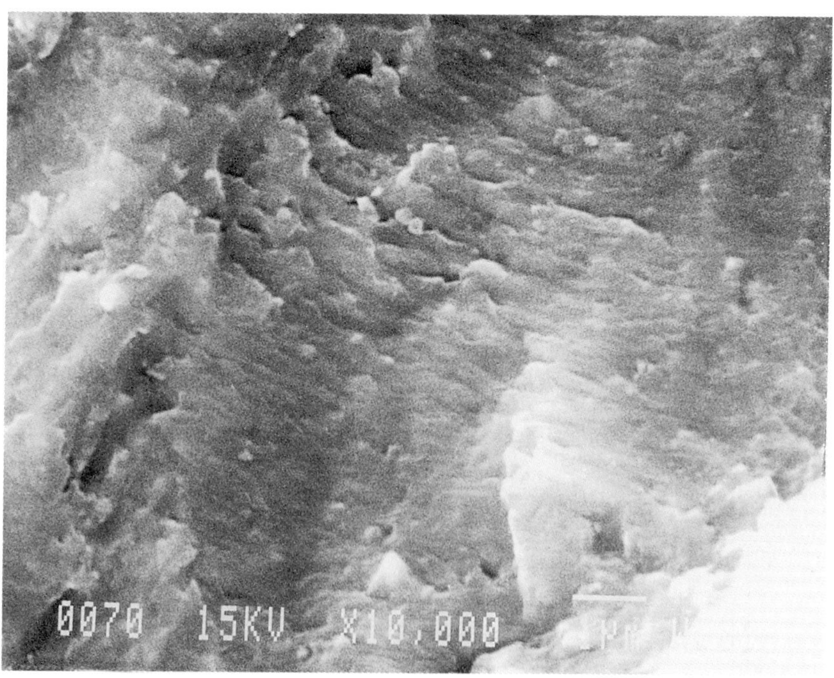

1-Open Hole, Alclad 2024-T3, Test: M7, W = 100 mm, t = 1.6 mm, D = 4.8 mm
$\sigma_{tension\ max}$ = 122 MPa, $\sigma_{bending\ max}$ = 89.6 MPa, $\sigma_{tension\ min}$ = 59.2 MPa,
$\sigma_{bending\ min}$ = 79.68 MPa, R = 0.656, c_1 = 0.313 mm

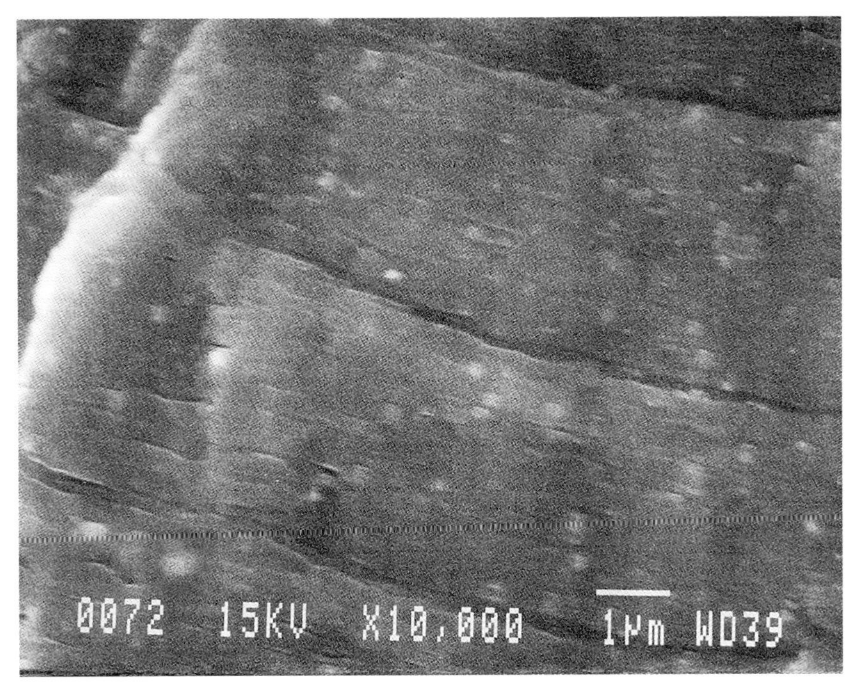

1-Open Hole, Alclad 2024-T3, Test: M7, W = 100 mm, t = 1.6 mm, D = 4.8 mm
$\sigma_{tension\ max}$ = 122 MPa, $\sigma_{bending\ max}$ = 89.6 MPa, $\sigma_{tension\ min}$ = 59.2 MPa,
$\sigma_{bending\ min}$ = 79.68 MPa, R = 0.656, c_1 = 2.0 mm

Figure D.2 Fatigue Striations and Marker Bands for 1-Open Hole Tension and Combined Tension and Bending Test Specimens

This page intentionally left blank

Appendix E

Stress Intensity Factor Solutions for Diametrically Opposed Part Elliptical, Oblique Through Cracks from a Centrally Located Hole in a Finite Width Sheet Subject to General Loading

Remote Tension

		a/c_1	0.2	0.3	0.4	0.6	1	2
a/t	r/t	z/t	β					
1.05	0.5	0.014	1.501	1.310	1.261	1.199	1.070	0.910
		0.058	1.368	1.354	1.324	1.273	1.151	0.998
		0.154	1.392	1.342	1.302	1.240	1.127	0.997
		0.360	1.477	1.372	1.305	1.205	1.084	0.975
		0.641	1.569	1.450	1.352	1.209	1.066	0.959
		0.846	1.724	1.679	1.531	1.336	1.145	0.993
		0.942	2.132	2.055	1.883	1.637	1.377	1.122
		0.986	5.062	3.991	3.431	2.787	2.145	1.467
	1	0.014	1.258	1.335	1.352	1.334	1.257	1.160
		0.058	1.303	1.396	1.420	1.416	1.356	1.263
		0.154	1.351	1.397	1.398	1.380	1.330	1.258
		0.360	1.472	1.450	1.409	1.349	1.288	1.232
		0.641	1.657	1.569	1.487	1.376	1.284	1.209
		0.846	2.068	1.874	1.732	1.559	1.405	1.244
		0.942	2.586	2.377	2.187	1.965	1.736	1.409
		0.986	5.699	4.676	4.118	3.450	2.715	1.780
	2	0.014	1.363	1.485	1.553	1.604	1.618	1.389
		0.058	1.410	1.542	1.624	1.700	1.730	1.529
		0.154	1.459	1.539	1.596	1.656	1.684	1.520
		0.360	1.598	1.609	1.618	1.624	1.622	1.483
		0.641	1.822	1.773	1.732	1.670	1.605	1.455
		0.846	2.324	2.169	2.061	1.915	1.750	1.498
		0.942	2.976	2.794	2.664	2.473	2.187	1.681
		0.986	6.821	5.884	5.265	4.471	3.324	2.085
1.09	0.5	0.014	1.527	1.429	1.276	1.203	1.063	0.899
		0.058	1.398	1.356	1.343	1.280	1.147	0.988
		0.154	1.422	1.341	1.321	1.249	1.126	0.989
		0.360	1.514	1.371	1.326	1.218	1.090	0.972
		0.641	1.638	1.446	1.386	1.232	1.081	0.963
		0.846	1.877	1.662	1.591	1.375	1.169	1.002
		0.942	2.472	2.171	2.016	1.713	1.411	1.126
		0.986	5.301	4.135	3.443	2.737	2.039	1.373
	1	0.014	1.296	1.362	1.367	1.332	1.244	1.135
		0.058	1.344	1.426	1.439	1.417	1.345	1.241
		0.154	1.388	1.425	1.417	1.385	1.324	1.241
		0.360	1.514	1.481	1.430	1.359	1.289	1.222
		0.641	1.727	1.614	1.519	1.397	1.297	1.211
		0.846	2.175	1.943	1.784	1.595	1.429	1.260
		0.942	2.845	2.539	2.311	2.044	1.767	1.421
		0.986	5.805	4.653	4.034	3.302	2.533	1.679
	2	0.014	1.401	1.510	1.566	1.595	1.584	1.377
		0.058	1.450	1.571	1.642	1.696	1.702	1.519
		0.154	1.497	1.567	1.615	1.656	1.665	1.516
		0.360	1.641	1.640	1.638	1.631	1.617	1.487
		0.641	1.892	1.819	1.765	1.693	1.622	1.473
		0.846	2.429	2.238	2.115	1.964	1.793	1.534
		0.942	3.258	2.992	2.817	2.576	2.236	1.711
		0.986	6.810	5.723	5.036	4.188	3.095	1.995

		a/c_1	0.2	0.3	0.4	0.6	1	2
a/t	r/t	z/t	β					
1.13	0.5	0.014	1.750	1.512	1.282	1.203	1.058	0.889
		0.058	1.380	1.540	1.352	1.282	1.143	0.979
		0.154	1.407	1.517	1.330	1.254	1.125	0.982
		0.360	1.508	1.539	1.337	1.226	1.093	0.968
		0.641	1.650	1.620	1.410	1.249	1.092	0.964
		0.846	1.977	1.851	1.638	1.401	1.185	1.004
		0.942	2.637	2.274	2.102	1.753	1.424	1.118
		0.986	5.275	4.104	3.437	2.684	1.960	1.306
	1	0.014	1.318	1.380	1.380	1.330	1.231	1.114
		0.058	1.369	1.446	1.455	1.418	1.335	1.222
		0.154	1.413	1.445	1.432	1.388	1.317	1.225
		0.360	1.548	1.501	1.446	1.365	1.288	1.211
		0.641	1.781	1.644	1.541	1.411	1.304	1.208
		0.846	2.243	1.993	1.820	1.618	1.441	1.263
		0.942	2.974	2.635	2.379	2.077	1.768	1.413
		0.986	5.799	4.609	3.956	3.183	2.400	1.606
	2	0.014	1.431	1.530	1.573	1.585	1.554	1.358
		0.058	1.482	1.594	1.653	1.690	1.676	1.503
		0.154	1.526	1.589	1.627	1.654	1.647	1.503
		0.360	1.674	1.662	1.651	1.633	1.609	1.481
		0.641	1.941	1.850	1.785	1.705	1.630	1.477
		0.846	2.505	2.285	2.149	1.990	1.813	1.545
		0.942	3.428	3.096	2.885	2.605	2.236	1.715
		0.986	6.752	5.571	4.852	3.965	2.931	1.915
1.17	0.5	0.014	1.776	1.456	1.310	1.196	1.053	0.880
		0.058	1.408	1.396	1.379	1.280	1.140	0.970
		0.154	1.432	1.381	1.354	1.251	1.125	0.975
		0.360	1.535	1.415	1.356	1.223	1.096	0.964
		0.641	1.696	1.518	1.427	1.247	1.100	0.962
		0.846	2.071	1.804	1.658	1.399	1.195	1.003
		0.942	2.760	2.365	2.148	1.735	1.427	1.107
		0.986	5.367	4.103	3.378	2.571	1.897	1.255
	1	0.014	1.533	1.385	1.389	1.325	1.220	1.095
		0.058	1.564	1.456	1.467	1.416	1.325	1.205
		0.154	1.587	1.450	1.445	1.388	1.310	1.211
		0.360	1.692	1.498	1.459	1.368	1.286	1.201
		0.641	1.893	1.633	1.559	1.420	1.307	1.204
		0.846	2.343	1.963	1.847	1.633	1.445	1.261
		0.942	2.969	2.532	2.421	2.090	1.756	1.399
		0.986	5.915	4.360	3.885	3.083	2.294	1.547
	2	0.014	1.457	1.543	1.579	1.576	1.528	1.344
		0.058	1.510	1.611	1.662	1.684	1.654	1.490
		0.154	1.552	1.605	1.636	1.651	1.631	1.494
		0.360	1.702	1.677	1.660	1.634	1.601	1.477
		0.641	1.981	1.872	1.798	1.713	1.632	1.480
		0.846	2.564	2.319	2.172	2.003	1.819	1.551
		0.942	3.541	3.156	2.916	2.605	2.218	1.708
		0.986	6.680	5.437	4.692	3.790	2.801	1.856

		a/c_1	0.2	0.3	0.4	0.6	1	2
a/t	r/t	z/t	β					
1.21	0.5	0.014	1.802	1.469	1.319	1.204	1.047	0.872
		0.058	1.434	1.414	1.392	1.288	1.136	0.962
		0.154	1.457	1.399	1.367	1.262	1.123	0.968
		0.360	1.560	1.434	1.370	1.239	1.098	0.959
		0.641	1.737	1.546	1.447	1.272	1.105	0.960
		0.846	2.148	1.854	1.689	1.440	1.202	1.000
		0.942	2.849	2.418	2.190	1.811	1.426	1.095
		0.986	5.418	4.070	3.342	2.565	1.843	1.213
	1	0.014	1.416	1.420	1.397	1.321	1.209	1.079
		0.058	1.434	1.485	1.478	1.414	1.316	1.190
		0.154	1.463	1.477	1.456	1.389	1.304	1.198
		0.360	1.588	1.532	1.470	1.371	1.283	1.191
		0.641	1.828	1.687	1.574	1.427	1.308	1.198
		0.846	2.308	2.055	1.869	1.643	1.446	1.255
		0.942	3.205	2.753	2.447	2.092	1.740	1.382
		0.986	5.830	4.517	3.818	2.994	2.206	1.498
	2	0.014	1.480	1.556	1.583	1.567	1.491	1.330
		0.058	1.535	1.627	1.670	1.678	1.627	1.477
		0.154	1.574	1.619	1.644	1.648	1.613	1.484
		0.360	1.725	1.690	1.668	1.634	1.590	1.472
		0.641	2.014	1.888	1.808	1.716	1.631	1.479
		0.846	2.611	2.344	2.187	2.008	1.821	1.551
		0.942	3.621	3.192	2.927	2.591	2.195	1.596
		0.986	6.609	5.314	4.550	3.643	2.725	1.806
2	0.5	0.014	2.156	1.779	1.432	1.204	0.985	0.779
		0.058	1.818	1.640	1.514	1.317	1.089	0.867
		0.154	1.802	1.620	1.501	1.315	1.098	0.882
		0.360	1.865	1.644	1.515	1.314	1.097	0.886
		0.641	2.109	1.791	1.620	1.366	1.122	0.893
		0.846	2.570	2.083	1.842	1.509	1.192	0.914
		0.942	3.039	2.381	2.092	1.731	1.302	0.944
		0.986	4.910	3.600	2.603	1.923	1.359	0.921
	1	0.014	2.083	1.863	1.442	1.271	1.085	0.916
		0.058	1.785	1.974	1.559	1.389	1.201	1.025
		0.154	1.782	1.953	1.547	1.389	1.212	1.045
		0.360	1.873	1.980	1.558	1.392	1.216	1.054
		0.641	2.162	2.157	1.669	1.456	1.249	1.067
		0.846	2.669	2.554	1.944	1.621	1.332	1.055
		0.942	3.170	3.092	2.372	1.883	1.458	1.130
		0.986	5.365	3.473	2.879	2.157	1.523	1.088
	2	0.014	1.767	1.660	1.564	1.427	1.279	1.191
		0.058	1.838	1.775	1.691	1.566	1.424	1.322
		0.154	1.835	1.760	1.679	1.566	1.439	1.339
		0.360	1.936	1.797	1.694	1.573	1.447	1.346
		0.641	2.237	1.978	1.823	1.653	1.492	1.362
		0.846	2.850	2.385	2.130	1.850	1.597	1.396
		0.942	3.771	3.035	2.606	2.140	1.737	1.434
		0.986	5.325	3.946	3.169	2.387	1.777	1.368

		a/c_1	0.2	0.3	0.4	0.6	1	2
a/t	r/t	z/t	β					
5	0.5	0.014	2.504	1.713	1.441	1.138	0.883	0.691
		0.058	2.400	1.850	1.620	1.347	1.052	0.778
		0.154	2.353	1.847	1.625	1.354	1.070	0.798
		0.360	2.319	1.863	1.639	1.365	1.084	0.807
		0.641	2.372	1.932	1.686	1.391	1.097	0.812
		0.846	2.523	2.043	1.755	1.425	1.110	0.814
		0.942	2.670	2.137	1.811	1.451	1.114	0.809
		0.986	3.219	2.432	1.921	1.395	0.989	0.738
	1	0.014	2.604	1.952	1.563	1.169	0.964	0.748
		0.058	2.154	1.842	1.622	1.375	1.090	0.843
		0.154	2.151	1.841	1.624	1.384	1.117	0.866
		0.360	2.185	1.859	1.639	1.398	1.130	0.877
		0.641	2.298	1.920	1.682	1.427	1.147	0.883
		0.846	2.458	2.005	1.741	1.466	1.170	0.886
		0.942	2.582	2.070	1.786	1.496	1.185	0.879
		0.986	4.168	2.860	2.212	1.407	1.099	0.801
	2	0.014	3.151	1.950	1.472	1.247	1.046	0.859
		0.058	2.067	1.882	1.672	1.419	1.183	0.976
		0.154	2.073	1.882	1.690	1.447	1.212	1.003
		0.360	2.113	1.905	1.713	1.467	1.228	1.017
		0.641	2.209	1.977	1.777	1.505	1.247	1.024
		0.846	2.331	2.077	1.876	1.566	1.272	1.027
		0.942	2.420	2.156	1.964	1.616	1.289	1.018
		0.986	5.342	2.835	1.971	1.531	1.194	0.916
10	0.5	0.014	2.612	1.721	1.533	1.089	0.832	0.617
		0.058	2.513	1.913	1.815	1.364	1.060	0.751
		0.154	2.504	1.916	1.819	1.376	1.071	0.768
		0.360	2.506	1.929	1.821	1.389	1.081	0.778
		0.641	2.545	1.959	1.832	1.402	1.087	0.780
		0.846	2.609	1.996	1.851	1.412	1.088	0.775
		0.942	2.661	2.023	1.864	1.416	1.084	0.762
		0.986	3.218	2.215	1.673	1.216	0.892	0.635
	1	0.014	2.724	1.987	1.430	1.113	0.834	0.685
		0.058	2.253	1.901	1.663	1.378	1.073	0.780
		0.154	2.255	1.903	1.667	1.383	1.085	0.803
		0.360	2.270	1.912	1.678	1.390	1.096	0.814
		0.641	2.310	1.933	1.698	1.401	1.102	0.817
		0.846	2.360	1.960	1.720	1.412	1.103	0.813
		0.942	2.395	1.978	1.735	1.418	1.099	0.797
		0.986	3.959	2.724	1.762	1.297	0.900	0.707
	2	0.014	*	*	1.562	1.142	0.950	0.743
		0.058	*	*	1.658	1.407	1.101	0.847
		0.154	*	*	1.660	1.412	1.131	0.873
		0.360	*	*	1.667	1.421	1.147	0.885
		0.641	*	*	1.683	1.434	1.156	0.889
		0.846	*	*	1.702	1.448	1.159	0.884
		0.942	*	*	1.715	1.456	1.147	0.866
		0.986	*	*	2.097	1.304	1.012	0.768

* Crack length outside FEM

Remote Bending

a/t	r/t	a/c₁ → z/t ↓	0.2	0.3	0.4	0.6	1	2
			β					
1.05	0.5	0.014	0.646	0.569	0.556	0.545	0.509	0.458
		0.058	0.549	0.568	0.562	0.550	0.512	0.457
		0.154	0.518	0.511	0.497	0.473	0.427	0.370
		0.360	0.443	0.401	0.371	0.329	0.272	0.218
		0.641	0.299	0.226	0.188	0.148	0.124	-0.131
		0.846	0.143	-0.130	-0.168	-0.209	-0.246	-0.272
		0.942	-0.052	-0.324	-0.407	-0.455	-0.461	-0.432
		0.986	-1.092	-1.124	-1.122	-1.051	-0.895	-0.660
	1	0.014	0.243	0.259	0.267	0.266	0.250	0.235
		0.058	0.243	0.262	0.269	0.268	0.251	0.233
		0.154	0.228	0.237	0.238	0.230	0.210	0.188
		0.360	0.191	0.186	0.178	0.160	0.134	0.110
		0.641	0.118	0.103	0.090	0.072	0.061	-0.066
		0.846	0.055	-0.069	-0.082	-0.101	-0.120	-0.134
		0.942	-0.105	-0.171	-0.196	-0.219	-0.227	-0.212
		0.986	-0.466	-0.529	-0.539	-0.513	-0.436	-0.310
	2	0.014	0.095	0.109	0.111	0.108	0.107	0.095
		0.058	0.095	0.110	0.111	0.108	0.106	0.094
		0.154	0.089	0.099	0.098	0.093	0.088	0.076
		0.360	0.074	0.077	0.073	0.064	0.056	0.045
		0.641	0.046	0.043	0.037	0.029	0.025	-0.026
		0.846	0.022	-0.026	-0.031	-0.039	-0.048	-0.053
		0.942	-0.043	-0.063	-0.075	-0.085	-0.090	-0.082
		0.986	-0.185	-0.215	-0.217	-0.204	-0.168	-0.120
1.09	0.5	0.014	0.658	0.635	0.562	0.547	0.510	0.455
		0.058	0.560	0.556	0.568	0.553	0.512	0.454
		0.154	0.527	0.500	0.500	0.474	0.425	0.367
		0.360	0.448	0.390	0.369	0.326	0.269	0.215
		0.641	0.297	0.220	0.184	0.147	-0.127	-0.136
		0.846	0.129	-0.110	-0.190	-0.231	-0.265	-0.284
		0.942	-0.154	-0.305	-0.470	-0.501	-0.493	-0.447
		0.986	-1.227	-1.264	-1.152	-1.056	-0.878	-0.639
	1	0.014	0.248	0.263	0.269	0.266	0.250	0.232
		0.058	0.249	0.266	0.272	0.268	0.251	0.230
		0.154	0.232	0.240	0.239	0.229	0.209	0.185
		0.360	0.193	0.187	0.177	0.157	0.132	0.108
		0.641	0.118	0.102	0.088	0.071	-0.063	-0.068
		0.846	0.058	-0.078	-0.092	-0.111	-0.130	-0.141
		0.942	-0.137	-0.203	-0.226	-0.243	-0.244	-0.221
		0.986	-0.501	-0.554	-0.552	-0.512	-0.427	-0.303
	2	0.014	0.097	0.110	0.111	0.108	0.106	0.119
		0.058	0.097	0.110	0.111	0.108	0.105	0.086
		0.154	0.091	0.099	0.098	0.092	0.087	0.055
		0.360	0.075	0.077	0.072	0.063	0.054	0.027
		0.641	0.046	0.042	0.035	0.028	-0.026	-0.044
		0.846	-0.024	-0.030	-0.036	-0.044	-0.053	-0.075
		0.942	-0.056	-0.078	-0.088	-0.096	-0.098	-0.093
		0.986	-0.199	-0.223	-0.222	-0.204	-0.165	-0.094

a/t	r/t	a/c₁ → z/t ↓	0.2	0.3	0.4	0.6	1	2
			β					
1.13	0.5	0.014	0.791	0.671	0.563	0.548	0.509	0.453
		0.058	0.564	0.651	0.569	0.553	0.511	0.451
		0.154	0.526	0.575	0.500	0.473	0.423	0.364
		0.360	0.448	0.437	0.366	0.322	0.265	0.212
		0.641	0.305	0.234	0.179	0.145	-0.131	-0.139
		0.846	0.130	-0.149	-0.210	-0.248	-0.279	-0.291
		0.942	-0.063	-0.369	-0.511	-0.529	-0.510	-0.454
		0.986	-1.416	-1.258	-1.165	-1.050	-0.860	-0.621
	1	0.014	0.251	0.265	0.270	0.265	0.249	0.229
		0.058	0.251	0.268	0.273	0.267	0.250	0.227
		0.154	0.234	0.241	0.240	0.228	0.207	0.183
		0.360	0.194	0.187	0.176	0.155	0.130	0.106
		0.641	0.118	0.100	0.086	0.070	-0.064	-0.069
		0.846	-0.062	-0.086	-0.101	-0.120	-0.138	-0.145
		0.942	-0.155	-0.224	-0.245	-0.258	-0.253	-0.224
		0.986	-0.518	-0.566	-0.557	-0.507	-0.417	-0.297
	2	0.014	0.098	0.110	0.111	0.108	0.105	0.093
		0.058	0.098	0.110	0.112	0.108	0.104	0.092
		0.154	0.092	0.099	0.098	0.092	0.086	0.074
		0.360	0.076	0.076	0.071	0.062	0.053	0.043
		0.641	0.045	0.041	0.035	0.028	-0.026	-0.028
		0.846	-0.026	-0.033	-0.040	-0.048	-0.056	-0.057
		0.942	-0.066	-0.087	-0.097	-0.103	-0.102	-0.088
		0.986	-0.208	-0.228	-0.224	-0.202	-0.162	-0.117
1.17	0.5	0.014	0.802	0.645	0.572	0.544	0.509	0.450
		0.058	0.573	0.567	0.574	0.550	0.510	0.448
		0.154	0.532	0.508	0.502	0.468	0.421	0.360
		0.360	0.450	0.389	0.364	0.316	0.262	0.209
		0.641	0.301	0.209	0.175	0.143	-0.134	-0.141
		0.846	0.118	-0.156	-0.224	-0.255	-0.290	-0.296
		0.942	-0.096	-0.407	-0.536	-0.528	-0.521	-0.457
		0.986	-1.496	-1.343	-1.165	-1.008	-0.842	-0.606
	1	0.014	0.293	0.264	0.271	0.265	0.249	0.227
		0.058	0.286	0.267	0.274	0.267	0.249	0.225
		0.154	0.262	0.240	0.240	0.227	0.206	0.180
		0.360	0.211	0.184	0.175	0.153	0.128	0.105
		0.641	0.125	0.098	0.084	0.070	-0.066	-0.071
		0.846	-0.066	-0.086	-0.109	-0.127	-0.143	-0.147
		0.942	-0.147	-0.215	-0.260	-0.268	-0.259	-0.226
		0.986	-0.561	-0.542	-0.559	-0.501	-0.408	-0.292
	2	0.014	0.099	0.110	0.111	0.108	0.104	0.092
		0.058	0.100	0.110	0.112	0.108	0.103	0.092
		0.154	0.093	0.099	0.097	0.092	0.085	0.074
		0.360	0.076	0.075	0.070	0.062	0.052	0.043
		0.641	0.045	0.040	0.034	0.028	-0.027	-0.028
		0.846	-0.028	-0.036	-0.043	-0.051	-0.059	-0.059
		0.942	-0.074	-0.094	-0.103	-0.107	-0.104	-0.089
		0.986	-0.214	-0.230	-0.224	-0.199	-0.159	-0.115

a/t	r/t	a/c₁ → z/t ↓	0.2	0.3	0.4	0.6	1	2
			β					
1.21	0.5	0.014	0.811	0.650	0.575	0.548	0.507	0.447
		0.058	0.580	0.572	0.577	0.552	0.508	0.445
		0.154	0.537	0.510	0.503	0.469	0.419	0.357
		0.360	0.451	0.387	0.361	0.314	0.259	0.206
		0.641	0.296	0.203	0.172	0.144	-0.136	-0.143
		0.846	0.110	-0.176	-0.241	-0.276	-0.299	-0.300
		0.942	-0.154	-0.441	-0.563	-0.570	-0.528	-0.457
		0.986	-1.557	-1.364	-1.167	-1.022	-0.826	-0.593
	1	0.014	0.267	0.270	0.272	0.265	0.248	0.224
		0.058	0.258	0.271	0.275	0.267	0.248	0.223
		0.154	0.238	0.241	0.240	0.226	0.205	0.179
		0.360	0.193	0.184	0.173	0.151	0.127	0.103
		0.641	0.114	0.096	0.083	0.069	-0.067	-0.071
		0.846	-0.067	-0.098	-0.116	-0.133	-0.148	-0.149
		0.942	-0.190	-0.251	-0.271	-0.274	-0.262	-0.227
		0.986	-0.564	-0.581	-0.559	-0.495	-0.400	-0.287
	2	0.014	0.100	0.110	0.111	0.107	0.102	0.092
		0.058	0.101	0.111	0.112	0.107	0.101	0.091
		0.154	0.093	0.098	0.097	0.091	0.084	0.073
		0.360	0.076	0.075	0.070	0.061	0.052	0.042
		0.641	0.045	0.039	0.033	0.028	-0.027	-0.029
		0.846	-0.030	-0.039	-0.046	-0.054	-0.060	-0.060
		0.942	-0.080	-0.100	-0.108	-0.110	-0.105	-0.090
		0.986	-0.220	-0.232	-0.223	-0.197	-0.159	-0.113
2	0.5	0.014	0.905	0.831	0.634	0.534	0.475	0.403
		0.058	0.641	0.598	0.565	0.530	0.472	0.400
		0.154	0.561	0.508	0.474	0.435	0.380	0.316
		0.360	0.410	0.344	0.303	0.264	0.220	0.176
		0.641	0.199	0.163	-0.167	-0.163	-0.159	-0.149
		0.846	-0.247	-0.295	-0.364	-0.366	-0.339	-0.297
		0.942	-0.481	-0.508	-0.593	-0.601	-0.516	-0.417
		0.986	-1.772	-1.521	-1.115	-0.810	-0.633	-0.471
	1	0.014	0.373	0.270	0.265	0.252	0.229	0.199
		0.058	0.245	0.264	0.265	0.252	0.228	0.197
		0.154	0.217	0.225	0.222	0.207	0.183	0.156
		0.360	0.159	0.152	0.144	0.127	0.106	0.087
		0.641	0.076	0.075	-0.075	-0.076	-0.077	-0.074
		0.846	-0.094	-0.159	-0.168	-0.171	-0.163	-0.147
		0.942	-0.183	-0.305	-0.308	-0.286	-0.248	-0.206
		0.986	-0.728	-0.517	-0.463	-0.392	-0.305	-0.231
	2	0.014	0.105	0.108	0.108	0.102	0.093	0.086
		0.058	0.102	0.109	0.108	0.102	0.092	0.084
		0.154	0.090	0.093	0.090	0.083	0.074	0.066
		0.360	0.066	0.063	0.059	0.051	0.043	0.037
		0.641	0.033	0.031	-0.030	-0.031	-0.031	-0.031
		0.846	-0.052	-0.063	-0.068	-0.069	-0.066	-0.062
		0.942	-0.114	-0.126	-0.124	-0.114	-0.100	-0.086
		0.986	-0.223	-0.208	-0.187	-0.154	-0.122	-0.096

a/t	r/t	a/c₁ → z/t ↓	0.2	0.3	0.4	0.6	1	2
			β					
5	0.5	0.014	1.164	0.855	0.766	0.671	0.457	0.336
		0.058	0.817	0.588	0.518	0.453	0.395	0.332
		0.154	0.671	0.471	0.412	0.356	0.311	0.259
		0.360	0.415	0.274	0.235	0.199	0.172	0.141
		0.641	-0.228	-0.197	-0.184	-0.170	-0.158	-0.133
		0.846	-0.465	-0.404	-0.370	-0.333	-0.303	-0.255
		0.942	-0.702	-0.574	-0.512	-0.451	-0.404	-0.339
		0.986	-1.640	-1.176	-0.986	-0.801	-0.515	-0.357
	1	0.014	0.552	0.490	0.455	0.292	0.323	0.161
		0.058	0.240	0.214	0.200	0.194	0.279	0.159
		0.154	0.197	0.172	0.159	0.153	0.220	0.124
		0.360	0.120	0.100	0.092	0.086	0.122	0.068
		0.641	-0.065	-0.067	-0.066	-0.073	-0.112	-0.064
		0.846	-0.130	-0.134	-0.131	-0.144	-0.214	-0.122
		0.942	-0.190	-0.187	-0.180	-0.196	-0.286	-0.162
		0.986	-0.735	-0.613	-0.546	-0.344	-0.364	-0.171
	2	0.014	0.292	0.176	0.090	0.084	0.077	0.065
		0.058	0.070	0.074	0.078	0.081	0.075	0.064
		0.154	0.059	0.060	0.063	0.064	0.059	0.050
		0.360	0.038	0.035	0.036	0.036	0.033	0.027
		0.641	-0.013	-0.024	-0.029	-0.030	-0.029	-0.026
		0.846	-0.020	-0.048	-0.059	-0.061	-0.058	-0.049
		0.942	-0.035	-0.067	-0.083	-0.086	-0.079	-0.065
		0.986	-0.320	-0.216	-0.119	-0.101	-0.086	-0.068
10	0.5	0.014	0.716	0.477	0.935	0.646	0.523	0.329
		0.058	0.311	0.204	0.627	0.430	0.345	0.282
		0.154	0.248	0.161	0.491	0.335	0.268	0.220
		0.360	0.142	0.090	0.272	0.183	0.146	0.120
		0.641	-0.101	-0.074	-0.234	-0.170	-0.139	-0.117
		0.846	-0.201	-0.143	-0.454	-0.324	-0.263	-0.219
		0.942	-0.276	-0.192	-0.610	-0.430	-0.345	-0.284
		0.986	-0.835	-0.539	-1.047	-0.708	-0.553	-0.339
	1	0.014	0.252	0.222	0.415	0.363	0.370	0.135
		0.058	0.059	0.053	0.175	0.155	0.244	0.133
		0.154	0.047	0.042	0.137	0.120	0.189	0.103
		0.360	0.028	0.024	0.076	0.066	0.103	0.056
		0.641	-0.016	-0.017	-0.066	-0.060	-0.099	-0.054
		0.846	-0.033	-0.034	-0.126	-0.114	-0.186	-0.103
		0.942	-0.045	-0.045	-0.167	-0.151	-0.244	-0.134
		0.986	-0.272	-0.235	-0.456	-0.388	-0.391	-0.139
	2	0.014	*	*	0.208	0.129	0.062	0.055
		0.058	*	*	0.050	0.054	0.059	0.054
		0.154	*	*	0.039	0.042	0.046	0.042
		0.360	*	*	0.022	0.023	0.025	0.023
		0.641	*	*	-0.017	-0.021	-0.024	-0.022
		0.846	*	*	-0.033	-0.041	-0.046	-0.042
		0.942	*	*	-0.044	-0.053	-0.061	-0.055
		0.986	*	*	-0.217	-0.137	-0.066	-0.057

* Crack length outside FEM

Biaxial Tension (B = 1.0)

		a/c_i	0.2	0.3	0.4	0.6	1	2
a/t	r/t	z/t	β					
1.05	0.5	0.014	0.745	0.652	0.626	0.590	0.510	0.386
		0.058	0.679	0.675	0.657	0.626	0.544	0.401
		0.154	0.692	0.669	0.646	0.608	0.523	0.366
		0.360	0.734	0.684	0.648	0.587	0.487	0.311
		0.641	0.779	0.721	0.667	0.578	0.452	0.257
		0.846	0.853	0.828	0.744	0.617	0.451	0.226
		0.942	1.047	1.000	0.894	0.722	0.499	0.222
		0.986	2.448	1.885	1.562	1.156	0.712	0.259
	1	0.014	0.616	0.649	0.642	0.596	0.475	0.260
		0.058	0.638	0.678	0.675	0.631	0.504	0.260
		0.154	0.662	0.679	0.665	0.612	0.481	0.227
		0.360	0.722	0.705	0.668	0.588	0.439	0.175
		0.641	0.810	0.754	0.687	0.567	0.384	0.113
		0.846	0.994	0.865	0.751	0.570	0.336	0.055
		0.942	1.215	1.036	0.864	0.613	0.310	-0.011
		0.986	2.516	1.841	1.414	0.875	0.348	-0.036
	2	0.014	0.639	0.654	0.626	0.522	0.317	0.027
		0.058	0.661	0.680	0.655	0.550	0.328	-0.007
		0.154	0.686	0.680	0.644	0.531	0.304	-0.022
		0.360	0.750	0.706	0.642	0.499	0.259	-0.030
		0.641	0.838	0.741	0.636	0.445	0.186	-0.088
		0.846	0.999	0.802	0.633	0.377	0.095	-0.152
		0.942	1.179	0.879	0.638	0.310	-0.038	-0.227
		0.986	2.286	1.407	0.861	0.271	-0.174	-0.330
1.09	0.5	0.014	0.760	0.713	0.635	0.594	0.509	0.386
		0.058	0.697	0.676	0.668	0.631	0.545	0.403
		0.154	0.709	0.669	0.658	0.614	0.526	0.370
		0.360	0.755	0.684	0.660	0.596	0.494	0.320
		0.641	0.817	0.720	0.686	0.594	0.467	0.274
		0.846	0.935	0.822	0.780	0.646	0.480	0.255
		0.942	1.228	1.063	0.974	0.784	0.554	0.271
		0.986	2.620	2.001	1.637	1.223	0.776	0.326
	1	0.014	0.640	0.666	0.655	0.601	0.479	0.267
		0.058	0.663	0.697	0.689	0.638	0.510	0.269
		0.154	0.686	0.697	0.678	0.620	0.488	0.237
		0.360	0.748	0.724	0.682	0.598	0.450	0.189
		0.641	0.850	0.780	0.708	0.585	0.404	0.133
		0.846	1.059	0.912	0.793	0.610	0.375	0.084
		0.942	1.363	1.148	0.966	0.704	0.386	0.052
		0.986	2.712	2.009	1.579	1.033	0.477	0.029
	2	0.014	0.664	0.674	0.643	0.535	0.329	0.386
		0.058	0.688	0.702	0.674	0.564	0.341	0.393
		0.154	0.711	0.701	0.661	0.545	0.318	0.348
		0.360	0.778	0.727	0.659	0.515	0.276	-0.015
		0.641	0.879	0.773	0.663	0.471	0.209	-0.073
		0.846	1.074	0.865	0.690	0.429	0.132	-0.133
		0.942	1.358	1.028	0.772	0.415	0.057	-0.196
		0.986	2.620	1.719	1.145	0.494	0.000	-0.261

		a/c_i	0.2	0.3	0.4	0.6	1	2
a/t	r/t	z/t	β					
1.13	0.5	0.014	0.874	0.756	0.639	0.595	0.508	0.386
		0.058	0.690	0.770	0.674	0.633	0.545	0.404
		0.154	0.703	0.759	0.663	0.618	0.529	0.375
		0.360	0.755	0.771	0.667	0.601	0.500	0.328
		0.641	0.826	0.810	0.701	0.605	0.480	0.288
		0.846	0.989	0.923	0.808	0.667	0.502	0.279
		0.942	1.316	1.128	1.027	0.821	0.588	0.306
		0.986	2.626	2.022	1.666	1.242	0.801	0.364
	1	0.014	0.654	0.678	0.665	0.606	0.483	0.273
		0.058	0.679	0.711	0.701	0.644	0.515	0.277
		0.154	0.701	0.710	0.689	0.627	0.495	0.247
		0.360	0.768	0.736	0.693	0.607	0.460	0.202
		0.641	0.880	0.799	0.724	0.601	0.422	0.151
		0.846	1.100	0.948	0.824	0.642	0.408	0.112
		0.942	1.443	1.220	1.033	0.767	0.443	0.095
		0.986	2.776	2.080	1.653	1.111	0.554	0.092
	2	0.014	0.684	0.690	0.656	0.545	0.339	0.055
		0.058	0.709	0.720	0.688	0.577	0.354	0.043
		0.154	0.730	0.717	0.675	0.558	0.332	0.019
		0.360	0.799	0.744	0.674	0.530	0.292	-0.012
		0.641	0.911	0.797	0.685	0.494	0.232	-0.053
		0.846	1.131	0.913	0.737	0.474	0.169	-0.105
		0.942	1.482	1.136	0.872	0.503	0.120	-0.155
		0.986	2.785	1.886	1.312	0.642	0.085	-0.195
1.17	0.5	0.014	0.888	0.729	0.654	0.592	0.508	0.386
		0.058	0.704	0.699	0.688	0.633	0.546	0.406
		0.154	0.717	0.691	0.676	0.618	0.531	0.378
		0.360	0.769	0.709	0.677	0.601	0.506	0.336
		0.641	0.850	0.760	0.710	0.607	0.491	0.301
		0.846	1.039	0.901	0.820	0.672	0.519	0.299
		0.942	1.383	1.177	1.056	0.825	0.611	0.333
		0.986	2.687	2.035	1.654	1.214	0.810	0.389
	1	0.014	0.763	0.682	0.673	0.608	0.485	0.278
		0.058	0.778	0.718	0.710	0.648	0.519	0.284
		0.154	0.790	0.714	0.699	0.632	0.501	0.256
		0.360	0.842	0.737	0.702	0.614	0.469	0.214
		0.641	0.940	0.797	0.738	0.613	0.437	0.168
		0.846	1.155	0.941	0.849	0.667	0.435	0.137
		0.942	1.454	1.192	1.079	0.810	0.486	0.131
		0.986	2.869	2.019	1.690	1.153	0.604	0.137
	2	0.014	0.701	0.703	0.667	0.554	0.348	0.067
		0.058	0.727	0.734	0.701	0.588	0.365	0.057
		0.154	0.747	0.731	0.688	0.569	0.344	0.035
		0.360	0.817	0.758	0.687	0.543	0.307	-0.011
		0.641	0.937	0.818	0.705	0.515	0.254	-0.036
		0.846	1.176	0.952	0.776	0.514	0.203	-0.082
		0.942	1.572	1.216	0.948	0.573	0.176	-0.122
		0.986	2.876	1.984	1.417	0.745	0.169	-0.149

		a/c_i	0.2	0.3	0.4	0.6	1	2
a/t	r/t	z/t	β					
1.21	0.5	0.014	0.902	0.736	0.659	0.596	0.506	0.385
		0.058	0.718	0.709	0.696	0.637	0.545	0.407
		0.154	0.730	0.701	0.684	0.624	0.533	0.381
		0.360	0.782	0.719	0.685	0.610	0.510	0.342
		0.641	0.872	0.776	0.722	0.621	0.499	0.312
		0.846	1.079	0.928	0.839	0.696	0.532	0.315
		0.942	1.432	1.208	1.083	0.870	0.627	0.353
		0.986	2.722	2.029	1.648	1.227	0.811	0.406
	1	0.014	0.706	0.701	0.679	0.610	0.488	0.283
		0.058	0.715	0.733	0.718	0.651	0.523	0.292
		0.154	0.730	0.729	0.706	0.636	0.507	0.265
		0.360	0.792	0.755	0.710	0.620	0.478	0.225
		0.641	0.909	0.825	0.750	0.625	0.452	0.183
		0.846	1.142	0.992	0.869	0.687	0.458	0.160
		0.942	1.576	1.310	1.112	0.842	0.519	0.161
		0.986	2.851	2.125	1.706	1.176	0.638	0.172
	2	0.014	0.716	0.714	0.677	0.562	0.356	0.078
		0.058	0.742	0.747	0.712	0.598	0.376	0.069
		0.154	0.762	0.743	0.699	0.580	0.358	0.048
		0.360	0.832	0.769	0.699	0.556	0.323	-0.015
		0.641	0.959	0.835	0.722	0.535	0.275	-0.016
		0.846	1.212	0.985	0.808	0.548	0.236	-0.059
		0.942	1.639	1.276	1.006	0.630	0.225	-0.092
		0.986	2.928	2.044	1.485	0.818	0.240	-0.109
2	0.5	0.014	1.079	0.891	0.720	0.604	0.489	0.371
		0.058	0.912	0.823	0.761	0.661	0.540	0.407
		0.154	0.904	0.814	0.755	0.660	0.543	0.406
		0.360	0.938	0.827	0.763	0.660	0.542	0.397
		0.641	1.064	0.904	0.817	0.686	0.553	0.396
		0.846	1.299	1.053	0.930	0.759	0.588	0.411
		0.942	1.538	1.205	1.057	0.870	0.643	0.433
		0.986	2.494	1.827	1.317	0.968	0.672	0.429
	1	0.014	1.047	0.800	0.720	0.623	0.501	0.341
		0.058	0.899	0.848	0.778	0.681	0.552	0.371
		0.154	0.898	0.838	0.773	0.680	0.554	0.366
		0.360	0.944	0.852	0.777	0.680	0.551	0.353
		0.641	1.090	0.931	0.832	0.709	0.560	0.347
		0.846	1.346	1.117	0.967	0.787	0.595	0.358
		0.942	1.599	1.388	1.180	0.913	0.652	0.377
		0.986	2.706	1.770	1.431	1.045	0.682	0.371
	2	0.014	0.883	0.815	0.745	0.628	0.458	0.218
		0.058	0.919	0.871	0.805	0.687	0.503	0.228
		0.154	0.917	0.863	0.798	0.685	0.501	0.216
		0.360	0.966	0.879	0.803	0.682	0.493	0.200
		0.641	1.114	0.963	0.858	0.708	0.495	0.188
		0.846	1.416	1.156	0.995	0.783	0.520	0.187
		0.942	1.870	1.467	1.212	0.900	0.562	0.195
		0.986	2.638	1.904	1.471	1.001	0.576	0.192

		a/c_i	0.2	0.3	0.4	0.6	1	2
a/t	r/t	z/t	β					
5	0.5	0.014	2.012	0.901	0.730	0.572	0.445	0.346
		0.058	1.965	0.973	0.822	0.678	0.530	0.389
		0.154	1.926	0.973	0.824	0.682	0.539	0.399
		0.360	1.886	0.981	0.832	0.688	0.546	0.403
		0.641	1.903	1.017	0.857	0.701	0.553	0.405
		0.846	1.995	1.075	0.893	0.719	0.560	0.407
		0.942	2.091	1.125	0.921	0.732	0.562	0.405
		0.986	2.477	1.282	0.979	0.705	0.499	0.369
	1	0.014	1.315	0.980	0.786	0.588	0.482	0.357
		0.058	1.092	0.927	0.817	0.692	0.545	0.402
		0.154	1.091	0.927	0.818	0.697	0.558	0.412
		0.360	1.109	0.937	0.826	0.704	0.565	0.416
		0.641	1.168	0.968	0.848	0.719	0.573	0.418
		0.846	1.250	1.012	0.878	0.738	0.584	0.420
		0.942	1.313	1.045	0.901	0.754	0.592	0.418
		0.986	2.126	1.447	1.118	0.709	0.549	0.382
	2	0.014	1.584	0.981	0.740	0.619	0.498	0.343
		0.058	1.041	0.948	0.840	0.704	0.563	0.386
		0.154	1.045	0.948	0.849	0.718	0.576	0.394
		0.360	1.065	0.960	0.861	0.728	0.582	0.396
		0.641	1.114	0.996	0.892	0.746	0.591	0.398
		0.846	1.175	1.047	0.942	0.776	0.603	0.401
		0.942	1.221	1.087	0.986	0.801	0.611	0.399
		0.986	2.696	1.425	0.990	0.759	0.567	0.362
10	0.5	0.014	2.097	0.907	1.254	0.575	0.420	0.311
		0.058	2.064	1.008	1.491	0.719	0.535	0.379
		0.154	2.057	1.010	1.494	0.726	0.541	0.387
		0.360	2.056	1.017	1.492	0.732	0.546	0.392
		0.641	2.081	1.033	1.495	0.739	0.549	0.393
		0.846	2.126	1.053	1.506	0.745	0.550	0.391
		0.942	2.162	1.068	1.513	0.747	0.548	0.384
		0.986	2.569	1.171	1.351	0.643	0.451	0.320
	1	0.014	1.376	0.999	0.725	0.560	0.420	0.342
		0.058	1.144	0.958	0.844	0.694	0.541	0.390
		0.154	1.145	0.959	0.847	0.697	0.547	0.401
		0.360	1.154	0.963	0.853	0.701	0.552	0.407
		0.641	1.174	0.975	0.863	0.706	0.556	0.408
		0.846	1.200	0.988	0.875	0.712	0.556	0.406
		0.942	1.218	0.997	0.882	0.715	0.555	0.398
		0.986	2.018	1.377	0.898	0.655	0.454	0.354
	2	0.014	*	*	0.787	0.575	0.475	0.354
		0.058	*	*	0.836	0.709	0.550	0.403
		0.154	*	*	0.837	0.711	0.565	0.415
		0.360	*	*	0.841	0.716	0.573	0.420
		0.641	*	*	0.849	0.722	0.578	0.422
		0.846	*	*	0.858	0.729	0.579	0.419
		0.942	*	*	0.865	0.733	0.573	0.411
		0.986	*	*	1.059	0.656	0.506	0.365

* Crack length outside FEM

Biaxial Tension (B = 0.5)

		a/c_1	0.2	0.3	0.4	0.6	1	2
a/t	r/t	z/t	β					
1.05	0.5	0.014	1.125	0.982	0.944	0.896	0.791	0.648
		0.058	1.025	1.016	0.992	0.950	0.848	0.700
		0.154	1.044	1.006	0.975	0.925	0.825	0.681
		0.360	1.107	1.029	0.977	0.897	0.786	0.643
		0.641	1.176	1.086	1.011	0.894	0.759	0.608
		0.846	1.291	1.255	1.139	0.977	0.798	0.610
		0.942	1.593	1.529	1.390	1.180	0.938	0.672
		0.986	3.761	2.941	2.498	1.973	1.429	0.863
	1	0.014	0.938	0.993	0.998	0.965	0.866	0.711
		0.058	0.972	1.038	1.048	1.024	0.930	0.762
		0.154	1.008	1.039	1.032	0.997	0.905	0.742
		0.360	1.098	1.079	1.039	0.969	0.863	0.703
		0.641	1.235	1.162	1.088	0.972	0.833	0.660
		0.846	1.533	1.371	1.242	1.065	0.870	0.649
		0.942	1.903	1.708	1.526	1.289	1.022	0.707
		0.986	4.113	3.261	2.767	2.163	1.531	0.868
	2	0.014	1.002	1.070	1.090	1.063	0.968	0.716
		0.058	1.036	1.112	1.140	1.126	1.030	0.778
		0.154	1.074	1.110	1.121	1.094	0.995	0.762
		0.360	1.175	1.158	1.130	1.061	0.940	0.726
		0.641	1.331	1.258	1.184	1.056	0.894	0.684
		0.846	1.663	1.486	1.346	1.144	0.918	0.672
		0.942	2.079	1.836	1.650	1.388	1.083	0.726
		0.986	4.557	3.646	3.062	2.367	1.576	0.877
1.09	0.5	0.014	1.145	1.073	0.956	0.899	0.787	0.643
		0.058	1.049	1.019	1.007	0.956	0.847	0.696
		0.154	1.067	1.007	0.990	0.932	0.827	0.680
		0.360	1.136	1.030	0.994	0.907	0.792	0.646
		0.641	1.229	1.086	1.037	0.914	0.774	0.618
		0.846	1.408	1.245	1.186	1.011	0.824	0.629
		0.942	1.852	1.621	1.496	1.249	0.983	0.699
		0.986	3.965	3.075	2.542	1.981	1.408	0.849
	1	0.014	0.969	1.015	1.012	0.967	0.862	0.701
		0.058	1.005	1.062	1.065	1.028	0.928	0.755
		0.154	1.038	1.062	1.048	1.003	0.906	0.739
		0.360	1.132	1.103	1.057	0.979	0.870	0.705
		0.641	1.290	1.198	1.114	0.991	0.850	0.671
		0.846	1.619	1.429	1.289	1.103	0.902	0.672
		0.942	2.106	1.845	1.639	1.374	1.076	0.738
		0.986	4.263	3.333	2.808	2.168	1.505	0.859
	2	0.014	1.033	1.093	1.105	1.065	0.957	-0.018
		0.058	1.070	1.137	1.159	1.131	1.022	-0.055
		0.154	1.105	1.135	1.139	1.101	0.992	-0.082
		0.360	1.211	1.184	1.149	1.073	0.946	-0.119
		0.641	1.386	1.297	1.214	1.081	0.915	-0.174
		0.846	1.753	1.552	1.402	1.195	0.960	-0.241
		0.942	2.309	2.011	1.794	1.494	1.145	-0.317
		0.986	4.719	3.723	3.091	2.340	1.537	-0.403

		a/c_1	0.2	0.3	0.4	0.6	1	2
a/t	r/t	z/t	β					
1.13	0.5	0.014	1.315	1.135	0.961	0.900	0.783	0.638
		0.058	1.037	1.157	1.014	0.958	0.845	0.692
		0.154	1.057	1.139	0.998	0.936	0.828	0.678
		0.360	1.134	1.156	1.003	0.914	0.797	0.648
		0.641	1.241	1.216	1.056	0.927	0.786	0.626
		0.846	1.486	1.388	1.224	1.035	0.843	0.642
		0.942	1.981	1.703	1.566	1.288	1.006	0.712
		0.986	3.960	3.066	2.553	1.964	1.381	0.835
	1	0.014	0.987	1.030	1.023	0.968	0.857	0.694
		0.058	1.025	1.079	1.078	1.031	0.925	0.750
		0.154	1.058	1.078	1.062	1.008	0.906	0.736
		0.360	1.159	1.120	1.070	0.986	0.874	0.706
		0.641	1.332	1.223	1.134	1.006	0.863	0.679
		0.846	1.673	1.472	1.323	1.130	0.924	0.688
		0.942	2.210	1.929	1.707	1.422	1.105	0.755
		0.986	4.292	3.346	2.806	2.148	1.477	0.851
	2	0.014	1.058	1.111	1.115	1.065	0.947	0.711
		0.058	1.096	1.157	1.171	1.134	1.016	0.777
		0.154	1.129	1.154	1.152	1.106	0.990	0.766
		0.360	1.237	1.203	1.163	1.082	0.951	0.738
		0.641	1.427	1.324	1.235	1.099	0.930	0.711
		0.846	1.820	1.600	1.443	1.231	0.990	0.718
		0.942	2.457	2.117	1.879	1.553	1.178	0.778
		0.986	4.772	3.730	3.083	2.303	1.509	0.859
1.17	0.5	0.014	1.335	1.094	0.983	0.895	0.781	0.633
		0.058	1.058	1.049	1.035	0.957	0.843	0.688
		0.154	1.077	1.038	1.016	0.935	0.828	0.677
		0.360	1.155	1.064	1.018	0.913	0.801	0.650
		0.641	1.276	1.141	1.070	0.927	0.796	0.631
		0.846	1.558	1.355	1.240	1.036	0.857	0.651
		0.942	2.076	1.774	1.604	1.280	1.019	0.720
		0.986	4.035	3.074	2.518	1.893	1.353	0.822
	1	0.014	1.149	1.034	1.032	0.967	0.853	0.687
		0.058	1.172	1.088	1.089	1.032	0.922	0.745
		0.154	1.190	1.083	1.073	1.010	0.906	0.734
		0.360	1.268	1.118	1.081	0.991	0.878	0.707
		0.641	1.418	1.216	1.149	1.017	0.872	0.686
		0.846	1.751	1.453	1.349	1.150	0.940	0.699
		0.942	2.213	1.863	1.751	1.450	1.121	0.765
		0.986	4.396	3.192	2.788	2.119	1.449	0.843
	2	0.014	1.080	1.124	1.123	1.065	0.939	0.709
		0.058	1.119	1.173	1.182	1.136	1.010	0.776
		0.154	1.151	1.168	1.162	1.110	0.988	0.767
		0.360	1.260	1.218	1.174	1.089	0.954	0.742
		0.641	1.460	1.345	1.252	1.114	0.942	0.721
		0.846	1.871	1.636	1.474	1.258	1.011	0.732
		0.942	2.558	2.187	1.932	1.589	1.197	0.791
		0.986	4.782	3.712	3.055	2.267	1.486	0.852

		a/c_1	0.2	0.3	0.4	0.6	1	2
a/t	r/t	z/t	β					
1.21	0.5	0.014	1.354	1.104	0.990	0.901	0.777	0.629
		0.058	1.078	1.063	1.044	0.963	0.841	0.685
		0.154	1.095	1.052	1.026	0.943	0.828	0.675
		0.360	1.174	1.078	1.028	0.925	0.804	0.651
		0.641	1.307	1.163	1.085	0.947	0.803	0.636
		0.846	1.616	1.393	1.265	1.069	0.868	0.658
		0.942	2.144	1.816	1.638	1.341	1.026	0.724
		0.986	4.077	3.054	2.496	1.897	1.327	0.810
	1	0.014	1.062	1.061	1.039	0.966	0.849	0.682
		0.058	1.075	1.110	1.099	1.033	0.920	0.741
		0.154	1.098	1.104	1.082	1.013	0.906	0.732
		0.360	1.191	1.144	1.091	0.996	0.881	0.708
		0.641	1.369	1.257	1.163	1.026	0.880	0.691
		0.846	1.727	1.525	1.369	1.165	0.952	0.708
		0.942	2.392	2.033	1.781	1.467	1.130	0.772
		0.986	4.344	3.323	2.763	2.085	1.422	0.836
	2	0.014	1.098	1.136	1.130	1.065	0.924	0.707
		0.058	1.139	1.187	1.192	1.138	1.002	0.775
		0.154	1.169	1.182	1.172	1.114	0.985	0.768
		0.360	1.280	1.230	1.184	1.095	0.957	0.746
		0.641	1.487	1.362	1.266	1.125	0.953	0.728
		0.846	1.913	1.665	1.498	1.278	1.028	0.743
		0.942	2.632	2.235	1.967	1.610	1.210	0.799
		0.986	4.772	3.681	3.018	2.230	1.483	0.846
2	0.5	0.014	1.619	1.337	1.077	0.905	0.738	0.575
		0.058	1.366	1.233	1.138	0.989	0.815	0.637
		0.154	1.354	1.218	1.129	0.988	0.821	0.644
		0.360	1.403	1.237	1.140	0.987	0.820	0.641
		0.641	1.588	1.349	1.219	1.026	0.838	0.645
		0.846	1.936	1.569	1.387	1.134	0.890	0.663
		0.942	2.291	1.794	1.575	1.301	0.973	0.688
		0.986	3.705	2.715	1.961	1.446	1.016	0.675
	1	0.014	1.566	1.197	1.082	0.947	0.794	0.629
		0.058	1.343	1.269	1.169	1.035	0.877	0.698
		0.154	1.341	1.254	1.161	1.035	0.884	0.706
		0.360	1.409	1.274	1.168	1.036	0.884	0.703
		0.641	1.627	1.394	1.251	1.083	0.905	0.707
		0.846	2.008	1.673	1.456	1.204	0.964	0.726
		0.942	2.385	2.079	1.777	1.398	1.055	0.753
		0.986	4.038	2.650	2.156	1.602	1.103	0.729
	2	0.014	1.326	1.238	1.155	1.028	0.868	0.705
		0.058	1.379	1.323	1.248	1.127	0.964	0.775
		0.154	1.377	1.312	1.239	1.126	0.970	0.778
		0.360	1.452	1.339	1.249	1.128	0.970	0.773
		0.641	1.676	1.471	1.341	1.181	0.993	0.775
		0.846	2.133	1.771	1.563	1.317	1.058	0.792
		0.942	2.822	2.252	1.909	1.520	1.150	0.815
		0.986	3.983	2.926	2.320	1.694	1.177	0.781

		a/c_1	0.2	0.3	0.4	0.6	1	2
a/t	r/t	z/t	β					
5	0.5	0.014	2.258	1.307	1.086	0.855	0.664	0.518
		0.058	2.183	1.412	1.222	1.013	0.792	0.583
		0.154	2.140	1.411	1.225	1.018	0.805	0.599
		0.360	2.103	1.422	1.237	1.027	0.815	0.605
		0.641	2.138	1.475	1.272	1.046	0.826	0.609
		0.846	2.259	1.560	1.325	1.073	0.835	0.611
		0.942	2.381	1.632	1.367	1.092	0.838	0.607
		0.986	2.848	1.858	1.451	1.051	0.745	0.554
	1	0.014	1.961	1.466	1.175	0.879	0.724	0.553
		0.058	1.624	1.385	1.220	1.034	0.818	0.623
		0.154	1.622	1.384	1.222	1.041	0.838	0.639
		0.360	1.648	1.399	1.233	1.052	0.848	0.647
		0.641	1.734	1.445	1.266	1.073	0.860	0.651
		0.846	1.855	1.509	1.310	1.103	0.877	0.653
		0.942	1.949	1.558	1.344	1.125	0.889	0.649
		0.986	3.149	2.154	1.665	1.059	0.824	0.592
	2	0.014	2.369	1.465	1.106	0.933	0.772	0.601
		0.058	1.555	1.415	1.256	1.062	0.873	0.681
		0.154	1.560	1.416	1.270	1.083	0.894	0.699
		0.360	1.590	1.433	1.288	1.098	0.905	0.707
		0.641	1.662	1.488	1.335	1.126	0.919	0.711
		0.846	1.754	1.563	1.409	1.171	0.938	0.714
		0.942	1.821	1.623	1.475	1.209	0.950	0.709
		0.986	4.021	2.129	1.481	1.145	0.881	0.639
10	0.5	0.014	2.355	1.314	1.394	0.832	0.626	0.464
		0.058	2.289	1.461	1.653	1.042	0.798	0.565
		0.154	2.281	1.464	1.657	1.052	0.807	0.577
		0.360	2.282	1.473	1.657	1.061	0.814	0.585
		0.641	2.314	1.497	1.664	1.071	0.819	0.587
		0.846	2.368	1.526	1.679	1.079	0.820	0.583
		0.942	2.412	1.546	1.689	1.082	0.816	0.574
		0.986	2.894	1.694	1.512	0.930	0.672	0.477
	1	0.014	2.051	1.493	1.078	0.837	0.628	0.514
		0.058	1.700	1.430	1.254	1.036	0.808	0.585
		0.154	1.701	1.431	1.258	1.040	0.816	0.602
		0.360	1.713	1.438	1.266	1.046	0.824	0.611
		0.641	1.743	1.454	1.281	1.054	0.829	0.613
		0.846	1.781	1.474	1.298	1.062	0.830	0.610
		0.942	1.808	1.488	1.309	1.067	0.827	0.598
		0.986	2.990	2.051	1.331	0.977	0.677	0.531
	2	0.014	*	*	1.175	0.859	0.713	0.549
		0.058	*	*	1.248	1.058	0.826	0.625
		0.154	*	*	1.249	1.062	0.848	0.644
		0.360	*	*	1.255	1.069	0.860	0.653
		0.641	*	*	1.267	1.079	0.867	0.655
		0.846	*	*	1.281	1.089	0.869	0.652
		0.942	*	*	1.290	1.095	0.860	0.639
		0.986	*	*	1.579	0.981	0.759	0.567

* Crack length outside FEM

Biaxial Tension (B = 0.25)

		a/c_1	0.2	0.3	0.4	0.6	1	2
a/t	r/t	z/t	β					
1.05	0.5	0.014	1.315	1.147	1.104	1.048	0.931	0.779
		0.058	1.198	1.186	1.159	1.112	1.000	0.849
		0.154	1.220	1.175	1.140	1.083	0.976	0.839
		0.360	1.294	1.202	1.142	1.051	0.935	0.809
		0.641	1.374	1.269	1.183	1.052	0.913	0.784
		0.846	1.510	1.468	1.336	1.157	0.972	0.801
		0.942	1.865	1.794	1.638	1.409	1.158	0.897
		0.986	4.418	3.469	2.967	2.381	1.788	1.165
	1	0.014	1.099	1.165	1.175	1.150	1.062	0.936
		0.058	1.139	1.218	1.235	1.221	1.144	1.013
		0.154	1.181	1.219	1.216	1.189	1.118	1.001
		0.360	1.287	1.265	1.225	1.160	1.076	0.967
		0.641	1.448	1.367	1.288	1.174	1.059	0.935
		0.846	1.803	1.624	1.488	1.312	1.137	0.947
		0.942	2.247	2.044	1.858	1.627	1.379	1.058
		0.986	4.911	3.972	3.444	2.807	2.123	1.324
	2	0.014	1.183	1.278	1.322	1.334	1.293	1.053
		0.058	1.224	1.328	1.383	1.413	1.380	1.154
		0.154	1.267	1.326	1.359	1.375	1.340	1.142
		0.360	1.387	1.384	1.374	1.343	1.281	1.105
		0.641	1.578	1.516	1.459	1.363	1.250	1.070
		0.846	1.995	1.829	1.704	1.529	1.334	1.085
		0.942	2.529	2.316	2.157	1.930	1.635	1.203
		0.986	5.693	4.767	4.164	3.419	2.450	1.481
1.09	0.5	0.014	1.338	1.254	1.117	1.052	0.926	0.771
		0.058	1.225	1.190	1.176	1.118	0.997	0.842
		0.154	1.246	1.176	1.157	1.091	0.977	0.835
		0.360	1.326	1.203	1.161	1.063	0.941	0.809
		0.641	1.435	1.268	1.212	1.074	0.928	0.791
		0.846	1.644	1.456	1.389	1.193	0.997	0.816
		0.942	2.164	1.899	1.757	1.482	1.197	0.912
		0.986	4.638	3.611	2.995	2.360	1.724	1.111
	1	0.014	1.134	1.189	1.190	1.150	1.053	0.919
		0.058	1.175	1.245	1.253	1.223	1.137	0.998
		0.154	1.215	1.245	1.233	1.194	1.115	0.990
		0.360	1.324	1.293	1.244	1.169	1.080	0.963
		0.641	1.510	1.407	1.317	1.195	1.074	0.941
		0.846	1.899	1.687	1.537	1.349	1.165	0.966
		0.942	2.478	2.193	1.976	1.709	1.422	1.079
		0.986	5.039	3.995	3.422	2.736	2.019	1.269
	2	0.014	1.218	1.302	1.336	1.330	1.271	1.023
		0.058	1.261	1.355	1.401	1.414	1.362	1.124
		0.154	1.302	1.352	1.377	1.379	1.329	1.116
		0.360	1.427	1.413	1.394	1.352	1.282	1.085
		0.641	1.640	1.559	1.490	1.387	1.268	1.061
		0.846	2.092	1.896	1.759	1.579	1.377	1.090
		0.942	2.785	2.503	2.306	2.035	1.691	1.203
		0.986	5.768	4.725	4.064	3.264	2.316	1.395

		a/c_1	0.2	0.3	0.4	0.6	1	2
a/t	r/t	z/t	β					
1.13	0.5	0.014	1.535	1.325	1.122	1.052	0.921	0.764
		0.058	1.211	1.350	1.184	1.121	0.994	0.836
		0.154	1.235	1.330	1.165	1.096	0.977	0.830
		0.360	1.324	1.349	1.171	1.071	0.946	0.808
		0.641	1.448	1.419	1.234	1.088	0.939	0.795
		0.846	1.735	1.621	1.432	1.218	1.014	0.823
		0.942	2.313	1.990	1.835	1.521	1.215	0.915
		0.986	4.627	3.588	2.997	2.325	1.671	1.070
	1	0.014	1.154	1.206	1.202	1.149	1.045	0.904
		0.058	1.198	1.264	1.267	1.225	1.130	0.986
		0.154	1.237	1.262	1.248	1.198	1.112	0.981
		0.360	1.355	1.311	1.259	1.176	1.081	0.959
		0.641	1.558	1.434	1.338	1.209	1.083	0.944
		0.846	1.960	1.734	1.572	1.375	1.182	0.976
		0.942	2.594	2.284	2.044	1.750	1.436	1.084
		0.986	5.050	3.980	3.383	2.666	1.938	1.228
	2	0.014	1.245	1.321	1.345	1.326	1.251	1.035
		0.058	1.290	1.376	1.413	1.412	1.346	1.140
		0.154	1.329	1.372	1.390	1.380	1.319	1.135
		0.360	1.456	1.433	1.407	1.358	1.280	1.110
		0.641	1.685	1.588	1.510	1.403	1.280	1.094
		0.846	2.164	1.943	1.797	1.610	1.401	1.132
		0.942	2.944	2.607	2.383	2.079	1.707	1.247
		0.986	5.766	4.652	3.968	3.134	2.220	1.387
1.17	0.5	0.014	1.558	1.277	1.147	1.046	0.917	0.757
		0.058	1.235	1.225	1.208	1.119	0.992	0.829
		0.154	1.257	1.211	1.186	1.094	0.977	0.826
		0.360	1.347	1.241	1.188	1.069	0.949	0.807
		0.641	1.488	1.331	1.250	1.088	0.948	0.797
		0.846	1.818	1.582	1.450	1.218	1.027	0.827
		0.942	2.422	2.073	1.877	1.508	1.224	0.913
		0.986	4.709	3.594	2.950	2.233	1.625	1.038
	1	0.014	1.343	1.210	1.211	1.146	1.037	0.891
		0.058	1.369	1.273	1.279	1.224	1.124	0.975
		0.154	1.390	1.267	1.259	1.199	1.108	0.973
		0.360	1.481	1.309	1.271	1.180	1.082	0.954
		0.641	1.657	1.426	1.355	1.219	1.090	0.945
		0.846	2.049	1.709	1.599	1.392	1.193	0.980
		0.942	2.593	2.199	2.087	1.770	1.439	1.082
		0.986	5.160	3.778	3.338	2.601	1.871	1.195
	2	0.014	1.269	1.334	1.351	1.321	1.233	1.027
		0.058	1.315	1.393	1.423	1.410	1.332	1.133
		0.154	1.352	1.387	1.400	1.381	1.310	1.131
		0.360	1.482	1.448	1.418	1.362	1.278	1.110
		0.641	1.722	1.609	1.525	1.413	1.287	1.100
		0.846	2.219	1.979	1.823	1.630	1.415	1.141
		0.942	3.051	2.673	2.425	2.097	1.707	1.249
		0.986	5.735	4.576	3.874	3.029	2.143	1.354

		a/c_1	0.2	0.3	0.4	0.6	1	2
a/t	r/t	z/t	β					
1.21	0.5	0.014	1.581	1.288	1.156	1.053	0.913	0.751
		0.058	1.258	1.241	1.219	1.126	0.989	0.824
		0.154	1.278	1.227	1.198	1.103	0.976	0.822
		0.360	1.369	1.258	1.200	1.083	0.951	0.805
		0.641	1.524	1.356	1.267	1.110	0.954	0.798
		0.846	1.885	1.626	1.478	1.255	1.035	0.829
		0.942	2.501	2.120	1.915	1.577	1.226	0.910
		0.986	4.755	3.567	2.921	2.232	1.585	1.012
	1	0.014	1.240	1.241	1.219	1.144	1.029	0.881
		0.058	1.256	1.298	1.289	1.224	1.118	0.966
		0.154	1.281	1.291	1.269	1.201	1.105	0.965
		0.360	1.391	1.339	1.281	1.184	1.082	0.950
		0.641	1.600	1.473	1.369	1.227	1.094	0.944
		0.846	2.019	1.791	1.620	1.405	1.199	0.981
		0.942	2.801	2.395	2.115	1.780	1.435	1.077
		0.986	5.091	3.922	3.292	2.540	1.814	1.167
	2	0.014	1.290	1.347	1.357	1.316	1.208	1.019
		0.058	1.338	1.408	1.431	1.408	1.315	1.127
		0.154	1.372	1.401	1.409	1.381	1.299	1.126
		0.360	1.503	1.461	1.426	1.364	1.274	1.109
		0.641	1.752	1.626	1.537	1.421	1.292	1.104
		0.846	2.263	2.005	1.843	1.643	1.424	1.147
		0.942	3.128	2.714	2.447	2.100	1.703	1.247
		0.986	5.694	4.499	3.785	2.936	2.104	1.326
2	0.5	0.014	1.889	1.559	1.255	1.055	0.862	0.677
		0.058	1.594	1.437	1.326	1.153	0.952	0.752
		0.154	1.579	1.420	1.316	1.152	0.960	0.763
		0.360	1.635	1.442	1.328	1.151	0.959	0.764
		0.641	1.850	1.571	1.420	1.196	0.980	0.769
		0.846	2.255	1.828	1.615	1.322	1.041	0.789
		0.942	2.667	2.089	1.834	1.516	1.138	0.816
		0.986	4.311	3.160	2.283	1.585	1.187	0.798
	1	0.014	1.826	1.396	1.262	1.109	0.940	0.772
		0.058	1.564	1.479	1.364	1.213	1.039	0.861
		0.154	1.562	1.462	1.354	1.212	1.048	0.876
		0.360	1.642	1.485	1.364	1.214	1.050	0.879
		0.641	1.895	1.625	1.460	1.270	1.077	0.887
		0.846	2.340	1.950	1.700	1.413	1.148	0.911
		0.942	2.779	2.424	2.075	1.641	1.256	0.941
		0.986	4.703	3.090	2.518	1.880	1.313	0.909
	2	0.014	1.547	1.449	1.360	1.228	1.074	0.948
		0.058	1.609	1.550	1.470	1.347	1.194	1.049
		0.154	1.607	1.537	1.459	1.346	1.205	1.058
		0.360	1.695	1.568	1.472	1.351	1.209	1.060
		0.641	1.957	1.725	1.582	1.417	1.243	1.069
		0.846	2.492	2.078	1.846	1.584	1.327	1.094
		0.942	3.298	2.644	2.258	1.830	1.444	1.124
		0.986	4.656	3.437	2.745	2.041	1.477	1.075

		a/c_1	0.2	0.3	0.4	0.6	1	2
a/t	r/t	z/t	β					
5	0.5	0.014	2.381	1.511	1.265	0.997	0.774	0.605
		0.058	2.292	1.632	1.422	1.180	0.922	0.681
		0.154	2.247	1.630	1.426	1.186	0.938	0.698
		0.360	2.211	1.643	1.439	1.196	0.950	0.706
		0.641	2.255	1.704	1.480	1.219	0.962	0.711
		0.846	2.391	1.802	1.541	1.249	0.973	0.713
		0.942	2.526	1.885	1.590	1.271	0.976	0.708
		0.986	3.034	2.146	1.687	1.223	0.867	0.646
	1	0.014	2.284	1.710	1.370	1.025	0.844	0.651
		0.058	1.890	1.614	1.422	1.205	0.954	0.733
		0.154	1.888	1.613	1.424	1.213	0.977	0.753
		0.360	1.918	1.629	1.437	1.225	0.989	0.762
		0.641	2.017	1.683	1.475	1.251	1.004	0.767
		0.846	2.157	1.758	1.526	1.285	1.024	0.769
		0.942	2.267	1.815	1.566	1.311	1.037	0.764
		0.986	3.660	2.508	1.939	1.233	0.961	0.697
	2	0.014	2.761	1.708	1.289	1.090	0.909	0.730
		0.058	1.812	1.649	1.465	1.241	1.029	0.829
		0.154	1.817	1.650	1.480	1.265	1.053	0.851
		0.360	1.852	1.670	1.501	1.282	1.067	0.862
		0.641	1.937	1.733	1.556	1.316	1.083	0.868
		0.846	2.043	1.822	1.643	1.369	1.105	0.870
		0.942	2.122	1.891	1.720	1.413	1.120	0.863
		0.986	4.683	2.481	1.726	1.338	1.038	0.778
10	0.5	0.014	2.484	1.518	1.463	0.961	0.730	0.541
		0.058	2.401	1.687	1.734	1.203	0.929	0.658
		0.154	2.392	1.691	1.738	1.214	0.939	0.673
		0.360	2.394	1.702	1.739	1.225	0.948	0.682
		0.641	2.430	1.728	1.748	1.237	0.954	0.684
		0.846	2.489	1.762	1.765	1.246	0.954	0.679
		0.942	2.536	1.785	1.776	1.249	0.951	0.668
		0.986	3.056	1.955	1.593	1.074	0.782	0.556
	1	0.014	2.389	1.741	1.255	0.975	0.731	0.599
		0.058	1.977	1.666	1.459	1.207	0.941	0.683
		0.154	1.979	1.667	1.463	1.212	0.951	0.703
		0.360	1.992	1.675	1.472	1.218	0.960	0.713
		0.641	2.028	1.694	1.490	1.228	0.966	0.715
		0.846	2.071	1.717	1.510	1.237	0.967	0.711
		0.942	2.103	1.733	1.523	1.242	0.964	0.697
		0.986	3.476	2.388	1.547	1.137	0.788	0.619
	2	0.014	*	*	1.370	1.001	0.832	0.646
		0.058	*	*	1.453	1.233	0.963	0.736
		0.154	*	*	1.455	1.238	0.990	0.758
		0.360	*	*	1.462	1.245	1.004	0.769
		0.641	*	*	1.476	1.257	1.012	0.772
		0.846	*	*	1.492	1.269	1.014	0.768
		0.942	*	*	1.503	1.276	1.003	0.752
		0.986	*	*	1.838	1.143	0.886	0.668

* Crack length outside FEM

Pin Loading (Concentrated Load)

		a/c_1	0.2	0.3	0.4	0.6	1	2
a/t	r/t	z/t	β					
1.05	0.5	0.014	0.064	0.069	0.079	0.097	0.119	0.144
		0.058	0.057	0.071	0.083	0.104	0.130	0.161
		0.154	0.058	0.070	0.082	0.102	0.129	0.165
		0.360	0.062	0.073	0.084	0.103	0.130	0.168
		0.641	0.071	0.085	0.096	0.114	0.140	0.175
		0.846	0.092	0.117	0.130	0.148	0.170	0.193
		0.942	0.137	0.170	0.188	0.209	0.227	0.228
		0.986	0.421	0.400	0.402	0.402	0.382	0.306
	1	0.014	0.081	0.111	0.135	0.169	0.204	0.234
		0.058	0.083	0.116	0.142	0.180	0.222	0.255
		0.154	0.086	0.116	0.140	0.177	0.220	0.255
		0.360	0.096	0.123	0.144	0.178	0.218	0.252
		0.641	0.118	0.146	0.167	0.196	0.229	0.252
		0.846	0.177	0.206	0.225	0.249	0.268	0.266
		0.942	0.255	0.301	0.323	0.345	0.349	0.307
		0.986	0.674	0.681	0.679	0.650	0.565	0.392
	2	0.014	0.137	0.188	0.229	0.280	0.325	0.304
		0.058	0.141	0.195	0.240	0.298	0.349	0.334
		0.154	0.145	0.195	0.236	0.291	0.340	0.332
		0.360	0.163	0.209	0.244	0.290	0.330	0.324
		0.641	0.203	0.249	0.279	0.311	0.334	0.318
		0.846	0.301	0.342	0.363	0.378	0.373	0.329
		0.942	0.429	0.481	0.503	0.508	0.473	0.370
		0.986	1.109	1.097	1.048	0.941	0.725	0.460
1.09	0.5	0.014	0.062	0.075	0.077	0.095	0.115	0.140
		0.058	0.056	0.070	0.082	0.102	0.126	0.157
		0.154	0.057	0.069	0.080	0.100	0.126	0.161
		0.360	0.061	0.072	0.083	0.101	0.128	0.165
		0.641	0.072	0.083	0.095	0.113	0.138	0.173
		0.846	0.095	0.113	0.127	0.144	0.167	0.190
		0.942	0.145	0.171	0.184	0.202	0.219	0.222
		0.986	0.365	0.379	0.349	0.350	0.334	0.276
	1	0.014	0.080	0.110	0.133	0.165	0.199	0.227
		0.058	0.083	0.115	0.140	0.176	0.216	0.249
		0.154	0.086	0.115	0.138	0.173	0.215	0.250
		0.360	0.095	0.122	0.143	0.175	0.215	0.248
		0.641	0.119	0.145	0.165	0.194	0.228	0.251
		0.846	0.175	0.202	0.221	0.246	0.267	0.268
		0.942	0.259	0.298	0.319	0.342	0.347	0.307
		0.986	0.596	0.603	0.605	0.584	0.511	0.367
	2	0.014	0.137	0.187	0.227	0.275	0.316	0.300
		0.058	0.141	0.195	0.238	0.293	0.340	0.331
		0.154	0.146	0.195	0.235	0.288	0.334	0.330
		0.360	0.163	0.208	0.243	0.288	0.327	0.323
		0.641	0.205	0.249	0.279	0.312	0.335	0.322
		0.846	0.300	0.340	0.363	0.381	0.380	0.336
		0.942	0.444	0.495	0.516	0.520	0.481	0.377
		0.986	1.016	1.008	0.964	0.865	0.671	0.440

		a/c_1	0.2	0.3	0.4	0.6	1	2
a/t	r/t	z/t	β					
1.13	0.5	0.014	0.071	0.074	0.076	0.093	0.112	0.136
		0.058	0.054	0.076	0.080	0.099	0.123	0.153
		0.154	0.055	0.074	0.079	0.098	0.123	0.158
		0.360	0.060	0.077	0.082	0.099	0.125	0.162
		0.641	0.071	0.088	0.094	0.111	0.136	0.170
		0.846	0.097	0.115	0.124	0.140	0.162	0.186
		0.942	0.144	0.158	0.178	0.192	0.209	0.214
		0.986	0.338	0.319	0.315	0.314	0.300	0.255
	1	0.014	0.079	0.108	0.130	0.161	0.193	0.221
		0.058	0.082	0.113	0.138	0.172	0.211	0.243
		0.154	0.085	0.113	0.136	0.170	0.211	0.245
		0.360	0.095	0.121	0.141	0.172	0.211	0.244
		0.641	0.119	0.144	0.163	0.191	0.225	0.249
		0.846	0.171	0.197	0.216	0.241	0.264	0.266
		0.942	0.252	0.289	0.310	0.332	0.338	0.303
		0.986	0.538	0.546	0.550	0.532	0.470	0.348
	2	0.014	0.136	0.186	0.224	0.270	0.307	0.295
		0.058	0.140	0.194	0.235	0.288	0.333	0.326
		0.154	0.145	0.193	0.232	0.284	0.328	0.326
		0.360	0.162	0.207	0.241	0.285	0.324	0.322
		0.641	0.204	0.247	0.277	0.310	0.334	0.322
		0.846	0.296	0.336	0.360	0.380	0.381	0.339
		0.942	0.443	0.491	0.512	0.516	0.477	0.377
		0.986	0.937	0.931	0.894	0.802	0.630	0.423
1.17	0.5	0.014	0.070	0.071	0.075	0.090	0.109	0.133
		0.058	0.053	0.067	0.080	0.097	0.120	0.149
		0.154	0.054	0.067	0.078	0.096	0.121	0.154
		0.360	0.059	0.070	0.081	0.097	0.123	0.158
		0.641	0.070	0.082	0.092	0.107	0.133	0.166
		0.846	0.097	0.110	0.121	0.133	0.157	0.182
		0.942	0.140	0.159	0.172	0.179	0.199	0.206
		0.986	0.311	0.302	0.287	0.279	0.274	0.237
	1	0.014	0.090	0.106	0.128	0.157	0.188	0.215
		0.058	0.091	0.111	0.136	0.168	0.206	0.238
		0.154	0.093	0.111	0.134	0.167	0.207	0.240
		0.360	0.101	0.118	0.139	0.169	0.208	0.240
		0.641	0.122	0.138	0.160	0.188	0.222	0.246
		0.846	0.170	0.186	0.210	0.235	0.259	0.263
		0.942	0.234	0.262	0.299	0.320	0.327	0.297
		0.986	0.509	0.480	0.505	0.491	0.436	0.332
	2	0.014	0.135	0.184	0.220	0.265	0.300	0.291
		0.058	0.140	0.192	0.232	0.284	0.326	0.323
		0.154	0.144	0.191	0.230	0.279	0.322	0.324
		0.360	0.161	0.205	0.238	0.281	0.320	0.320
		0.641	0.202	0.244	0.273	0.307	0.332	0.322
		0.846	0.291	0.330	0.354	0.377	0.379	0.339
		0.942	0.435	0.481	0.502	0.506	0.469	0.375
		0.986	0.871	0.866	0.834	0.751	0.597	0.409

		a/c_1	0.2	0.3	0.4	0.6	1	2
a/t	r/t	z/t	β					
1.21	0.5	0.014	0.069	0.069	0.074	0.088	0.106	0.129
		0.058	0.053	0.067	0.078	0.095	0.117	0.146
		0.154	0.054	0.066	0.077	0.094	0.118	0.151
		0.360	0.059	0.069	0.080	0.096	0.120	0.155
		0.641	0.070	0.081	0.091	0.106	0.130	0.163
		0.846	0.096	0.108	0.118	0.132	0.153	0.177
		0.942	0.137	0.153	0.165	0.178	0.190	0.199
		0.986	0.290	0.278	0.265	0.263	0.253	0.223
	1	0.014	0.081	0.106	0.126	0.153	0.184	0.210
		0.058	0.082	0.111	0.133	0.165	0.202	0.233
		0.154	0.083	0.111	0.132	0.163	0.202	0.236
		0.360	0.093	0.117	0.137	0.166	0.204	0.237
		0.641	0.115	0.139	0.157	0.184	0.218	0.243
		0.846	0.161	0.188	0.205	0.229	0.254	0.260
		0.942	0.243	0.272	0.288	0.308	0.316	0.291
		0.986	0.470	0.470	0.470	0.456	0.408	0.318
	2	0.014	0.134	0.181	0.217	0.260	0.290	0.287
		0.058	0.139	0.190	0.230	0.279	0.318	0.319
		0.154	0.143	0.189	0.227	0.275	0.316	0.321
		0.360	0.160	0.202	0.235	0.278	0.315	0.318
		0.641	0.200	0.240	0.269	0.303	0.330	0.321
		0.846	0.285	0.324	0.348	0.371	0.376	0.339
		0.942	0.425	0.469	0.489	0.494	0.460	0.372
		0.986	0.815	0.811	0.783	0.707	0.575	0.397
2	0.5	0.014	0.058	0.060	0.057	0.063	0.072	0.088
		0.058	0.048	0.055	0.061	0.069	0.081	0.100
		0.154	0.048	0.054	0.060	0.069	0.082	0.104
		0.360	0.050	0.056	0.062	0.070	0.084	0.107
		0.641	0.058	0.062	0.068	0.075	0.088	0.111
		0.846	0.072	0.074	0.079	0.085	0.095	0.114
		0.942	0.086	0.086	0.091	0.098	0.105	0.117
		0.986	0.142	0.133	0.114	0.110	0.109	0.113
	1	0.014	0.084	0.086	0.092	0.107	0.126	0.151
		0.058	0.072	0.091	0.100	0.117	0.140	0.170
		0.154	0.072	0.091	0.099	0.118	0.143	0.176
		0.360	0.076	0.093	0.101	0.120	0.146	0.179
		0.641	0.091	0.104	0.111	0.129	0.153	0.184
		0.846	0.115	0.128	0.133	0.146	0.166	0.191
		0.942	0.138	0.161	0.165	0.172	0.183	0.197
		0.986	0.240	0.203	0.202	0.198	0.191	0.189
	2	0.014	0.113	0.140	0.159	0.185	0.212	0.241
		0.058	0.118	0.150	0.173	0.204	0.237	0.268
		0.154	0.118	0.150	0.172	0.206	0.241	0.273
		0.360	0.126	0.154	0.176	0.209	0.245	0.275
		0.641	0.149	0.174	0.194	0.224	0.256	0.280
		0.846	0.195	0.215	0.232	0.255	0.277	0.289
		0.942	0.263	0.278	0.287	0.298	0.303	0.297
		0.986	0.376	0.365	0.351	0.334	0.310	0.283

		a/c_1	0.2	0.3	0.4	0.6	1	2
a/t	r/t	z/t	β					
5	0.5	0.014	0.048	0.036	0.034	0.033	0.035	0.043
		0.058	0.046	0.039	0.038	0.039	0.042	0.049
		0.154	0.045	0.039	0.038	0.039	0.043	0.051
		0.360	0.044	0.040	0.039	0.040	0.044	0.052
		0.641	0.045	0.041	0.040	0.041	0.044	0.053
		0.846	0.048	0.043	0.042	0.042	0.045	0.053
		0.942	0.051	0.046	0.043	0.043	0.045	0.052
		0.986	0.062	0.052	0.046	0.041	0.039	0.047
	1	0.014	0.062	0.057	0.054	0.052	0.061	0.075
		0.058	0.051	0.054	0.057	0.062	0.070	0.086
		0.154	0.051	0.054	0.057	0.063	0.072	0.089
		0.360	0.052	0.054	0.057	0.064	0.073	0.091
		0.641	0.055	0.056	0.059	0.065	0.074	0.092
		0.846	0.059	0.059	0.061	0.067	0.076	0.092
		0.942	0.062	0.061	0.063	0.069	0.077	0.090
		0.986	0.100	0.085	0.078	0.064	0.071	0.082
	2	0.014	0.111	0.088	0.080	0.090	0.106	0.129
		0.058	0.072	0.085	0.092	0.103	0.121	0.148
		0.154	0.072	0.085	0.093	0.106	0.124	0.153
		0.360	0.074	0.087	0.094	0.107	0.127	0.156
		0.641	0.077	0.090	0.098	0.111	0.129	0.157
		0.846	0.082	0.095	0.104	0.115	0.132	0.157
		0.942	0.085	0.099	0.109	0.119	0.133	0.155
		0.986	0.192	0.131	0.110	0.113	0.123	0.139
10	0.5	0.014	0.050	0.036	0.029	0.023	0.022	0.024
		0.058	0.048	0.041	0.034	0.029	0.028	0.030
		0.154	0.048	0.041	0.035	0.029	0.028	0.031
		0.360	0.048	0.041	0.035	0.029	0.029	0.031
		0.641	0.048	0.042	0.035	0.030	0.029	0.031
		0.846	0.050	0.042	0.035	0.030	0.029	0.031
		0.942	0.051	0.043	0.035	0.030	0.029	0.030
		0.986	0.061	0.047	0.032	0.026	0.023	0.025
	1	0.014	0.065	0.058	0.034	0.032	0.033	0.043
		0.058	0.053	0.055	0.039	0.040	0.043	0.050
		0.154	0.053	0.055	0.039	0.040	0.043	0.051
		0.360	0.054	0.056	0.040	0.041	0.044	0.052
		0.641	0.055	0.056	0.040	0.041	0.044	0.053
		0.846	0.056	0.057	0.041	0.041	0.044	0.052
		0.942	0.057	0.058	0.041	0.041	0.044	0.051
		0.986	0.094	0.080	0.042	0.038	0.035	0.045
	2	0.014	*	*	0.054	0.051	0.061	0.076
		0.058	*	*	0.058	0.064	0.070	0.087
		0.154	*	*	0.058	0.064	0.072	0.089
		0.360	*	*	0.058	0.065	0.074	0.091
		0.641	*	*	0.059	0.065	0.074	0.092
		0.846	*	*	0.059	0.066	0.074	0.091
		0.942	*	*	0.060	0.066	0.074	0.089
		0.986	*	*	0.074	0.059	0.065	0.078

* Crack length outside FEM

Pin Loading (Cosθ Load Distribution)

		a/c₁	0.2	0.3	0.4	0.6	1	2
a/t	r/t	z/t	β					
1.05	0.5	0.014	0.063	0.068	0.078	0.095	0.114	0.135
		0.058	0.056	0.070	0.081	0.101	0.125	0.152
		0.154	0.057	0.069	0.080	0.100	0.124	0.156
		0.360	0.061	0.072	0.082	0.100	0.125	0.159
		0.641	0.070	0.083	0.094	0.111	0.134	0.166
		0.846	0.090	0.115	0.126	0.142	0.161	0.184
		0.942	0.134	0.165	0.181	0.199	0.214	0.220
		0.986	0.410	0.385	0.384	0.379	0.359	0.296
	1	0.014	0.079	0.108	0.130	0.161	0.192	0.221
		0.058	0.081	0.112	0.137	0.172	0.209	0.244
		0.154	0.084	0.112	0.135	0.168	0.208	0.247
		0.360	0.093	0.119	0.139	0.169	0.207	0.248
		0.641	0.115	0.141	0.160	0.186	0.219	0.252
		0.846	0.171	0.197	0.214	0.236	0.258	0.270
		0.942	0.246	0.287	0.306	0.327	0.339	0.315
		0.986	0.644	0.644	0.642	0.620	0.554	0.404
	2	0.014	0.132	0.180	0.218	0.266	0.313	0.304
		0.058	0.136	0.187	0.228	0.283	0.337	0.338
		0.154	0.140	0.187	0.225	0.277	0.330	0.339
		0.360	0.157	0.199	0.232	0.277	0.323	0.337
		0.641	0.195	0.236	0.265	0.299	0.332	0.338
		0.846	0.287	0.325	0.347	0.368	0.378	0.356
		0.942	0.409	0.460	0.486	0.504	0.490	0.407
		0.986	1.062	1.061	1.029	0.953	0.763	0.510
1.09	0.5	0.014	0.061	0.073	0.076	0.093	0.111	0.132
		0.058	0.055	0.069	0.080	0.099	0.121	0.148
		0.154	0.056	0.068	0.079	0.098	0.121	0.152
		0.360	0.061	0.071	0.082	0.099	0.122	0.156
		0.641	0.071	0.082	0.093	0.109	0.132	0.164
		0.846	0.093	0.110	0.124	0.139	0.158	0.181
		0.942	0.143	0.166	0.178	0.193	0.206	0.212
		0.986	0.358	0.368	0.337	0.333	0.314	0.264
	1	0.014	0.079	0.107	0.128	0.157	0.187	0.214
		0.058	0.081	0.112	0.135	0.168	0.204	0.237
		0.154	0.084	0.112	0.133	0.166	0.203	0.241
		0.360	0.093	0.119	0.138	0.167	0.204	0.243
		0.641	0.116	0.141	0.159	0.185	0.217	0.250
		0.846	0.170	0.194	0.211	0.233	0.256	0.269
		0.942	0.250	0.285	0.303	0.323	0.333	0.310
		0.986	0.573	0.573	0.571	0.551	0.492	0.371
	2	0.014	0.132	0.179	0.216	0.261	0.304	0.300
		0.058	0.136	0.187	0.227	0.278	0.328	0.334
		0.154	0.141	0.186	0.223	0.273	0.323	0.337
		0.360	0.158	0.199	0.231	0.274	0.319	0.336
		0.641	0.197	0.237	0.265	0.298	0.332	0.340
		0.846	0.286	0.323	0.346	0.369	0.382	0.361
		0.942	0.423	0.470	0.494	0.509	0.490	0.408
		0.986	0.966	0.960	0.929	0.856	0.692	0.479

		a/c₁	0.2	0.3	0.4	0.6	1	2
a/t	r/t	z/t	β					
1.13	0.5	0.014	0.070	0.073	0.007	0.091	0.108	0.128
		0.058	0.053	0.074	0.030	0.097	0.118	0.144
		0.154	0.054	0.073	0.078	0.096	0.119	0.149
		0.360	0.059	0.076	0.183	0.097	0.120	0.153
		0.641	0.070	0.087	0.326	0.108	0.130	0.161
		0.846	0.095	0.113	0.431	0.135	0.154	0.177
		0.942	0.142	0.155	0.480	0.185	0.198	0.203
		0.986	0.333	0.311	0.502	0.300	0.283	0.241
	1	0.014	0.078	0.105	0.007	0.154	0.182	0.208
		0.058	0.080	0.110	0.030	0.165	0.199	0.231
		0.154	0.083	0.110	0.078	0.163	0.199	0.236
		0.360	0.093	0.118	0.183	0.164	0.201	0.239
		0.641	0.116	0.139	0.326	0.182	0.214	0.246
		0.846	0.166	0.190	0.431	0.228	0.252	0.265
		0.942	0.245	0.278	0.480	0.313	0.322	0.303
		0.986	0.520	0.521	0.502	0.501	0.448	0.346
	2	0.014	0.131	0.178	0.007	0.256	0.295	0.295
		0.058	0.136	0.185	0.030	0.274	0.320	0.329
		0.154	0.140	0.185	0.078	0.269	0.317	0.332
		0.360	0.157	0.198	0.183	0.271	0.315	0.332
		0.641	0.196	0.235	0.326	0.296	0.329	0.338
		0.846	0.283	0.319	0.431	0.366	0.380	0.360
		0.942	0.422	0.465	0.480	0.501	0.481	0.404
		0.986	0.890	0.882	0.502	0.783	0.640	0.453
1.17	0.5	0.014	0.069	0.070	0.007	0.088	0.105	0.125
		0.058	0.053	0.067	0.030	0.095	0.115	0.141
		0.154	0.054	0.066	0.078	0.094	0.116	0.145
		0.360	0.059	0.069	0.183	0.095	0.118	0.150
		0.641	0.070	0.081	0.326	0.104	0.127	0.157
		0.846	0.095	0.108	0.431	0.129	0.150	0.172
		0.942	0.139	0.156	0.480	0.173	0.189	0.195
		0.986	0.307	0.296	0.502	0.269	0.259	0.223
	1	0.014	0.088	0.103	0.007	0.150	0.178	0.202
		0.058	0.089	0.109	0.030	0.161	0.195	0.226
		0.154	0.091	0.108	0.078	0.159	0.195	0.231
		0.360	0.099	0.115	0.183	0.162	0.197	0.235
		0.641	0.120	0.135	0.326	0.179	0.211	0.243
		0.846	0.166	0.180	0.431	0.223	0.246	0.261
		0.942	0.228	0.252	0.480	0.302	0.311	0.294
		0.986	0.494	0.461	0.502	0.462	0.414	0.327
	2	0.014	0.131	0.176	0.007	0.251	0.287	0.290
		0.058	0.135	0.184	0.030	0.269	0.313	0.324
		0.154	0.139	0.183	0.078	0.265	0.311	0.329
		0.360	0.156	0.196	0.183	0.267	0.310	0.330
		0.641	0.195	0.233	0.326	0.293	0.326	0.336
		0.846	0.279	0.314	0.431	0.361	0.376	0.358
		0.942	0.416	0.456	0.480	0.488	0.469	0.398
		0.986	0.829	0.820	0.502	0.726	0.599	0.434

		a/c₁	0.2	0.3	0.4	0.6	1	2
a/t	r/t	z/t	β					
1.21	0.5	0.014	0.068	0.069	0.073	0.087	0.102	0.122
		0.058	0.052	0.066	0.077	0.093	0.113	0.137
		0.154	0.053	0.065	0.075	0.092	0.114	0.142
		0.360	0.058	0.068	0.073	0.094	0.115	0.146
		0.641	0.069	0.080	0.08[illegible]	0.103	0.124	0.154
		0.846	0.095	0.106	0.11[illegible]	0.128	0.145	0.167
		0.942	0.135	0.150	0.162	0.172	0.181	0.188
		0.986	0.286	0.273	0.259	0.254	0.240	0.209
	1	0.014	0.079	0.104	0.122	0.147	0.173	0.197
		0.058	0.080	0.108	0.130	0.158	0.190	0.221
		0.154	0.082	0.108	0.128	0.156	0.191	0.226
		0.360	0.091	0.115	0.133	0.159	0.194	0.230
		0.641	0.113	0.135	0.152	0.175	0.207	0.239
		0.846	0.158	0.182	0.197	0.217	0.241	0.256
		0.942	0.237	0.262	0.275	0.292	0.299	0.286
		0.986	0.458	0.452	0.448	0.430	0.386	0.311
	2	0.014	0.130	0.174	0.207	0.246	0.277	0.286
		0.058	0.135	0.182	0.219	0.264	0.305	0.320
		0.154	0.138	0.182	0.216	0.261	0.304	0.325
		0.360	0.155	0.194	0.224	0.264	0.305	0.327
		0.641	0.193	0.229	0.255	0.289	0.322	0.334
		0.846	0.274	0.308	0.330	0.355	0.371	0.355
		0.942	0.407	0.444	0.463	0.474	0.456	0.391
		0.986	0.778	0.767	0.742	0.679	0.572	0.417
2	0.5	0.014	0.057	0.059	0.056	0.062	0.071	0.084
		0.058	0.048	0.054	0.060	0.068	0.079	0.096
		0.154	0.048	0.054	0.060	0.068	0.081	0.100
		0.360	0.050	0.055	0.061	0.069	0.082	0.103
		0.641	0.057	0.062	0.067	0.074	0.086	0.106
		0.846	0.071	0.074	0.078	0.084	0.093	0.109
		0.942	0.085	0.085	0.090	0.097	0.102	0.112
		0.986	0.141	0.132	0.114	0.108	0.107	0.108
	1	0.014	0.084	0.085	0.090	0.104	0.120	0.141
		0.058	0.071	0.090	0.098	0.114	0.134	0.160
		0.154	0.071	0.090	0.098	0.115	0.137	0.166
		0.360	0.076	0.092	0.100	0.117	0.140	0.170
		0.641	0.090	0.103	0.109	0.125	0.147	0.175
		0.846	0.114	0.127	0.131	0.142	0.159	0.180
		0.942	0.137	0.159	0.162	0.167	0.175	0.186
		0.986	0.238	0.200	0.197	0.192	0.182	0.178
	2	0.014	0.111	0.137	0.154	0.[illegible]77	0.200	0.232
		0.058	0.116	0.146	0.167	0.[illegible]95	0.225	0.260
		0.154	0.116	0.146	0.166	0.[illegible]96	0.229	0.265
		0.360	0.124	0.150	0.170	0.199	0.233	0.269
		0.641	0.146	0.169	0.187	0.213	0.243	0.275
		0.846	0.192	0.209	0.223	0.243	0.263	0.283
		0.942	0.258	0.270	0.276	0.283	0.287	0.291
		0.986	0.368	0.353	0.337	0.347	0.294	0.276

		a/c₁	0.2	0.3	0.4	0.6	1	2
a/t	r/t	z/t	β					
5	0.5	0.014	0.048	0.036	0.034	0.032	0.034	0.043
		0.058	0.046	0.039	0.038	0.039	0.042	0.049
		0.154	0.045	0.039	0.038	0.039	0.043	0.050
		0.360	0.044	0.039	0.039	0.040	0.043	0.051
		0.641	0.045	0.041	0.040	0.041	0.044	0.052
		0.846	0.048	0.043	0.042	0.042	0.045	0.052
		0.942	0.051	0.045	0.043	0.042	0.045	0.051
		0.986	0.061	0.052	0.046	0.041	0.039	0.046
	1	0.014	0.062	0.057	0.054	0.052	0.060	0.073
		0.058	0.051	0.053	0.056	0.062	0.068	0.083
		0.154	0.051	0.053	0.056	0.062	0.070	0.086
		0.360	0.052	0.054	0.057	0.063	0.072	0.087
		0.641	0.054	0.056	0.059	0.065	0.073	0.088
		0.846	0.058	0.059	0.061	0.067	0.074	0.088
		0.942	0.061	0.061	0.063	0.068	0.075	0.087
		0.986	0.100	0.084	0.078	0.063	0.070	0.079
	2	0.014	0.111	0.087	0.079	0.088	0.103	0.122
		0.058	0.071	0.085	0.091	0.101	0.117	0.140
		0.154	0.072	0.085	0.092	0.103	0.120	0.145
		0.360	0.073	0.086	0.093	0.105	0.122	0.147
		0.641	0.077	0.089	0.097	0.108	0.124	0.149
		0.846	0.081	0.094	0.103	0.113	0.127	0.149
		0.942	0.085	0.098	0.108	0.117	0.128	0.147
		0.986	0.191	0.130	0.108	0.110	0.119	0.132
10	0.5	0.014	0.050	0.036	0.029	0.023	0.021	0.024
		0.058	0.048	0.040	0.034	0.029	0.028	0.030
		0.154	0.048	0.040	0.035	0.029	0.028	0.031
		0.360	0.048	0.041	0.035	0.029	0.028	0.031
		0.641	0.048	0.041	0.035	0.030	0.029	0.031
		0.846	0.050	0.042	0.035	0.030	0.029	0.031
		0.942	0.051	0.043	0.035	0.030	0.028	0.030
		0.986	0.061	0.047	0.032	0.026	0.023	0.025
	1	0.014	0.064	0.057	0.033	0.032	0.032	0.043
		0.058	0.053	0.055	0.039	0.040	0.043	0.049
		0.154	0.053	0.055	0.039	0.040	0.043	0.051
		0.360	0.053	0.055	0.040	0.040	0.044	0.051
		0.641	0.054	0.056	0.040	0.041	0.044	0.052
		0.846	0.056	0.057	0.041	0.041	0.044	0.051
		0.942	0.056	0.057	0.041	0.041	0.044	0.050
		0.986	0.094	0.080	0.042	0.037	0.035	0.044
	2	0.014	*	*	0.053	0.050	0.060	0.073
		0.058	*	*	0.057	0.063	0.069	0.083
		0.154	*	*	0.058	0.063	0.071	0.086
		0.360	*	*	0.058	0.064	0.072	0.088
		0.641	*	*	0.058	0.065	0.073	0.088
		0.846	*	*	0.059	0.065	0.073	0.088
		0.942	*	*	0.060	0.066	0.072	0.086
		0.986	*	*	0.073	0.058	0.064	0.076

* Crack length outside FEM

Pin Loading ($Cos^2\theta$ Load Distribution)

		a/c_i	0.2	0.3	0.4	0.6	1	2
a/t	r/t	z/t	β					
1.05	0.5	0.014	0.063	0.068	0.078	0.096	0.115	0.138
		0.058	0.057	0.070	0.082	0.102	0.126	0.155
		0.154	0.057	0.069	0.081	0.100	0.126	0.159
		0.360	0.061	0.072	0.083	0.101	0.126	0.164
		0.641	0.071	0.083	0.095	0.112	0.136	0.172
		0.846	0.090	0.116	0.127	0.144	0.165	0.190
		0.942	0.135	0.167	0.183	0.202	0.219	0.226
		0.986	0.413	0.389	0.389	0.387	0.368	0.305
	1	0.014	0.079	0.109	0.131	0.164	0.197	0.229
		0.058	0.082	0.113	0.138	0.174	0.214	0.252
		0.154	0.084	0.113	0.136	0.171	0.213	0.255
		0.360	0.094	0.120	0.141	0.172	0.212	0.256
		0.641	0.116	0.143	0.162	0.190	0.225	0.260
		0.846	0.173	0.200	0.218	0.242	0.267	0.278
		0.942	0.248	0.292	0.312	0.337	0.351	0.323
		0.986	0.654	0.659	0.660	0.640	0.573	0.415
	2	0.014	0.133	0.183	0.223	0.274	0.324	0.315
		0.058	0.137	0.190	0.233	0.291	0.348	0.349
		0.154	0.142	0.190	0.229	0.285	0.341	0.349
		0.360	0.159	0.203	0.237	0.284	0.334	0.344
		0.641	0.198	0.241	0.272	0.308	0.342	0.344
		0.846	0.292	0.333	0.357	0.380	0.389	0.361
		0.942	0.417	0.473	0.501	0.520	0.503	0.411
		0.986	1.089	1.094	1.063	0.982	0.781	0.514
1.09	0.5	0.014	0.062	0.074	0.076	0.093	0.112	0.134
		0.058	0.055	0.069	0.081	0.100	0.122	0.151
		0.154	0.056	0.068	0.079	0.098	0.123	0.156
		0.360	0.061	0.071	0.082	0.099	0.124	0.160
		0.641	0.071	0.082	0.094	0.110	0.134	0.169
		0.846	0.094	0.111	0.125	0.140	0.161	0.187
		0.942	0.143	0.167	0.180	0.195	0.211	0.218
		0.986	0.360	0.371	0.340	0.338	0.321	0.271
	1	0.014	0.079	0.108	0.129	0.160	0.191	0.222
		0.058	0.082	0.112	0.136	0.171	0.209	0.246
		0.154	0.084	0.113	0.135	0.168	0.208	0.249
		0.360	0.094	0.120	0.140	0.170	0.209	0.251
		0.641	0.117	0.142	0.161	0.188	0.223	0.258
		0.846	0.171	0.197	0.214	0.238	0.264	0.277
		0.942	0.253	0.289	0.309	0.332	0.344	0.320
		0.986	0.580	0.583	0.584	0.568	0.510	0.383
	2	0.014	0.133	0.182	0.220	0.268	0.314	0.311
		0.058	0.138	0.189	0.231	0.286	0.339	0.345
		0.154	0.142	0.189	0.228	0.281	0.334	0.347
		0.360	0.159	0.202	0.236	0.282	0.330	0.344
		0.641	0.199	0.241	0.271	0.307	0.342	0.346
		0.846	0.291	0.330	0.355	0.381	0.393	0.366
		0.942	0.431	0.482	0.509	0.526	0.505	0.414
		0.986	0.988	0.989	0.959	0.885	0.711	0.486

		a/c_i	0.2	0.3	0.4	0.6	1	2
a/t	r/t	z/t	β					
1.13	0.5	0.014	0.071	0.074	0.075	0.091	0.109	0.130
		0.058	0.054	0.075	0.079	0.098	0.119	0.147
		0.154	0.055	0.074	0.078	0.097	0.120	0.152
		0.360	0.060	0.076	0.081	0.098	0.122	0.157
		0.641	0.070	0.087	0.093	0.108	0.132	0.166
		0.846	0.096	0.113	0.122	0.136	0.157	0.182
		0.942	0.142	0.156	0.174	0.187	0.201	0.209
		0.986	0.334	0.313	0.308	0.304	0.288	0.248
	1	0.014	0.078	0.106	0.127	0.156	0.186	0.215
		0.058	0.081	0.111	0.134	0.167	0.204	0.239
		0.154	0.083	0.111	0.133	0.165	0.204	0.244
		0.360	0.094	0.118	0.138	0.167	0.206	0.247
		0.641	0.117	0.141	0.159	0.185	0.220	0.254
		0.846	0.168	0.192	0.209	0.233	0.259	0.274
		0.942	0.247	0.281	0.300	0.321	0.333	0.312
		0.986	0.525	0.529	0.530	0.515	0.464	0.358
	2	0.014	0.133	0.181	0.217	0.263	0.305	0.305
		0.058	0.137	0.188	0.228	0.281	0.331	0.340
		0.154	0.141	0.188	0.226	0.276	0.328	0.342
		0.360	0.159	0.201	0.234	0.278	0.325	0.341
		0.641	0.199	0.239	0.268	0.305	0.340	0.345
		0.846	0.287	0.326	0.351	0.378	0.392	0.367
		0.942	0.429	0.477	0.502	0.517	0.496	0.411
		0.986	0.908	0.906	0.880	0.810	0.660	0.462
1.17	0.5	0.014	0.069	0.070	0.075	0.088	0.106	0.127
		0.058	0.053	0.067	0.079	0.095	0.116	0.143
		0.154	0.054	0.066	0.078	0.094	0.117	0.149
		0.360	0.059	0.069	0.080	0.095	0.119	0.154
		0.641	0.070	0.081	0.091	0.105	0.129	0.162
		0.846	0.096	0.109	0.119	0.130	0.152	0.177
		0.942	0.139	0.157	0.169	0.175	0.192	0.201
		0.986	0.308	0.298	0.282	0.272	0.263	0.229
	1	0.014	0.088	0.104	0.125	0.152	0.182	0.210
		0.058	0.090	0.109	0.133	0.163	0.199	0.234
		0.154	0.091	0.109	0.131	0.162	0.200	0.239
		0.360	0.099	0.115	0.136	0.164	0.202	0.242
		0.641	0.120	0.136	0.156	0.182	0.216	0.250
		0.846	0.167	0.182	0.204	0.227	0.253	0.270
		0.942	0.230	0.255	0.290	0.309	0.321	0.304
		0.986	0.498	0.467	0.489	0.474	0.428	0.339
	2	0.014	0.132	0.178	0.214	0.257	0.297	0.301
		0.058	0.137	0.186	0.225	0.276	0.323	0.335
		0.154	0.141	0.186	0.223	0.272	0.321	0.338
		0.360	0.158	0.199	0.231	0.275	0.320	0.338
		0.641	0.197	0.237	0.265	0.301	0.337	0.344
		0.846	0.283	0.320	0.345	0.372	0.388	0.366
		0.942	0.422	0.466	0.490	0.504	0.484	0.406
		0.986	0.844	0.840	0.816	0.751	0.619	0.444

		a/c_i	0.2	0.3	0.4	0.6	1	2
a/t	r/t	z/t	β					
1.21	0.5	0.014	0.068	0.069	0.073	0.087	0.103	0.124
		0.058	0.053	0.066	0.078	0.094	0.114	0.140
		0.154	0.053	0.065	0.077	0.093	0.115	0.145
		0.360	0.058	0.069	0.079	0.094	0.117	0.150
		0.641	0.069	0.080	0.090	0.104	0.126	0.158
		0.846	0.095	0.107	0.116	0.129	0.148	0.172
		0.942	0.136	0.151	0.163	0.174	0.184	0.193
		0.986	0.287	0.274	0.261	0.256	0.243	0.215
	1	0.014	0.080	0.104	0.123	0.149	0.177	0.205
		0.058	0.081	0.109	0.131	0.160	0.195	0.228
		0.154	0.082	0.109	0.129	0.159	0.196	0.234
		0.360	0.091	0.115	0.134	0.161	0.198	0.238
		0.641	0.113	0.136	0.153	0.178	0.212	0.246
		0.846	0.159	0.183	0.199	0.221	0.248	0.265
		0.942	0.239	0.265	0.279	0.298	0.309	0.296
		0.986	0.461	0.457	0.455	0.440	0.398	0.322
	2	0.014	0.131	0.176	0.211	0.252	0.287	0.296
		0.058	0.136	0.184	0.223	0.271	0.315	0.331
		0.154	0.140	0.184	0.220	0.268	0.314	0.335
		0.360	0.156	0.196	0.228	0.271	0.315	0.335
		0.641	0.195	0.233	0.261	0.297	0.333	0.342
		0.846	0.277	0.313	0.338	0.366	0.383	0.363
		0.942	0.413	0.454	0.476	0.489	0.471	0.401
		0.986	0.790	0.785	0.762	0.702	0.592	0.428
2	0.5	0.014	0.057	0.059	0.057	0.062	0.071	0.085
		0.058	0.048	0.055	0.060	0.068	0.079	0.097
		0.154	0.048	0.054	0.060	0.069	0.081	0.101
		0.360	0.050	0.055	0.061	0.070	0.083	0.104
		0.641	0.057	0.062	0.067	0.074	0.087	0.107
		0.846	0.071	0.074	0.078	0.084	0.094	0.111
		0.942	0.085	0.086	0.090	0.097	0.103	0.113
		0.986	0.141	0.133	0.114	0.109	0.107	0.109
	1	0.014	0.084	0.085	0.091	0.105	0.122	0.145
		0.058	0.072	0.091	0.098	0.115	0.136	0.164
		0.154	0.072	0.090	0.098	0.116	0.139	0.170
		0.360	0.076	0.092	0.100	0.118	0.142	0.175
		0.641	0.090	0.103	0.110	0.126	0.149	0.180
		0.846	0.114	0.127	0.131	0.143	0.161	0.185
		0.942	0.137	0.159	0.162	0.168	0.177	0.191
		0.986	0.238	0.201	0.199	0.194	0.185	0.183
	2	0.014	0.112	0.137	0.155	0.180	0.205	0.241
		0.058	0.116	0.147	0.169	0.198	0.230	0.269
		0.154	0.116	0.147	0.168	0.199	0.235	0.275
		0.360	0.124	0.151	0.172	0.203	0.239	0.278
		0.641	0.147	0.170	0.189	0.217	0.250	0.284
		0.846	0.193	0.211	0.225	0.247	0.270	0.293
		0.942	0.259	0.272	0.279	0.288	0.295	0.301
		0.986	0.370	0.356	0.342	0.323	0.302	0.286

		a/c_i	0.2	0.3	0.4	0.6	1	2
a/t	r/t	z/t	β					
5	0.5	0.014	0.048	0.036	0.034	0.033	0.034	0.043
		0.058	0.046	0.039	0.038	0.039	0.042	0.049
		0.154	0.045	0.039	0.038	0.039	0.043	0.050
		0.360	0.044	0.039	0.039	0.040	0.043	0.052
		0.641	0.045	0.041	0.040	0.041	0.044	0.052
		0.846	0.048	0.043	0.042	0.042	0.045	0.052
		0.942	0.051	0.045	0.043	0.043	0.045	0.051
		0.986	0.061	0.052	0.046	0.041	0.039	0.046
	1	0.014	0.062	0.057	0.054	0.052	0.060	0.074
		0.058	0.051	0.053	0.056	0.062	0.069	0.084
		0.154	0.051	0.054	0.056	0.062	0.071	0.087
		0.360	0.052	0.054	0.057	0.063	0.072	0.088
		0.641	0.054	0.056	0.059	0.065	0.073	0.089
		0.846	0.058	0.059	0.061	0.067	0.075	0.089
		0.942	0.061	0.061	0.063	0.068	0.076	0.088
		0.986	0.100	0.085	0.078	0.064	0.070	0.080
	2	0.014	0.111	0.087	0.080	0.089	0.104	0.125
		0.058	0.072	0.085	0.091	0.102	0.118	0.143
		0.154	0.072	0.085	0.092	0.104	0.121	0.148
		0.360	0.073	0.086	0.094	0.106	0.123	0.151
		0.641	0.077	0.090	0.097	0.109	0.126	0.152
		0.846	0.081	0.095	0.103	0.113	0.128	0.152
		0.942	0.085	0.098	0.108	0.117	0.130	0.150
		0.986	0.191	0.130	0.108	0.111	0.120	0.134
10	0.5	0.014	0.050	0.036	0.029	0.023	0.022	0.024
		0.058	0.048	0.040	0.034	0.029	0.028	0.030
		0.154	0.048	0.040	0.035	0.029	0.028	0.031
		0.360	0.048	0.041	0.035	0.029	0.028	0.031
		0.641	0.048	0.041	0.035	0.030	0.029	0.031
		0.846	0.050	0.042	0.035	0.030	0.029	0.031
		0.942	0.051	0.043	0.035	0.030	0.029	0.030
		0.986	0.061	0.047	0.032	0.026	0.023	0.025
	1	0.014	0.064	0.058	0.034	0.032	0.032	0.043
		0.058	0.053	0.055	0.039	0.040	0.043	0.049
		0.154	0.053	0.055	0.039	0.040	0.043	0.051
		0.360	0.054	0.056	0.040	0.040	0.044	0.052
		0.641	0.055	0.056	0.040	0.041	0.044	0.052
		0.846	0.056	0.057	0.041	0.041	0.044	0.052
		0.942	0.057	0.058	0.041	0.041	0.044	0.050
		0.986	0.094	0.080	0.042	0.037	0.035	0.044
	2	0.014	*	*	0.054	0.050	0.060	0.074
		0.058	*	*	0.058	0.063	0.069	0.084
		0.154	*	*	0.058	0.064	0.072	0.087
		0.360	*	*	0.058	0.064	0.073	0.089
		0.641	*	*	0.058	0.065	0.073	0.089
		0.846	*	*	0.059	0.065	0.074	0.088
		0.942	*	*	0.060	0.066	0.073	0.086
		0.986	*	*	0.073	0.058	0.064	0.076

* Crack length outside FEM

Appendix F

Crack Growth Prediction Model Program Flow Chart

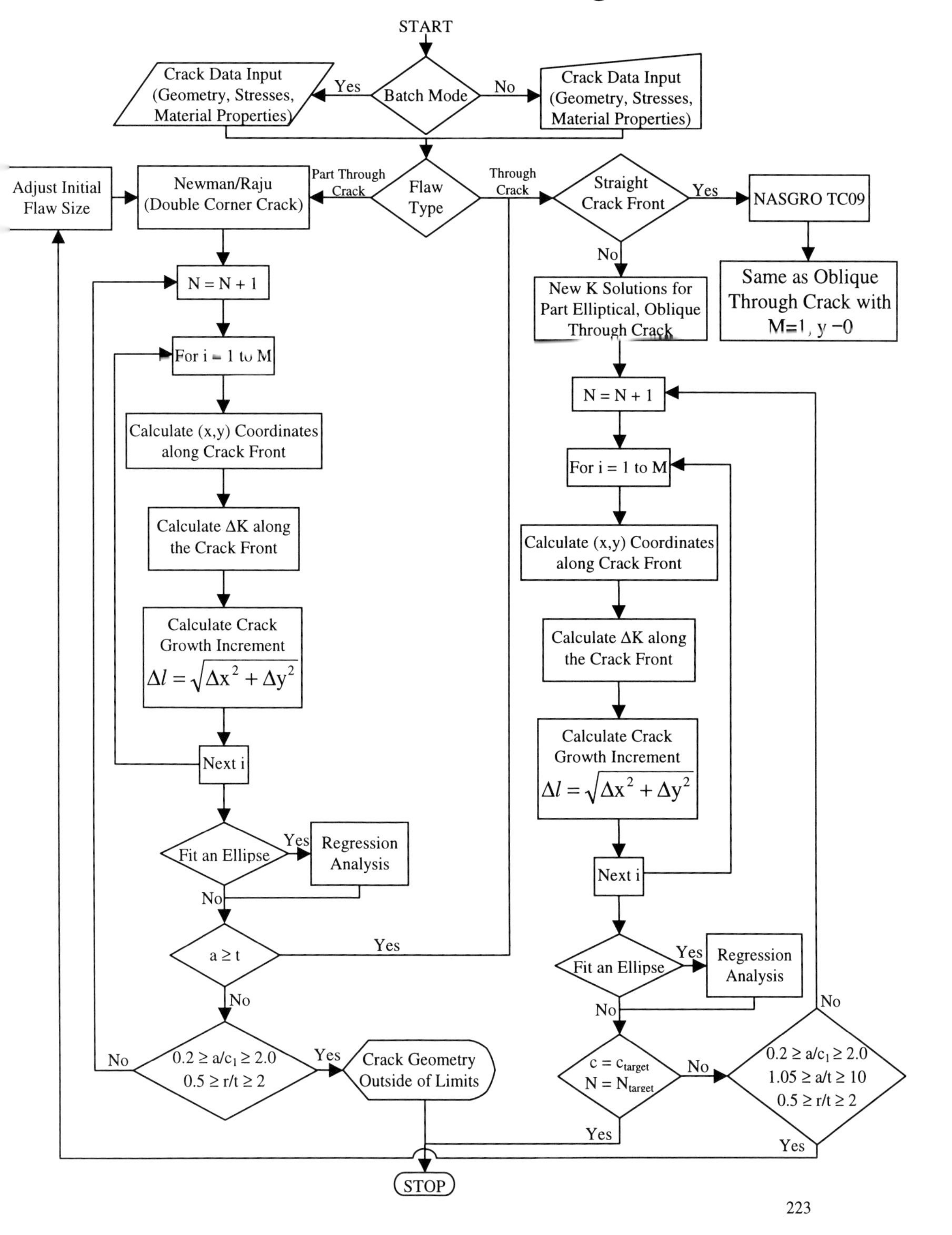

This page intentionally left blank

Appendix G

Newman/Raju Corner Crack and NASGRO TC09 Stress Intensity Factor Solutions[1-4]

G.1 General Stress Intensity Factor Equation

$$K_I = (\sigma_o F_o + \sigma_1 F_1 + \sigma_3 F_3)\sqrt{\pi c}$$

σ_o = Remote tensile stress

σ_1 = Remote bending stress

σ_3 = Bearing stress (pin loading)

F_0 = Stress intensity magnification factor for tension

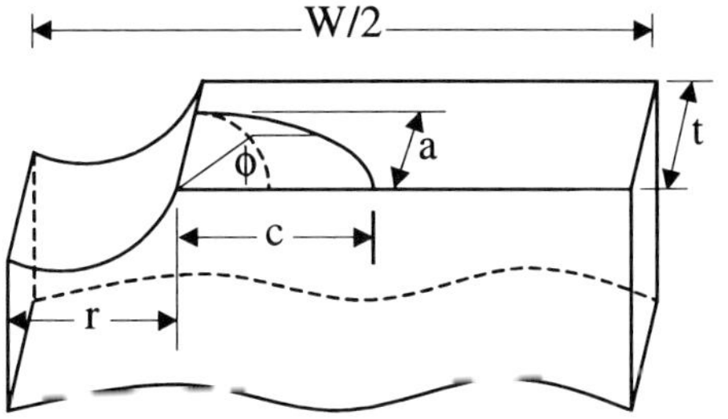

Figure G.1 Crack and Geometric Parameters

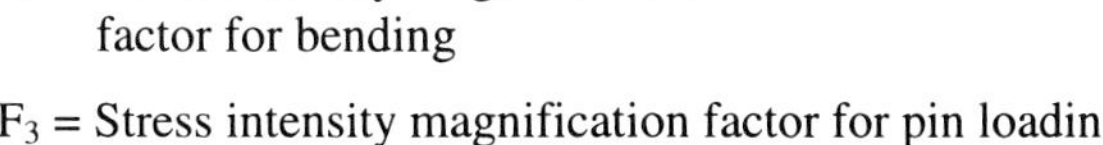

F_1 = Stress intensity magnification factor for bending

F_3 = Stress intensity magnification factor for pin loading

c = Crack length

G.2 Common Symbols

a	=	Crack Depth
t	=	Plate Thickness
D	=	Hole Diameter
W	=	Plate Width

Geometric Ratios

$$u = \frac{c}{D} \quad v = \frac{a}{t} \quad w = \frac{c}{W} \quad x = \frac{a}{c} \quad y = \frac{D}{W}$$

Approximation to the Square of the Complete Elliptical Integral of the Second Kind (Shape Parameter)

$$f_x = \left[1 + 1.464x^{1.65}\right]^{-\frac{1}{2}} \quad \text{for } x \le 1$$

$$= \left[1 + 1.464x^{-1.65}\right]^{-\frac{1}{2}} \quad \text{for } x > 1$$

Fitting Functions for Tension $f_o(z)$ and Bending $f_1(z)$

$$f_o(z_i) = 0.7071 + 0.7548z_i + 0.3415z_i^2 + 0.642z_i^3 + 0.9196z_i^4$$

$$f_1(z_i) = 0.078 + 0.7588z_i + 0.4293z_i^2 + 0.0644z_i^3 + 0.651z_i^4$$

Boundary Correction Factor for Embedded Elliptical Crack in an Infinite Solid

$$f_\phi = \left[(x\cos\phi)^2 + \sin^2\phi\right]^{\frac{1}{4}} \quad \text{for } x \le 1$$

$$= \left[\cos^2\phi + \left(x^{-1}\sin\phi^2\right)\right]^{\frac{1}{4}} \text{ for } x > 1$$

Bending Multiplier for Corner Crack at a Hole in a Plate

$$H_{ch} = H_1 + (H_2 - H_1)\sin^p\phi$$

Curve Fitting Functions

$$M_o = M_1 + M_2 v^2 + M_3 v^4$$

Functions to Simplify Algebraic Equations

$$I = 1 - \sin\phi \qquad J = 1 - \cos\phi$$

G.3 Newman/Raju Boundary Correction Factor Equations

$$F_o = G_o G_w \qquad F_3 = \left[\frac{G_o y}{2} + G_3\right] G_w \qquad F_1 = C_r C_f G_1 G_w H_{ch}$$

$$G_o = \frac{f_o(z_o)}{d_o} \quad G_3 = \frac{f_1(z_o)g_p}{d_o} \qquad G_1 = \frac{f_o(z_2)}{d_2} \quad G_w = M_o g_1 g_3 g_4 f_w f_\phi f_x$$

where

$$z_{o,2} = \left[1 + 2\frac{c}{D}\cos(\mu_{o,2}\phi)\right]^{-1} \quad d_{o,2} = 1 + 0.13 z_{o,2}^2 \quad \mu_o = 0.85 \text{ (Tension)}$$

$$\mu_2 = 0.85 - 0.25 v^{\frac{1}{4}} \text{ (Bending)} \qquad g_p = \sqrt{\frac{1+y}{1-y}}$$

Finite Width Correction Factor

$$f_w = \sqrt{\frac{1}{\beta}\sin\beta\sec\lambda\sec\left(\frac{\pi y}{2}\right)}$$

$$\lambda = \frac{\pi}{2}\sqrt{v}\,\frac{D+c}{W-c}$$

Bending Multipliers

$$H_1 = 1 + G_{11}v + G_{12}v^2 + G_{13}v^3 \qquad H_2 = 1 + G_{21}v + G_{22}v^2 + G_{23}v^3$$

C_r = Stress Concentration Factor from Reissner[5]

C_f = Empirical Correction Factor from Forman[4]

$$C_f = 0.637 - \frac{0.24\frac{D}{t}}{\sqrt{19.51 + \left(\frac{D}{t}\right)^2}}$$

G.3.1 Parameters Depending on the a/c Ratio

G.3.1.1 For $a \leq c$:

Boundary Correction Factor at Maximum Depth Point

$$M_o = [1.13 - 0.09x] + \left[-0.54 + \frac{0.89}{0.2 + x}\right]v^2 + \left[0.5 - \frac{1}{0.65 + x} + 14(1 - x)^{24}\right]v^4$$

Products of Functions for Curve Fitting

$$g_1 = 1 + (0.1 + 0.35v^2)I^2 \qquad g_3 = (1 + 0.0.4x)(1 + 0.1J^2)\left(0.85 + 0.15v^{\frac{1}{4}}\right)$$

$$g_4 = 1 - 0.71(1 - v)(x - 0.2)(1 - x) \qquad p = 0.1 + 1.13v + 1.1x - 0.7xv$$

$$G_{11} = -0.43 - 0.74x - 0.84x^2 \qquad G_{12} = 1.25 - 1.19x - 4.39x^2$$

$$G_{13} = -1.94 + 4.22x - 5.51x^2 \qquad G_{21} = -1.5 - 0.04x - 1.73x^2$$

$$G_{22} = 1.71 - 3.17x + 6.84x^2 \qquad G_{23} = -1.28 + 2.71x - 5.22x^2$$

G.3.1.2 For $a > c$:

$$M_o = \frac{1 + \frac{0.04}{x}}{\sqrt{x}} + \frac{0.2v^2}{x^4} - \frac{0.11v^4}{x^4}$$

$$g_1 = 1 + \left(0.1 + \frac{0.35v^2}{x}\right)I^2 \qquad g_3 = \left(1.13 - \frac{0.09}{x}\right)(1 + 0.1J^2)\left(0.85 + 0.15v^{\frac{1}{4}}\right)$$

$$g_4 = 1 \quad p = 0.2 + \frac{1}{x} + 0.6v$$

$$G_{11} = -2.07 + \frac{0.06}{x} \qquad G_{12} = 4.35 + \frac{0.16}{x} \qquad G_{13} = -2.93 - \frac{0.3}{x}$$

$$G_{21} = -3.64 + \frac{0.37}{x} \qquad G_{22} = 5.87 - \frac{0.49}{x} \qquad G_{23} = -4.32 + \frac{0.53}{x}$$

$$\phi = 0^\circ \text{ for } \frac{dc}{dN} \quad \text{and} \quad \phi = 80^\circ \text{ for } \frac{da}{dN}$$

Using $\phi = 80^\circ$ is recommended by reference [1] to better correlate actual and predicted crack shapes and fatigue lives.

G.4 Boundary Correction Factors Equations for NASGRO TC09 Through Crack from Hole in a Plate under Combined Loading [1,6,7,8]

$$F_o = \sum_{n=0}^{4} [A_n(1+B) + B_n] b^n$$

$$F_1 = [(1+u)b]^{\frac{3}{2}} \frac{(1+\nu)}{(3+\nu)}$$

$$F_2 = \left(\frac{D}{3X}\right) \sum_{n=0}^{5} C_n b^n$$

$$F_2 = \sum_{n=0}^{5} D_n b^n$$

$$b = \frac{1}{(1+2u)}$$

Table G.1 Values of Coefficients A_n, B_n, C_n, D_n

n	A_n	B_n	C_n	D_n
0	-0.00074	0.70920	0.7968	0
1	0.06391	0.68902	0.5326	0.0780
2	-0.10113	0.52270	0.2767	0.7588
3	-0.29411	0.65768	0.0630	-0.4293
4	-0.79179	1.91920	-0.0166	0.0644
5	-	-	1.7197	0.6510

---

[1] NASGRO Fatigue Crack Growth Computer Program, Version 2.01, NASA JSC-22267A, 1994.

[2] Shivakumar, V. and Y. C. Hsu. Stress Intensity Factors for Cracks Emanating from the Loaded Fastener Hole. Proc. of the International Conference on Fracture Mechanics and Technology, Hong Kong, March 1977.

[3] Newman, Jr., J. C. and I. S. Raju. Stress Intensity Factor Equations for Cracks in Three-Dimensional Finite Bodies Subjected to Tension and Bending Loads. NASA-TP-85793, 1985.

[4] Forman, R. G., and S. R. Mettu. Behavior of Surface and Corner Cracks Subjected to Tensile and Bending Loads in Ti-6Al-4V Alloy. Fracture Mechanics: Twenty-Second Symposium, Vol. 1, ASTM STP 1131, H. A. Ernst, A. Saxena, and D. L. McDowell, Eds., American Society for Testing and Materials, Philadelphia, 1992. 519-546.

[5] Reissner, Eric. "The Effect of Transverse Shear Deformation on the Bending of Elastic Plates." Journal of Applied Mechanics. (1945): A69-A77.

[6] Shivakumar, V. and R. G. Forman. "Green's Function for a Crack Emanating from a Circular Hole in an Infinite Sheet." International Journal of Fracture, 16 (1980): 305-316.

[7] Roberts, R. and T. Rich, "Stress Intensity Factors for Plate Bending," Transactions of ASME, Journal of Applied Mechanics, 34 (1967): 777-779.

[8] Roberts, Richard, and John J. Kibler. "Some Aspects of Fatigue Crack Propagation," Engineering Fracture Mechanics. 2 (1971): 243-260.

Published Papers

Based on the research presented in this thesis, the following papers have been published or will be published shortly.

- Fawaz, S. A. "Application of the Virtual Crack Closure Technique to Calculate Stress Intensity Factors for Through Cracks with an Elliptical Crack Front." Engineering Fracture Mechanics, *accepted for publication,* 1997.
- Fawaz, S. A., J. Schijve, and A. U. de Koning. Fatigue Crack Growth in Riveted Lap-Splice Joints, Proc. of the 19th Symposium of the International Committee on Aeronautical Fatigue, 16-20 June 1997, Edinburgh, Scot. Scotland, UK: EMAS/SoMat Systems International Ltd, 1997.
- Fawaz, S. A. Application of the Virtual Crack Closure Technique to Calculate Stress Intensity Factors for Through Cracks with an Oblique Elliptical Crack Front, Report LR-805. Delft, NL: Delft University of Technology UP, 1996.
- Fawaz, S. A. Application of the Gel Electrode Method in Thin Sheet Fatigue Specimens, B2-97-04. Delft, NL: Faculty of Aerospace Engineering, Delft University of Technology, 1997.
- Fawaz, S. A. and J. Schijve. Fatigue Crack Growth in Riveted Lap Joints. Proc. of the 1995 USAF Structural Integrity Program Conference, 28-30 Nov 1995, San Antonio, TX, WL-TR-96-4093.
- Fawaz, S. A. and J. Schijve. Multiple Site Damage (MSD) in a Pressurized Fuselage Riveted Lap Joint. Proc. of the 1994 USAF Structural Integrity Program Conference, 6-8 Dec 1994, San Antonio, TX, WL-TR-96-4030.

This page intentionally left blank

Curriculum Vitae

The author was born on June 30, 1965 in Harbor City, California. After graduating from high school in 1983, he entered the United States Air Force Academy, Colorado Springs, Colorado. He graduated with academic distinction with a Bachelor of Science degree in Engineering Mechanics and was commissioned a Regular Officer in the United States Air Force in May 1987.

As a Second Lieutenant, he earned a Master of Science degree in Aeronautical Engineering at The Air Force Institute of Technology, Dayton, Ohio. In 1989 he was transferred to the San Antonio Air Logistics Center at Kelly Air Force Base, San Antonio, Texas. As the lead damage tolerance analysis engineer, he was responsible for the structural integrity programs of the Lockheed C-5, Cessna T-37, Cessna O/A-37, Northrop T-38, Northrop F-5, and McDonnell Douglas C-17. In addition, he was the lead aircraft battle damage repair engineer for the Lockheed C-5.

After promotion to Captain, the author was assigned to the United States Air Force Academy in Colorado Springs, Colorado. He served as an Instructor and later an Assistant Professor in the Department of Engineering Mechanics. He had several extra curricular activities with the cadets as the officer-in-charge of both the cadet chapter of the American Society of Mechanical Engineers and Model Engineering Club.

In 1994 he moved to Delft, The Netherlands to begin his doctoral studies under the sponsorship of the Air Force Institute of Technology and supervision of Prof. ir. L. B. Vogelesang and Prof. dr. ir. J. Schijve. As part of the damage tolerance group, the author investigated the fatigue behavior of riveted joints

The author is married to the former Jennifer Lynn Abbott of Pueblo, Colorado and they have two children Sarah Elizabeth and Dylan Anthony.

This page intentionally left blank

Samenvatting

Het huidige onderzoek bestrijkt een experimenteel deel en een analytisch deel gebaseerd op breukmechanica. Het doel is meer kennis te verwerven over de groei van kleine en grotere scheuren in geklonken lapnaden, zoals die worden toegepast in operationele transportvliegtuigen. Voorts is het onderzoek gericht op de mogelijkheden voor het voorspellen van vermoeiingsscheuren, zowel voor kleine scheuren als voor grotere scheuren. Kleine scheuren met een kwart-elliptisch scheurfront zijn nog niet door de gehele plaatdikte heen gegroeid. Bij grotere scheuren is dat wel het geval, maar staat het scheurfront nog steeds scheef op het plaatoppervlak. Een karakteristiek aspect van vermoeiing van geklonken lapnaden is het optreden van scheurgroei onder een complex spanningssysteem, dat in de eenvoudigste vorm bestaat uit wisselende trek met daaraan toegevoegd een wisselende buiging als gevolg van de excentriciteit in de lapnaad. Bovendien leidt het stuiken van de klinknagels tot het plastisch oprekken van de gaten en tot ingebouwde restspanningen. Voorts wordt het contactgedrag van de nagel in het nagelgat gecompliceerd door het scheeftrekken van de nagels.

In het empirische deel van het onderzoek is eerst een eenvoudiger probleem geanalyseerd, namelijk vermoeiingsscheurgroei in een plaat met meerdere open gaten onder gecombineerde trek- en buigspanning. Het bleek dat de scheurgroeiontwikkeling voor kleine scheuren (nog niet door de dikte heen gegroeid) gevolgd kon worden door fractografische metingen gebruikmakend van zgn. markerloads, d.w.z. kleinere belastingscycli tussen de constante amplitude belastingen door. Dezelfde markeringstechniek is gebruikt voor lapnaden met twee klinknagelrijen met vier klinknagels in elke rij. Hiermee kon de scheurgroeigeschiedenis worden gereconstrueerd van een scheurlengte van 75 μm tot uiteindelijke breuk bij 4.5 mm.

Voor het analytische gedeelte waren K-oplossingen nodig, d.w.z. spanningsintensiteitsfactoren langs elliptische scheurfronten. Voor kleine scheuren met een kwart-elliptische vorm zijn de bekende Newman- Raju oplossingen beschikbaar. Die zijn van toepassing op bovengenoemde proefstukken met open gaten. Nadat grotere scheuren door de dikte heen zijn verkregen blijven deze groeien met een scheef elliptisch scheurfront door de gecombineerde trek en buiging. Omdat hiervoor geen K-oplossingen beschikbaar zijn, zijn voor deze scheuren met de eindige elementen methode middels een drie dimensionale virtuele scheursluit techniek (virtual crack closure technique) K-oplossingen berekend. Aangetoond werd dat de resultaten verkregen met deze techniek nauwelijks afhankelijk zijn van de oriëntatie van de elementgrenzen in het scheurvlak, waardoor met een minimale hoeveelheid pre-processing kan worden volstaan om nieuwe K-oplossingen te genereren. De techniek is geverifiëerd door resultaten te vergelijken met K-waardes voor scheurvormen waarvoor oplossingen aanwezig zijn in de literatuur. Verkregen K-waardes voor scheurvormen in de open gat plaatproefstukken werden berekend en gebruikt voor de voorspelling van de groei van deze scheuren. Een goede overeenstemming werd gevonden. Voorts zijn K-waardes berekend voor "door de dikte scheuren" met scheefstaande scheurfronten voor verschillende scheurdiepte tot scheurlengte verhoudingen (a/c = 0,2, 0,3, 0,4, 0,6 ,1,0, 2,0), scheurdiepte tot plaatdikte verhoudingen (a/t = 1,05, 1,09, 1,13, 1,17, 1,21, 2,0, 5,0, 10,0) en gatstraal tot plaatdikte verhoudingen (r/t = 0,5, 1,0, 2,0). De Newman/Raju K-oplossingen voor kleine scheuren (niet door de dikte heen) en de nieuw berekende K=oplossingen zijn gebruikt in een voorspellingsmodel voor scheurgroei onder een wisselende trek en buigbelasting. Het algoritme voorspelt niet alleen het vermoeiingsleven binnen 6% van het empirische resultaat, maar ook nauwkeurig de scheurgroeigeschiedenis tot net voor de uiteindelijke breuk optrad.

Het klinkproces heeft grote invloed op het vermoeiingsgedrag en de levensduur van geklonken lapnaden zoals Müller heeft aangetoond. Dat geldt met name voor de klinkkracht met daaraan gerelateerd de gatexpansie en de restspanning rond het nagelgat. Om dit nader te onderzoeken is gebruik gemaakt van een 3D eindige elementen model van een geklonken lapnaad met drie rijen nagels. Hiermee is een analyse gemaakt van de spanningen, inclusief de optredende restsspanningen t.g.v. het klinken, die optraden rondom de nagelgaten en in de kritieke doorsnede als gevolg van de trekbelasting op de lapnaad en daardoor veroorzaakte secundaire buiging. Door toepassing van contact-elementen kon ook het contact tussen de platen en tussen de klinknagel en het nagelgat worden onderzocht. Berekeningen zijn gemaakt voor een lage klinkkracht en een hoge klinkkracht. De lage klinkkracht leidt tot een schuifpassing, waarbij geen restspanningen optreden. Onder belasting op de lapnaad wordt dan plaatselijk het contact tussen nagel en gatwand verbroken, en de spanningsconcentratie bij het gat is aanzienlijk. In het tweede geval ontstaat een perspassing tussen nagel en nagelgat en zijn er hoge inwendige spanningen door gatoprekking. De hoogste spanning in de plaat treedt niet meer op naast het gat in de minimum doorsnede, maar aan de bovenzijde van het gat. De verschuiving van de plaats waar de hoogste spanning optreedt stemt overeen met de verschuiving van het scheurinitiatiepunt bij de vermoeiingsproeven op lapnaden geklonken met een lage, respectievelijk een hoge klinkkracht. Voorts zijn berekeningen gemaakt met kleine scheuren in het eindige elementen model. Daarbij bleek dat de amplitude van de spanningsintensiteitsfactor bij een hoge klinggracht ca. 3x lager was dan voor een lage klinkkracht. Ook dat stemt overeen met de veel langere levensduur van de lapnaden geklonken met een hoge klinkkracht.

Scott Anthony FAWAZ